INSECTICIDE RESISTANCE
From Mechanisms to Management

INSECTICIDE RESISTANCE
From Mechanisms to Management

Amelia Lewis
Hunter Coleman

Kruger Brentt
Publishers
2026

Kruger Brentt Publishers UK. LTD.
Company Number 9728962

Regd. Office: 68 St Margarets Road, Edgware, Middlesex HA8 9UU

ISBN: 978-1-78715-616-6

For information on all our publications visit our website at http://krugerbrentt.com/

PREFACE

Insecticide resistance is one of the most significant challenges in modern agriculture, public health, and vector control. The widespread and often indiscriminate use of chemical insecticides has led to the evolution of resistance in many pest populations, threatening crop yields, food security, and the control of disease vectors. Understanding the biological mechanisms of resistance and developing effective management strategies are critical for sustainable pest control and environmental protection.

This book, Insecticide Resistance: From Mechanisms to Management, provides a comprehensive and detailed overview of the science, mechanisms, and management of insecticide resistance. It is designed for entomologists, agricultural scientists, pest management professionals, and students who seek a thorough understanding of how resistance develops and how it can be mitigated in practice.

The opening chapters introduce the fundamental concepts of insecticide action, resistance evolution, and population genetics. Subsequent sections explore the biochemical, physiological, and molecular mechanisms that contribute to resistance, including target site mutations, metabolic detoxification, behavioral changes, and cuticular modifications. Detailed discussions help readers understand the interplay between pest biology, insecticide chemistry, and environmental factors that drive resistance.

The book also addresses practical management strategies, emphasizing integrated pest management (IPM) approaches, rotation and mixture of insecticides, biological control methods, and regulatory policies. Case studies highlight successful resistance management programs in agricultural crops, livestock systems, and public health initiatives, demonstrating the real-world application of scientific principles.

Additionally, emerging technologies such as molecular diagnostics, genetic tools, and computational modeling are discussed for monitoring resistance and predicting its spread. Ethical and environmental considerations, including minimizing non-target effects and promoting sustainable pest control practices, are also covered to provide a holistic perspective.

Insecticide Resistance: From Mechanisms to Management is an essential resource for anyone involved in pest management, agriculture, or vector control. By combining rigorous scientific explanation with practical strategies and case studies, this book equips readers with the knowledge needed to understand, monitor, and manage insecticide resistance effectively.

Amelia Lewis
Hunter Coleman

CONTENTS

1 Chapter

INSECTICIDE RESISTANCE IN COFFEE LEAFMINER POPULATIONS

1. INTRODUCTION: AN OVERVIEW

The coffee leafminer Leucoptera coffeella (Guérin-Méneville, 1842) (Lepidoptera: Lyonetiidae) is indigenous to Africa and has emerged as a significant pest in numerous coffee-producing nations (Coffea arabica and Coffea canephora) [1,2]. The highly varied life cycle of this species and their detrimental impact on coffee harvests render them a pest with significant destructive potential [3-5]. Insecticides represent the most effective means of managing this pest, with over 30 distinct active insecticides authorized for application against L. coffeella in Brazil [6]. The excessive application of pesticides by farmers has resulted in insect resistance, despite the presence of multiple active chemicals. [7]. The inaugural recorded instance of resistance occurred in 1914, involving the San Jose Scale (Quadraspidiotus perniciosus) (Comstock, 1881) (Hemiptera: Diaspididae) subjected to successive applications of sulfur powder [8]. Insect resistance reports began to rise in the 1940s with the advent of insecticides and miticides. There are more than 7,740 documented instances of resistance, encompassing 331 chemicals and over 540 species of insect and mite pests [9]. Between 1914 and 2007, the predominant instances of resistance were observed in Lepidoptera, totaling 1,799 confirmed occurrences. Lepidopteran species, including Alabama argillacea (Lepidoptera: Noctuidae), Plutella xylostella (Lepidoptera: Plutellidae), and Tuta absoluta (Lepidoptera: Gelechiidae), have demonstrated resistance to various classes of insecticides. The scientists investigated insect populations across various locales, employing distinct classes of pesticides with diverse modes of action. Research on L. coffeella has only concentrated on the organophosphate category, with no investigations into other chemical groups [13,14]. Consequently, investigations involving diverse populations and many pesticide categories are essential. Most pesticides employed are neurotoxins, and this category poses the greatest challenges regarding insect resistance [9]. Neurotoxic insecticides, such as organophosphates and pyrethroids, induce quick mortality in vulnerable insects, whereas abamectin, neonicotinoids, and diamides result in a more protracted demise of insects. Resistance to a specific active component can be identified by comparing the mortality time of each population against several neurotoxic insecticides. Comparable tests have been conducted with other species, including the mosquito Culex tarsalis (Diptera: Culicidae) [16]. Prolonged fatalities may suggest that the population is starting to develop resistance. Delayed mortality may be likened to the impact of sublethal doses, which induce stress in insects and diminish their metabolism prior to death [17]. One method to assess resistance through lethal time of death (LT) involves gathering geographically disparate groups to acquire more accurate data and facilitate comparisons among regional populations, as resistance is relative. Consequently, by examining the mechanism of action of each pesticide group, one can assess resistance by evaluating the speed at which the insecticides affect a population. Two investigations examine the identification of insecticide resistance in L. coffeella populations, specifically concerning organophosphate insecticides. We propose to examine various groupings and regions. This study sought to identify populations of L. coffeella throughout several regions of Brazil that exhibited resistance to neurotoxic pesticides by analyzing lethal time comparisons.

2. MATERIALS AND METHODS

2.1 Insect populations

This study was conducted at the Laboratory of Integrated Pests Management at Universidade Federal de Viçosa, Rio Paranaíba Campus (UFV-CRP). We selected six municipalities with coffee cultivation of the species C. arabica and C. canephora, located in producing regions of the Brazilian states of Minas Gerais, Espírito Santo, São Paulo, and Pernambuco. These areas were selected because they are the largest coffee producing regions in Brazil. In these regions, we collected leaves from the middle third of plants randomly selected in commercial crops during the 2012-2013 crop season, with active mines (live caterpillars) of L. coffeella. These crops were georeferenced with the help of a portable Garmin E-trex Summit Hc GPS (Figure 1). The leaves collected in each region were transported to the laboratory in separate plastic bags for visual selection of mines that did not present any harm (e.g., open or with signs of parasitism/predation). Selected mined leaves were combined for insect rearing in a greenhouse (20 × 10 m). These leaves were placed in vials with water (25 mL) inside wooden cages covered with organza. The larvae were fed seedlings coffee of Catuaí cultivar grown in a greenhouse without insecticide application. Only larvae with at least one generation in the laboratory were used in bioassays to prevent the expression of insecticide tolerance due to differing environmental conditions at the different sampling sites (i.e., differences without any genetic basis).

Regions City-State	Geographic coordinates	Average temperature (°C)	Precipitation (mm/year)	Altitude (m)
Rio Paranaiba-MG	-19.21,-46.14	21.33	1478	1071
Abaeté dos Mendes-MG	-19.17,-46.09	21.31	1475	849
Carmo do Paranaiba-MG	-19.04,-46.22	21.25	1472	1120
Guaraciaba-MG	-20.34,-43.00	21.11	1611	1120
Franca-SP	-20.58,-47.43	21.05	1644	578
Santa Teresa-ES	-19.93,-40.59	21.54	1162	174
Guaranhuns-PE	-08.89,-36.49	22.05	857	900

Figure 1. Location and characterization of Leucoptera coffeella collection in coffee-producing regions. Dark spots represent coffee-producing regions. White spherical symbols within the dark spots represent collection sites of leafminer populations.

2.2 Insecticides

Six neurotoxic insecticides were selected for bioassays of L. coffeella resistance to the concentrated active ingredients abamectin 18 g l-1 EC (emulsifiable concentrate) (Syngenta, São Paulo, Brazil), chlorantraniliprole 350 g l-1 WG (water-dispersible granules) (DuPont, Paulínia, Brazil), chlorpyrifos 480 g l-1 EC (Fersol, Mairinque, Brazil), deltamethrin 25 g l-1 EC (Bayer SA, São Paulo, Brazil), profenofos 550 g l-1 EC (Syngenta, São Paulo, Brazil), and thiamethoxam 250 g l-1 WG (Syngenta, São Paulo, Brazil) (Table 1).

Table 1. Time and mortality curves (LT50) of Brazilian populations of Leucoptera coffeella under the effect of seven insecticides at the recommended doses.

Insecticide	Population	LT50 [a] (CI95%)[b]	n	RT_{50} [c]	χ^2 [d](df)[e]	pf
Abamectin	Rio Paranaíba-MG	13.29 (11.29–15.34)	40	1.74	2.87 (3)	0.59
	Abaeté dos Mendes-MG	14.70 (12.59–16.53)	40	1.92	6.95 (3)	0.07
	Carmo do Paranaíba-MG	36.75 (33.63–40.97)	40	4.80	4.05 (3)	0.26
	Santa Teresa-ES	9.11 (6.03–11.52)	40	1.19	2.89 (3)	0.59
	Guaranhuns-PE	17.85 (15.87–19.71)	40	2.33	4.97 (3)	0.17
	Franca-SP	12.41 (10.99–15.12)	40	1.62	3.42 (3)	1.12
	Guaraciaba-MG	7.65 (6.85–10.11)	40	1.00	5.63 (3)	3.11
Chlorpyrifos	Rio Paranaíba-MG	8.16 (7.02–9.20)	40	8.08	7.35 (4)	0.12
	Abaeté dos Mendes-MG	17.18 (15.68–18.75)	40	17.01	9.07 (4)	0.06
	Carmo do Paranaíba-MG	16.39 (15.12–17.76)	40	16.23	1.66 (3)	0.65
	Santa Teresa-ES	4.58 (3.62–5.54)	40	4.53	7.56 (5)	0.18
	Guaranhuns-PE	8.59 (6.70–10.21)	40	8.50	2.39 (3)	0.50
	Franca-SP	18.82 (17.54–20.15)	40	18.63	8.20 (4)	0.08
	Guaraciaba-MG	1.01 (0.35–2.07)	40	1.00	6.32 (7)	0.06
Chlorantraniliprole	Rio Paranaíba-MG	27.70 (24.70–31.56)	40	1.98	3.66 (3)	0.30
	Abaeté dos Mendes-MG	26.30 (22.15–34.79)	40	1.88	1.57 (2)	0.54
	Carmo do Paranaíba-MG Santa Teresa-ES	14.01 (11.87–16.47)	40	1.00	7.51 (5)	0.18
	Santa Teresa-ES	31.53 (28.44–35.74)	40	2.25	5.50 (3)	0.14
	Guaranhuns-PE	18.82 (17.54–20.15)	40	3.23	8.20 (4)	0.08
	Franca-SP	14.28 (11.00–18.23)	40	1.02	6.30 (5)	1.22
	Guaraciaba-MG	8.59 (6.70–10.21)	40	1.88	2.39 (3)	0.50
Deltamethrin	Rio Paranaíba-MG	31.12 (27.59–36.20)	40	5.35	4.96 (4)	0.17
	Abaeté dos Mendes-MG	25.73 (23.34–28.56)	40	4.42	3.83 (3)	0.28
	Carmo do Paranaíba-MG	28.18 (24.46–34.29)	40	4.84	2.22 (3)	0.53
	Santa Teresa-ES	5.82 (4.23–7.65)	40	1.00	5.99 (4)	0.07
	Guaranhuns-PE	20.38 (17.53–23.23)	40	3.50	6.04 (3)	0.11
	Franca-SP	18.82 (17.54–20.15)	40	3.23	8.20 (4)	0.08
	Guaraciaba-MG	6.11 (5.03–7.84)	40	1.05	5.81 (4)	0.06
Profenofos	Rio Paranaíba-MG	15.66 (13.96–17.17)	40	522	0.85 (5)	0.66
	Abaeté dos Mendes-MG	12.25 (11.10–13.19)	40	408	1.35 (6)	0.51
	Carmo do Paranaíba-MG	6.96 (4.28–9.00)	40	232	3.85 (5)	0.28
	Santa Teresa-ES	1.96 (0.08–3.00)	40	65.3	3.71 (3)	2.32
	Guaranhuns-PE	10.96 (8.50–11.00)	40	365	1.36 (4)	0.44

	Franca-SP	12.96 (8.24–14.53)	40	432	4.12 (3)	0.21
	Guaraciaba-MG	0.03 (0.01–0.50)	40	1.00	1.58 (3)	0.23
Thiamethoxam	Rio Paranaíba-MG	37.29 (33.32–43.21)	40	4.41	2.54 (3)	0.53
	Abaeté dos Mendes-MG	23.10 (21.11–25.27)	40	2.73	0.43 (3)	0.93
	Carmo do Paranaíba-MG	89.93 (61.70–180.00)	40	10.61	6.54 (4)	0.16
	Santa Teresa-ES	10.49 (9.13–11.78)	40	1.24	8.65 (4)	0.07
	Guaranhuns-PE	13.57 (12.07–14.87)	40	1.61	7.69 (3)	0.06
	Franca-SP	8.45 (7.07–10.95)	40	1.00	5.66 (3)	1.05
	Guaraciaba-MG	9.36 (7.01–10.34)	40	1.11	6.71 (3)	0.06

aLT50 = time (h) lethal to kill 50% of the population.

bCI = confidence interval of 95%.

cRT50 = ratio of lethal time to kill 50% of the population.

dχ2 = chi-square.

edf = degrees of freedom.

fP = probability.

The registered label rates of the respective active ingredients in Brazil were 0.18 mg mL-1 (0.026 mg a.i. mL-1) for abamectin, 0.072 mg mL-1 and 0.078 mg a.i. mL-1 for chlorantraniliprole, 0.05 mg mL-1 (4.800 mg a.i. mL-1) for chlorpyrifos, 0.032 mg mL-1 (0.013 mg a.i. mL-1) for deltamethrin, 0.4 mg mL-1 (1.100 mg a.i. mL-1) for profenofos, and 0.024 mg mL-1 (2.000 mg a.i. mL-1) for thiamethoxam.

2.3 Time-mortality bioassay

For time-mortality analysis, circular discs (diameter 90 mm) of filter paper were dipped into the insecticide solutions diluted in distilled water, using the recommended doses to control L. coffeella. The control used embedded disks with distilled water. The discs containing the insecticides and the water were fixed on a clothesline to dry in the shade and then placed separately into Petri dishes (9.0 × 1.5 cm). Ten larvae of L. coffeella reared in the lab were transferred to each Petri dish using a fine-tipped brush. The Petri dishes with the larvae were kept in the BOD incubator (model SP-500) at 25°C ± 1°C until the time of evaluation. The experiments were conducted in a completely randomized design with four replications.

Preliminary tests using only discs soaked in water were carried out to observe caterpillar mortality over a 48-h period. This was necessary to estimate the maximum evaluation time after bioassay assembly that causes 20% lower mortality in the control [18]. Thus, to have a mortality range from 0% to 100%, evaluations were made at 2, 6, 12, 16, 24, 32 and 48 h (treatments) after bioassay assembly. The time intervals were assessed in independent experimental units, to avoid pseudoreplicates. We considered insects dead when they did not move after being touched with the fine-tipped brush.

2.4 Spatial dependence of insecticide resistance

To determine the spatial dependence of L. coffeella insecticide resistance, the semivariance statistical model of LT50 values to L. coffeella populations for each insecticide and the distance between sampling locations of each population were used. The distance between the sampling sites of each insect population was determined using geographic coordinates with a global positioning system (GPS 12, Garmin International, Olathe, KS). The semivariograms were estimated from the semivariance data of the LTs50 of each population for each insecticide and used as dependent variables in regression analysis, with the distance between the sampling sites as an independent variable. The first inflection point of the semivariogram curve represents the maximum distance of interference between the populations of L. coffeella in relation to susceptibility to a given insecticide.

3. RESULTS AND DISCUSSION

Resistance to neurotoxic insecticides varied generally among the different populations of L. coffeella in Brazil. RT50 varied from 1.02 to 522. Low resistance levels were observed for chlorantraniliprole insecticides (1.02-3.23 times), abamectin (1.19-4.80 times), and deltamethrin (1.05-5.35 times).

On the other hand, intermediate resistance was observed for thiamethoxam (1.11-10.61 times) and chlorpyrifos (4.53-18.63 times), while resistance was high for profenofos (65.3-522 times) (Table 1). Higher levels of organophosphate resistance were observed in Minas Gerais (Abaeté dos Mendes, Rio Paranaíba and Carmo do Paranaíba), Pernambuco (Guaranhuns), and São Paulo (Franca).

The RT50 values are supported by the LT50 values, which were variable among populations and insecticides. The population from Carmo do Paranaíba-MG was noteworthy as it took 89.93 h for 50% of the population to die after contact with the insecticide thiamethoxam. The organo-phosphate and pyrethroid insecticides had lower lethal times. Chlorantraniliprole showed lower LT50 of 8.59 h.

Two canonical axes were significant among the five canonical axes identified, showing linear associations between LT50 of the insecticides with the geographical regions of the population origins of L. coffeella, which showed that the four canonical axes were significant, with the first three axes explaining 90% of the total variance data (Table 1 and Figure 2). The highest absolute values of the canonical coefficients show which insecticides most contributed to the standard deviation of resistance among the different localities. For the first canonical axis of greater importance in the analysis, the insecticides chlorpyrifos, profenofos, and deltamethrin showed positive correlations and higher values of coefficients and thus higher contributions to the differences between the resistant populations (Table 2). Profenofos and deltamethrin, with a positive relationship, contributed to the pattern of divergence on the second axis.

The opposite relationship was observed for assistance with the chlorpyrifos insecticide on the third and fourth axes. On the fifth and sixth axes, a positive relationship was observed between the profenofos insecticide and the standard deviation. It is important to highlight that the new insecticide chlorantraniliprole did not contribute to the resistance of populations (Table 2). Graphs of this analysis done with the first two axes explained 92% of the total variance of the data to show the grouping between locations (Table 2 and Figure 2).

The weight of organophosphate (profenofos and chlorpyrifos) and pyrethroid (deltamethrin) insecticides on the first two axes enhanced the resistance process since they are among the main groups with examples of insect resistance (quotation). Two grouping patterns were observed, with one group for the populations of L. coffeella Rio Parnaíba-MG, Carmo do Paranaíba-MG, and Abaeté dos Mendes-MG and a second group for the populations of Santa Teresa-ES and Guaraciaba-MG, but these patterns did not occur in the other populations (Figure 2).

Table 2. Canonical axes and coefficients (grouped in the canonical structure) of mortalities of Leucoptera coffeella caused by six neurotoxic insecticides.

Variables/mortality	Canonical axes					
	1	2	3	4	5	6
Abamectin	−0.21	−0.10	−0.21	−0.46	−0.38	−0.05
Chlorpyrifos	0.77	−0.65	−0.33	−0.45	−0.10	−0.35
Chlorantraniliprole	−0.10	0.00	−0.21	0.00	0.04	0.00
Deltamethrin	0.51	0.68	0.17	0.05	0.39	0.55
Profenofos	0.64	−0.59	0.38	0.55	0.64	0.64
Thiamethoxam	0.40	0.42	0.11	0.19	0.40	0.38
F	31.02	25.51	20.36	15.02	13.55	9.11
df_xa	68; 181	54; 181	46; 181	32; 181	20; 181	16; 181
$R^2{}_x$b	0.90	0.89	0.78	0.66	0.61	0.53

adfx = degrees of freedom (numerator/denominator).
bR2xb = canonical correlation square.

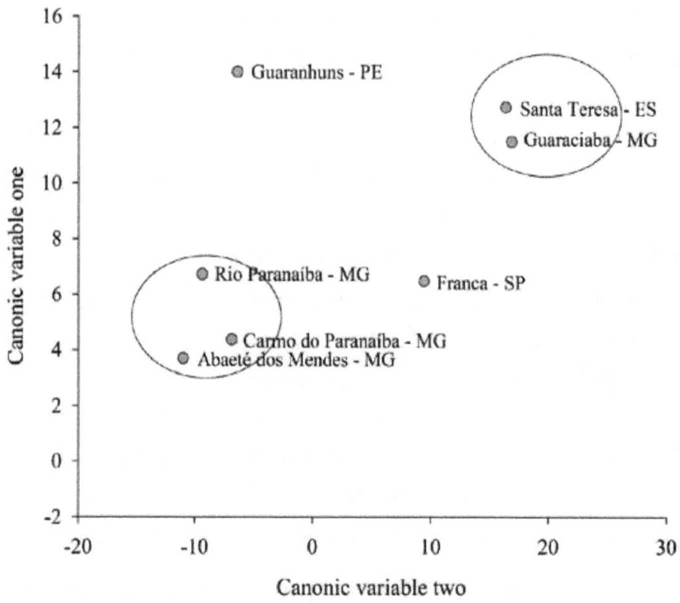

Figure 2. Ordination diagram showing discrimination of insecticide resistance between Brazilian populations of Leucoptera coffeella. Spherical gray symbols are centroids of treatments and represent the average of canonical variable classes. Large circles indicates treatment groups with no significant difference between them (approximate F test, P < 0. 05), based on the Mahalanobis distance (D2) between averages.

The semivariogram models related to the LT50 values of L. coffeella with the distance between the sampling sites obtained for only two insecticides, the organophosphates chlorpyrifos and the pirimiphos. The first inflection points for the models were lengths of 169 and 1,956 km for the insecticides chlorpyrifos and pirimiphos (Figure 3). Therefore, these were the maximum distances between the interference resistance levels of the L. coffeella sampling sites.

Our study reported high variations in the resistance ratio (RT50) of the organophosphates profenofos (522 times) and chlorpyrifos (19 times) compared to the susceptible population of L. coffeella. This large variation represented by RT50 indicates that populations show differences in susceptibility and greater or lesser sensitivity to the enzyme acetylcholinesterase since variations were observed between populations that died the fastest and those that died more slowly.

This shows that this group of insecticides is extremely important in managing resistance because of its intense use, with this group being highly toxic and presenting higher neurotoxic action [19]. Many studies on resistance to the organophosphate insecticide group showed high variation in the mortality of the resistant population compared to other lepidopteran populations [20,21]. Extensive insecticide use in coffee crops and high death speed are among the main factors of resistance [22]. Fragoso et al. [13] observed up to 22 applications of organophosphate insecticides, detecting high levels of resistance when larvae were kept exposed to the discriminating concentration. These concentrations were higher than those tested for profenofos and chlorpyrifos in our study.

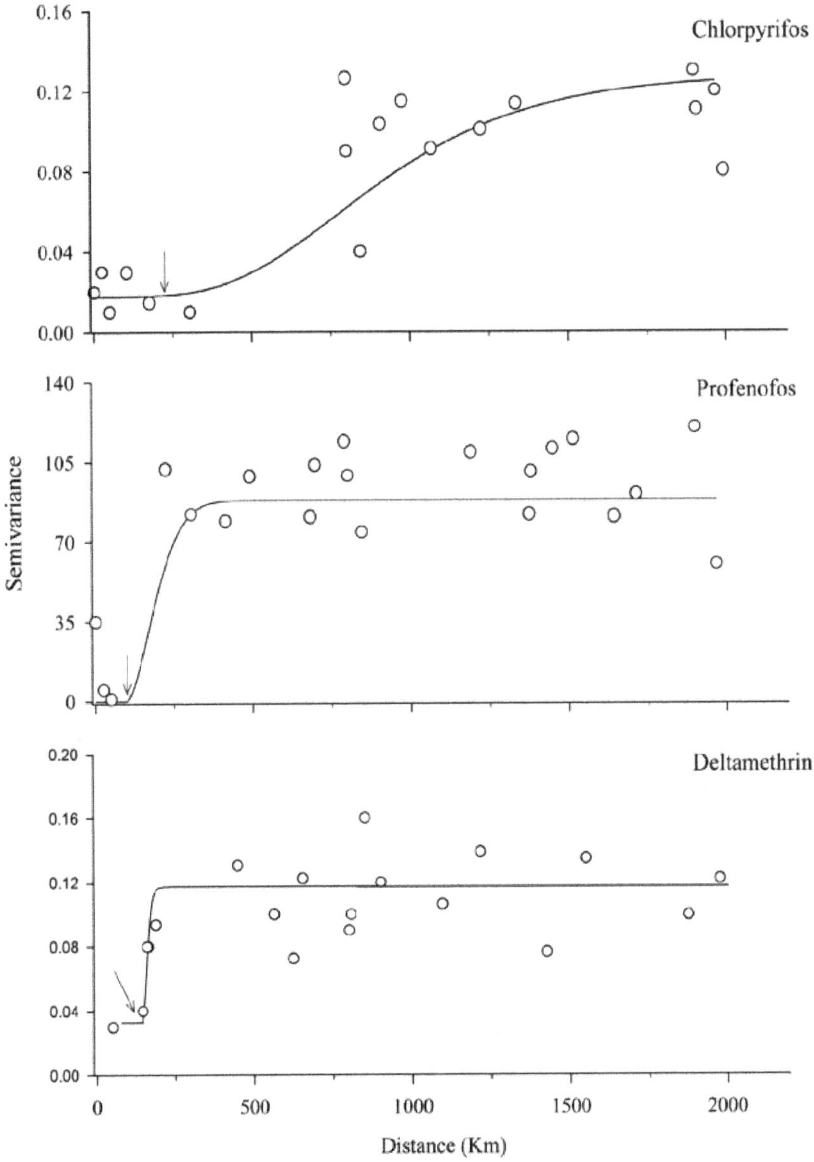

Figure 3. Semivariogram of the LT50 of chlorpyrifos, profenofos, and deltamethrin according to the distance between sampled points from populations of Leucoptera coffeella. The first inflection point of the semivariogram curve, represented by a down ward-pointing arrow, represents the maximum distance of interference of the resistance to the insecticides.

On the other hand, chlorantraniliprole, abamectin, and deltamethrin insecticides showed low levels of RT50 variation. The result with the chlorantraniliprole insecticide was as expected since this insecticide has only recently been commercialized [23-25] and has a highly efficient molecule since low doses of this insecticide (31.5 g a.i. ha-1) cause high mortality to L. coffeella; moreover, it is selective for wasps [26].

Selectivity is an important factor in managing resistance in pest insects [27]. Many studies with basic lines of susceptibility have been done with chlorantraniliprole insecticide and Lepidoptera, and the observations are that populations show susceptibility with low variation in mortality [28,29]. The insecticide abamectin is not considered old and has been effective in controlling this pest insect, with no flaws detected in its control of L. coffeella as of yet. Despite the abamectin insecticide not being among those at risk of resistance in L. coffeella, this insecticide has not been studied. However, many arthropod pests have been classified as being at risk for resistance to this group. Among them are Leptinotarsa decemlineata (Say) [30], Musca domestica [31], P. xylostella [32], Frankliniella occidentalis [33], and Tetranychus urticae [34]. Abamectin resistance has been observed in populations of F. occidentalis [35] and Liriomyza trifolii [36]. Deltamethrin had surprising results, with low discrepancy between the resistant and the susceptible populations (5 times) compared to their insecticides such as thiamethoxam (10 times) that are less used in coffee plantation. In recent years, however, the number of pyrethroid applications in coffee production has been greatly reduced. Despite the low resistance to pyrethroids, however, the variation has been observed in Brazil for the moth P. xylostella [37] as well as with other pyrethroids (cypermethrin, β-cypermethrin, deltamethrin, and esfenvalerate) in Pakistan, India, China, and Korea [38,39]. Although deltamethrin has affected fewer Brazilian populations of L. coffeella, a difference of 5 times is cause for concern since it should have been more effective.

Insects usually have a resistance mechanism that confers nerve insensitivity, known as knockdown resistance (Kdr), as first reported in M. domestica (L.) (Diptera: Muscidae) [40]. This type of resistance is found in other agricultural pests based on patterns of cross-resistance and the absence of compound synergism that inhibits the activity of cytochrome P450 and esterase enzymes [41].

The insecticide thiamethoxam has been frequently used and can be applied as a spray or via the soil [42]. There are no studies of lepidopteran resistance to this insecticide. Control failures were observed depending on the time of application, however, for example [43] observed effectiveness of 4.1%, 50.6%, 62.1%, and 69.0%.

The grouping of populations from Rio Paranaíba, Carmo do Paranaíba, and Abaeté (Group I) and Santa Teresa with Guaraciaba (Group II), coupled with the significant response of the effect of distance on the LT50 of the chlorpyrifos, profenofos, and deltamethrin insecticides, showed that resistance was affected by the collection distance of these populations since more closely connected populations had similar resistance responses.

Studies have shown a strong relationship between collection distance and resistance patterns [44,10,45,12]. All of these studies showed significant association of resistance with distance, and nearby populations tended to show more similar responses, as is the case for P. xylostella (L.) (Lepidoptera: Plutellidae). Chen et al. [46] studied the resistance of pyrethroids to Culex pipiens (Diptera: Culicidae) and found different frequencies of resistance at different locations, ranging from 21.4% to 79.8%. Moreover, this type of response may be associated with the large dispersal capacity of adult L. coffeella and the sampling characteristics.

Adults of L. coffeella disperse easily between coffee crops and have different densities in different environments [47-49]. Moreover, there is a geographic corridor between the largest-producing Brazilian states (Figure 1). Isaaks and Srivastava [50] determined that to identify disparities in geostatistical analyses of geographical distribution, it was essential to gather samples from both proximal and distal locations.

We find that Brazilian populations of L. coffeella exhibited enhanced resistance to organophosphates. Moreover, resistance may correlate with the proximity of producing regions, and local selection influenced by dispersal appears crucial for the development of pesticide resistance in Brazilian populations of L. coffeella, warranting consideration in the formulation of pest management strategies. Insecticides that do not induce death in L. coffeella should be used under specific conditions, and a diverse array of insecticides (beyond the cross-resistance and multiple-resistance spectra) should be utilized in rotation to mitigate the risk of resistance evolution..

REFERENCES

1. Lomeli-Flores RJ, Barrera JF, Bernal JS. Impacts of weather, shade cover and elevation on coffee leafminer Leucoptera coffeella (Lepidoptera: Lyonetiidae) population dynamics and natural enemies. Crop Protection. 2010; 29: 1039-1048. DOI: 10.1016/ j.cropro.2010.03.007

2. Righi CA, Campoe OC, Bernardes MS, Lunz AMP, Piedade SMS, Pereira CR. Influence of rubber trees on leaf-miner damage to coffee plants in an agroforestry system. Agroforestry Systems. 2013; 87: 1351-1362. DOI: 10.1007/ s10457-013-9642-9.

3. Ramiro DA, Guerreiro-Filho O, Queiroz-Voltan RB, Matthiesen SC. Caracterização anatômica de folhas de cafeeiros resistentes e suscetíveis ao bicho-mineiro. Bragantia. 2004; 63: 363-372, DOI: 10.1590/S0006-87052004000300006

4. Vega FE, Posada F, Infante F. Coffee insect: ecology and control. In: Pimentel D, editor. Encyclopedia of Pest Management. 1st ed. London: Taylor & Francis; 2007. p. 1-4. DOI: 10.1081/E-EPM-120042132

5. Michereff MFF, Michereff-Filho M, Vilela EF. Comportamento de acasalamento do bicho-mineiro-do-cafeeiro, Leucoptera coffeella (Guérin-Mèneville) (Lepidoptera: Lyonetiidae). Neotropical Entomology. 2007; 36: 376-382. DOI: 10.1590/ S1519-566X2007000300005

6. MAPA-Ministério da Agricultura, Pecuária e Abastecimento [Internet]. 2014. Available from: http://extranet.agricultura.gov.br/agrofit_cons/principal_agrofit_cons. Accessed: 2014-12-22.

7. Fragoso DB, Guedes RNC, Rezende ST. Glutathione S-transferase detoxification as a potential pyrethroid resistance mechanism in the weevil, Sithophilus zeamais. Entomologia Experimentalis et Applicata. 2003; 109: 21-29. DOI: 10.1046/j. 1570-7458.2003.00085.x

8. Messing R, Croft BA. Insecticide in pest management. In: Metcalf RL, Luckman WH, editors. Introduction to Insect Pest Management. 1st ed. New York: John Wiley & Sons; 1994. p. 217-277.

9. Whalon M, Mota-Sanchez D, Hollingworth RM. Global Pesticide Resistance in Arthropods. 7th ed. London: CABI; 2009. 192 p. DOI: 10.1079/9781845933531.0000

10. Silva TBM, Siqueira HAA, Oliveira AC, Torres, JB, Oliveira JV, Montarroyos PAV, Farias M. Insecticide resistance in Brazilian populations of the cotton leaf worm, Alabama argillacea. Crop Protection. 2011; 30: 1156-1161. DOI: 10.1016/j. cropro.2011.05.022

11. Zago HB, Siqueira HAA, Pereira EJG, Picanço MC, Barros R. Resistance and behavioral response of Plutella xylostella (Lepidoptera: Plutellidae) populations to Bacillus thuringiensis formulations. Pest Management Science. 2014; 70: 488-495. DOI: 10.1002/ps.3600

12. Gontijo PC, Picanço MC, Pereira EJG, Martins JC, Chediak M, Guedes RNC. Spatial and temporal variation in the control failure likelihood of the tomato leafminer, Tuta absoluta. Annals of Applied Biology. 2012; 162: 50-59. DOI: 10.1111/aab.12000

13. Fragoso DB, Guedes RNC, Picanço MC, Zambolim L. Insecticide use and organophosphate resistance in the coffee leafminer Leucoptera coffeella (Lepidoptera: Lyonetidae). Bulletin of Entomological Research. 2002; 92: 203-212. DOI: 10.1079/BER2002156

14. Ribeiro BM, Guedes RNC, Oliveira EE, Santos JP. Insecticide resistance and synergism in Brazilian populations of Sitophilus zeamais (Coleoptera:

Curculionidae). Journal of Stored Products Research. 2003; 39: 21-31. DOI: 10.1016/S0022-474X(02)00014-0

15. Nauen R. Insecticide mode of action: return of the ryanodine receptor. Pest Management Science. 2006; 62: 690-692. DOI: 10.1002/ps.1254
16. Strong AC, Kondratieff BC, Doyle MS, Black WC. Resistance to permethrin in Culex tarsalis in northeastern Colorado. Journal of the American Mosquito Control Association. 2008; 24: 281-288. DOI: 10.2987/5593.1
17. Piiroinen S, Lyytinen A, Lindström L. Stress for success? Temperature stress of preceding generations modifies the response to insecticide stress in an invasive pest insect. Evolutionary Applications. 2013; 6: 313-323. DOI: 10.1111/eva.12001
18. Szendrei Z, Grafius E, Byrne A, Ziegler A. Resistance to neonicotinoid insecticides in field populations of the Colorado potato beetle (Coleoptera: Chrysomelidae). Pest Management Science. 2012; 68: 941-946. DOI: 10.1002/ps.3258
19. Baron RL. Delayed neurotoxicity and other consequences of organophosphate esters. Annual Review of Entomology. 1981; 26: 29-48. DOI: 10.1146/annurev.en. 26.010181.000333
20. Zibaee A, Sendi JJ, Ghadamyari M, Alinia F, Etebari K. Diazinon resistance in different selected strains of Chilo suppressalis (Lepidoptera: Crambidae) in northern Iran. Journal of Economic Entomology. 2009; 102: 1189-1196. DOI: 10.1603/029.102.0343
21. Saeed Q, Saleem MA, Ahmad M. Toxicity of some commonly used synthetic insecticides against Spodoptera exigua (Fab) (Lepidoptera: Noctuidae). Pakistan Journal of Zoology. 2012; 44: 1197-1201.
22. Alves PMP, Lima JOG, Oliveira LM. Monitoramento da resistência do bicho-mineiro- do-café, Leucoptera coffeella (Lepidoptera: Lyonetiidae), a inseticidas em Minas Gerais. Anais da Sociedade Entomológica do Brasil. 1992; 21: 77-91.
23. Jeanguenat A. The story of a new insecticidal chemistry class: the diamides. Pest Management Science. 2013; 69: 7-14. DOI: 10.1002/ps.3406
24. Sattelle DB, Cordova D, Cheek TR. Insect ryanodine receptors: molecular targets for novel pest control chemicals. Invertebrate Neuroscience. 2008; 8: 107-119. DOI: 10.1007/s10158-008-0076-4
25. Lahm GP, Cordova D, Barry JD. New and selective ryanodine receptor activators for insect control. Bioorganic & Medicinal Chemistry. 2009; 17: 4127-4133. DOI: doi: 10.1016/j.bmc.2009.01.018
26. Fernandes FL, Silva PR, Gorri JER, Pucci LF, Silva IW. Selectivity of old and new insecticides and behaviour of Vespidae predators in coffee crop. Sociobiology. 2014; 60: 471-476. DOI: 10.13102/sociobiology.v60i4.471-476

27. Casida JE, Durkin KA. Neuroactive insecticides: targets, selectivity, resistance, and secondary effects. Annual Review of Entomology. 2013; 58: 99-117. DOI: 10.1146/ annurev-ento-120811-153645

28. Silva EJ, Siqueira HAA, Silva TBM, Campos RM. Baseline susceptibility to chlorantraniliprole of Brazilian populations of Plutella xylostella. Crop Protection. 2012; 35: 97-101. DOI: 10.1016/j.cropro.2012.01.013

29. Gao C, Yao R, Zhang Z, Wu M, Zhang S, Su J. Susceptibility baseline and chlorantraniliprole resistance monitoring in Chilo suppressalis (Lepidoptera: Pyralidae). Journal of Economic Entomology 106: 2190-2194. DOI: 10.1603/ EC13058

30. Argentine JA, Clark JM. Selection for abamectin resistance in Colorado potato beetle (Coleoptera: Chrysonelidae). Pesticide Science. 1990; 28: 17-24. DOI: 10.1002/ps. 2780280104

31. Scott JG, Roush RT, Liu N. Selection of high-level abamectin resistance from field- collected house flies, Musca domestica. Experientia. 1991; 47: 288-291. DOI: 10.1007/ BF01958163

32. Liang P, Gao X, Zheng B. Genetic basis of resistance and studies on cross-resistance in a population of diamondback moth, Plutella xylostella (Lepidoptera: Plutellidae).Pest Management Science. 2003; 59: 1232-1236. DOI: 10.1002/ps.760

33. Chen X, Yuan L, Du Y, Zhang Y, Wang J. Cross-resistance and biochemical mechanisms of abamectin resistance in the western flower thrips, Frankliniella occidentalis. Pesticide Biochemistry and Physiology. 2011; 101: 34-38. DOI: 10.1016/j.pestbp. 2011.07.001

34. Kwon DH, Seong GM, Kang TJ, Lee SH. Multiple resistance mechanisms to abamectin in the two spotted spider mite. Journal of Asia-Pacific Entomology. 2010; 13: 229-232. DOI: 10.1016/j.aspen.2010.02.002

35. Immaraju JA, Paine TD, Bethke JA, Robb KL, Newman JP. Western flower thrips (Thysanoptera: Thripidae) resistance to insecticides in coastal California greenhouse. Journal of Economic Entomology. 1992; 85: 9-14. DOI: 10.1093/ jee/85.1.9

36. Ferguson SJ. Development and stability of insecticide resistance in the leafminer Liriomyza trifolii (Diptera: Agromyzidae) to cyromazine, abamectin and spinosad. Journal of Economic Entomology. 2004; 97: 112-119. DOI: 10.1093/ jee/97.1.112

37. Oliveira AC, Siqueira HA, Oliveira JV, Silva JE, Michereff-Filho M. Resistance of Brazilian diamondback moth populations to insecticides. Scientia Agricola. 2011; 68: 154-159. DOI: 10.1590/S0103-90162011000200004

38. Kwon DH, Choi BR, Park HM, Lee SH, Miyata T, Clark JM, Lee SH. Knockdown resistance allele frequency in field populations of Plutella xylostella in Korea. Pesticide Biochemistry and Physiology. 2004; 80: 21-30. DOI: 10.1016/j.pestbp.2004.06.001

39. Balasubramani V, Sayyed AH, Crickmore N. Genetic characterization of resistance to deltamethrin in Plutella xylostella (Lepidoptera: Plutellidae) from India. Journal of Economic Entomology. 2008; 101: 1911-1918. DOI: 10.1603/0022-0493-101.6.1911

40. Busvine JR. Mechanism of resistance to insecticide in houseflies. Nature. 1951; 168: 193-195. DOI: 10.1038/168193a0

41. Soderlund DM. Molecular mechanisms of insecticide resistance. In: Sjut V, editor, Molecular Mechanisms of Resistance to Agrochemicals. Berlin: Springer; 1997. p. 21-56. DOI: 10.1007/978-1-4684-6429-0_4

42. Martins LD, Rodrigues WN, Tomaz MA, De Souza AF, De Jesus Junior WC. Função de crescimento vegetativo de mudas de cafeeiro conilon a níveis de ciproconazol+tiametoxam e nitrogênio. Revista de Ciências Agrárias. 2012; 35: 173-183.

43. Jones AK, Raymond-Delpech V, Thany SH, Gauthier M, Sattelle DB. The nicotinic acetylcholine receptor gene family of the honey bee, Apis mellifera. Genome Research. 2006; 16: 1422-1430. DOI: 10.1101/gr.4549206

44. Scott JG, Alefantis TG, Kaufman PE, Rutz DA. Insecticide resistance in house flies from caged-layer poultry facilities. Pest Management Science. 2000; 56: 147-153. DOI: 10.1007/s00436-009-1425-x

45. Shah R, Worner SP, Chapman RB. Determination of the influence of dispersion pattern of pesticide-resistant individuals on the reliability of resistance estimates using different sampling plans. Bulletin of Entomological Research. 2012; 102: 531-538. DOI: 10.1017/S0007485312000065

46. Chen L, Zhong D, Zhang D, Shi L, Zhou G, Gong M, Zhou H, Sun Y, Ma L, He J, Hong S, Zhou D, Xiong C, Yan G. Molecular ecology of pyrethroid knockdown resistance in Culex pipiens pallens mosquitoes. PloS One. 2010; 5: 1-9. DOI: 10.1371/journal.pone.0011681

47. Bacca T, Lima ER, Picanço MC, Guedes RN, Viana JHM. Optimum spacing of pheromone traps for monitoring the coffee leafminer Leucoptera coffeella. Entomoloia Experimentalis et Applicata. 2006; 119: 39-45. DOI: 10.1111/j.1570-7458.2006.00389.x

48. Bacca T, Lima ER, Picanço MC, Guedes RN, Viana JHM. Sampling plan for the coffee leafminer Leucoptera coffeella with sex pheromone traps. Journal of Applied Entomology. 2008; 132: 430-438. DOI: 10.1111/j.1439-0418.2007.01264.x

49. Fernandes FL, Mantovani, EC, Neto HB, Nunes VV. Effects of irrigation, environmental variability and predatory wasp on Leucoptera coffeella (Guerin-Meneville) (Lepidoptera: Lyonetiidae), in coffee plants. Neotropical Entomology. 2009. 38: 410-417. DOI: 10.1590/S1519-566X2009000300018

50. Isaaks EH, Srivastava RM. An Introduction to Applied Geostatistics. 1st ed. Oxford: University Press; 1989. 592 p.

Chapter 2

THE FUNCTION OF ACHE IN THE RESISTANCE OF COLORADO POTATO BEETLE (LEPTINOTARSA DECEMLINEATA SAY) TO CARBAMATES AND ORGANOPHOSPHATES

1. INTRODUCTION: AN OVERVIEW

Colorado potato beetle – CPB (Leptinotarsa decemlineata) The Coleoptera: Chrysomelidae is an oligophagous species that infests many cultivated and wild plants within the Solanaceae family across North America, Europe, and certain regions of Asia [1, 2, 3]. The Colorado potato beetle, belonging to the genus Leptinotarsa, exhibits the most extensive host range, consuming at least 10 varieties of both wild and farmed Solanum. Defoliation of potatoes, particularly pronounced during the flowering phase, results in significant reductions in tuber yield, contingent upon the growth stage. Botanically, the potato, Solanum tuberosum L., is a perennial species; nevertheless, it is commonly cultivated as an annual due to its vegetative propagation strategy. The potato is a cultivated plant native to the Western Hemisphere. It initially emerged in Europe in 1573 in Spain, and by 1586, it had arrived in England and Ireland. The potato was introduced to the Western Balkans in 1759, specifically in the Banat region. It is the third-largest crop in our nation by cultivated area, following maize and wheat. Nonetheless, the overall acreage devoted to potato cultivation is diminishing, reflecting a prevailing trend in Europe [6]. The potato is vulnerable to pests during all phases of its growth, with the Colorado potato beetle (CPB) being the most significant threat. This pest is also present on tomatoes, aubergines, and peppers, as well as buttercups, within the Balkan region's spontaneous flora. The Colorado potato beetle is a constraint on potato production both locally and globally, and in certain areas, it also poses a significant threat to tomatoes and eggplants. Under our conditions, CPB generates two generations annually (along with an incomplete third generation), hence necessitating 3-4 control treatments each year to manage substantial populations.

Despite the existence of alternate approaches [7], traditional pesticides remain crucial in the management of CPB. Adults can hibernate in the soil at a depth of 7.5–12.5 cm. Upon emerging from the earth in spring, they disseminate by traversing and flying to neighboring fields, where they promptly commence feeding on host plants. Females deposit eggs in clusters of 20–60 on the undersides of leaves after a period of 5–10 days. Under laboratory conditions, females have a lifespan of 120 days, with a maximum fecundity exceeding 4,000 eggs [1]. The eggs hatch concurrently, and the newly emerged larvae promptly commence feeding. Larval development, comprising four stages, endures for 10 to 20 days, contingent upon temperature variations. Feed consumption is contingent upon plant hosts, being essentially constant across all larval stages for a single plant host. Defoliation of potatoes, particularly pronounced during the flowering phase, results in significant reductions in tuber yields [8, 9, 5]. Significantly reduced yields were attributable to substantial defoliation (20%) occurring weeks prior to harvest [10]. Upon cessation of feeding, the fourth instar larvae descend from the plant, excavate into the soil, and undergo pupation. Imagoes eclose from the ground, locate the nearest plant host, and commence feeding. Imagoes mate and generate a new generation of beetles, migrate to different fields, or cease eating and enter diapause, contingent upon several parameters such as temperature, photoperiod, and the condition of the plant host [1]. The exceptional capacity of CPB to adapt to unfavorable environmental conditions is seen in its pesticide resistance, leading to heightened potato production costs, environmental contamination, and disruption of biocenotic equilibrium. A comprehensive array of resistance mechanisms, including reduced permeability, increased pesticide metabolism, alterations in target-site sensitivity, and behavioral modifications, renders the Colorado potato beetle (CPB) exceedingly challenging to control [11]. It has acquired resistance to all categories of insecticides, significantly reducing the duration of efficacy for new pesticides. Numerous writers worldwide have explored the resistance of CPB to toxicological, biochemical, and genetic approaches. An effectively structured CPB management program can successfully safeguard crops and prolong the duration of insecticide application. Resistance monitoring is a fundamental component of integrated pest management. [6, 12].

2. MANAGEMENT OF COLORADO POTATO BEETLE

Rapid adjustment to new biotic and abiotic conditions is highly present within CPB populations. As a result, it is considered to be the most serious defoliator pest of potatoes in the world [11]. Nontreated pest populations can defoliate and completely destroy yields in the period of tuber formation [1]. Before CPB, we did not have a pest able to seriously harm potato crops. After coming to a new habitat, this pest finds conditions favorable for life and reproduction, which enables them to occur permanently and at a massive scale. The destructive activity of this pest is such that potato crops would be destroyed without efficient protection [13].

In most regions with dense CPB populations, chemical treatments are still the number one method in potato protection, considering that alternative measures are not efficient enough [11]. Most potato protection programs require decreasing CPB populations at the beginning and middle of the vegetation period, and tolerating higher populations at the end of the vegetation period. In general, the goal of such programs is to limit total defoliation to 10–25% during the most critical periods of potato development. Without chemical treatments, yield losses would amount to 74% [6].

In Balkan region, potatoes are treated with insecticides 3–4 times per year to ensure normal yields [14]. CPB management costs potato growers hundreds of million dollars a year [2]. Large numbers of pesticides from different chemical classes have been used so far to manage CPB, such as insecticides from the class of chlorinated hydrocarbons, organophosphates, carbamates, pyrethroids, nereistoxins, insect growth regulators, neonicotinoids, as well as bacterial products, GMOs, products of plant origin (azadirachtin), etc. Over the last few decades, only a few insecticides in our country have been satisfactorily efficient [15, 16, 17, 18].

3. BRIEF OVERVIEW OF CPB RESISTANCE

CPB has an amazing ability to adjust to toxicants from different chemical classes [19], by developing a range of different resistance levels to all classes of insecticides applied to manage this pest [1, 20, 21, 22, 23, 24]

The use of DDT and other organochlorine insecticides was efficient after World War II. The first information on DDT resistance dated back to early 1950s, specifically to 1952, reported by Quinton [19] and Hoffmaster and Waterfield [25]. Since then, CPB has developed resistance to a wide range of insecticides, including arsenic compounds, organochlorine compounds, carbamates, organophosphates, and pyrethroids [1, 19] and more recently to neonicotinoids [13, 23, 26, 27, 28, 29]. Experimental proofs on development of resistance to Bacillus thuringiensis products and to transgenic plants, were also obtained [19, 30, 31]

In Balkan region, a significant level of CPB resistance was detected in 1967 to insecticides from the class of chlorinated hydrocarbons [32], which was proved for most localities of ex- Yugoslavia [33], where the resistance to organophosphorus insecticides and carbamates was detected in some CPB populations. Studying CPB resistance was continued in the years to follow [34, 35]. Remarkably high levels of resistance of the fourth instar larvae to quinalphos and carbaryl were recorded [36, 37]. Research on insecticide resistance level of CPB to most commonly used insecticides is ongoing [6, 17, 18, 38].

The rate of resistance development increases progressively with the introduction of new, synthetic insecticides. When it comes to pyrethroids, this resistance occurred 2–4 years after pyrethroids were put into practice and widely used. Physiological and genetic mechanisms of CPB resistance have been little studied. It is expected that in the future CPB will develop resistance to all newly introduced insecticides [1].

4. INSECT RESISTANCE TO INSECTICIDES

According to IRAC (Insecticide Research Action Committee), resistance may be defined as "a heritable change in the sensitivity of a pest population that is reflected in the repeated failure of a product to achieve the expected level of control when used according to the label recommendation for that pest species."

Cross-resistance may be described as an ability to simultaneously develop tolerance to substances they have never been exposed to (resistance to one insecticide confers resistance to another, newly introduced insecticide). Due to large populations and numerous descendants (they breed quickly), there is always a risk that insecticide selection pressure will ultimately result in insecticide resistance. Insecticide resistance may evolve rapidly, especially when field application of insecticides is misused or overused.

The resistance occurs with behavior change, when the insects avoid contacts with the toxicant [39], or physiological change [40], where they survive toxicant exposure. Most important mechanisms of resistance are reduced cuticle penetration, increased excretion, increased metabolic detoxification and altered target-site sensitivity. Knowing molecular basis for emergence and development of resistance is very important for developing appropriate measures and strategies to slow down the resistance [41].

There is never only one cause of resistance. Hence, regardless of the type, the cause of resistance to different pesticides can vary from substance to substance.

According to Hassall [42], factors that lead to resistance are:

1) morphological, 2) physiological and biochemical, and 3) behavioral.

4.1 Morphological changes

Such changes lead to a reduction of the amount of a pesticide that comes into an insect's body in a unit of time, compared to the amount that comes into the body of a susceptible insect. The change in lipid content of the insect's cuticle can result in reduced absorption, as a factor for overall resistance development. This so-called penetration resistance is quite significant since it often goes along with other resistance mechanisms, enhancing their effect. In its simplest form, this can be explained by the fact that different speed of intake is a factor that, combined with excretion, provides a high level of resistance, even during usual enzyme activity [43].

4.2 Metabolic changes

Populations of resistant insects may detoxify or destroy the toxin faster than susceptible insects, or quickly rid their bodies of the toxic molecules. This type of resistance is the common mechanism and often presents the greatest challenge. Resistant populations may possess higher levels or more efficient forms of these enzymes. In addition to being more efficient, these enzyme systems also may have a broad spectrum of activity (i.e., they can degrade many different insecticides).

Changes in insect metabolism could be manifested as:

- Change in metabolic rate of a certain insecticide. Resistant strains increase the amounts of enzymes or their activity (a wide range of impact). These changes come as a result of the organism's reaction to stimulants. When it comes to transcription control, the metabolism is reformulated in order to increase the level of transcription, which leads to better supply of enzymes (gene amplification, P450 for example).
- Loss of sensitivity of the target-site, receptors, or enzymes as a responsive reaction to a stimulant, or a change in transcription control, i.e., change in encoding gene expression (in structural genes, direct change occurs in the structure of the enzymes, i.e., there are different forms of P450, ALiE, GST, AChE) [44].

4.2.1 Gene amplification

Gene amplification is the multiple copying of structural genes that manage the synthesis of enzymes, thus ensuring hundreds of copies of structural genes. Increased detoxification can be a result of better supply of enzymes, and this quantitative increase is caused by gene amplification [45]. Gene amplification has been determined as a key factor for increased esterase production or change in sensitivity of AChE and Na+ channels (the target-site of 90% of insecticides), but not GST [46].

In the first stage, the metabolism of insecticides is manifested through many reactions, most important of which are oxidation, reduction, and hydrolysis. In the second stage, conjugates are formed, which are practically nontoxic. Selective toxicity of insecticides mostly comes from the balance of the reactions included in activation and detoxification [6].

4.2.1.1 Oxidative processes

Oxidative processes have a dominant role in the metabolism of insecticides present in living organisms. These reactions are catalyzed by enzymes of multifunctional oxidases (MFO) or mono-oxygenises, located in the endoplasmic reticulum, i.e., in microsoms. Cyt P450 is the active center of MFO. MFO mechanism has evolved due to the need of living organisms to protect themselves from many natural toxicants they are constantly exposed to [47, 48]. It is clear that several forms of Cyt P450 exist both in resistant and susceptible strains, while there is a quality difference in different strains [49].

MFO activity of CPB (epoxidation, N- and O-demethylation) is 2–3 times higher in resistant species than in susceptible species. Resistant species have two different types of mfo. Type 1 mfo provides resistance to permethrin, and weak cross-resistance to azinphosmethyl and carbofuran. Type 2 provides resistance azinphosmethyl and carbofuran, but not to permethrin [50]. This mechanism is comprised in the resistance of CPB larvae and adults to imidaclopride [29].

4.2.1.2 Hydrolytic processes

Insecticide detoxification primarily unfolds through molecule hydrolysis on different sites, thereby breaking ester, carboxyl-ester, amide, and other chemical bonds. Pyrethrins, pyrethroids, organophosphates, carbamates, and other insecticides are degraded by hydrolysis. This is the basis for the selective effect of insecticides and for insects' resistance mechanisms. The most important hydrolytic enzymes are phosphoric triesters and carboxylesterases (ALiE esterases, nonspecific or B-esterases) [50].

Esterase-related insect resistance is based on the following:

- Increase in the total amount of esterase – by altering regulatory genes or regulatory loci combined with structural genes, which results in change in enzyme synthesis in the organism [51] or amplification of genes responsible for DNA methylation.
- Change in their activity – by altering structural genes that directly determine the nature of enzymes.

The impact of nonspecific esterase on the level of resistance to carbamates has not been confirmed [52], which was also [36] indicated in the case of CPB. The role of esterase in CPB resistance was confirmed [29, 36, 53].

4.2.1.3 Conjugation processes

Forming of conjugates almost always implies detoxification, but sometimes there are some cases of toxicant reactivation. The most important conjugation reactions are: glutathione conjugation, glucoside or glucuronide conjugation and amino acids conjugation. A change in GST activities depends on modifying a series of enzymes, rarely on only one enzyme, as in esterase. In this class of enzymes, there is no proof that enzyme amplification and resistance are related [54]. When determining the amount of GST-metabolite of azinphosmethyl in resistant CPB strains, no direct impacts of GST to resistance were found. The authors think that the share of GST in total resistance is manifested through the transformation of toxic oxidative metabolites of azinphosmethyl, whose levels are higher in resistant strains [46].

4.2.2 Change in target-site sensitivity

There are four basic groups of macromolecules, depending on the neurotoxic insecticide target- site:

Target-site	Active ingredient
Acetylcholinesterase (AChE)	Organophosphates, carbamates
Na+ channels	Pyrethroids, DDT
GABA receptors	Lindane, cyclodienes, fipronil
ACh receptors	Neonikotinoides, nereistoxins

4.2.2.1 Acetylcholinesterase (AChE)

AChE is a target-site for organophosphate and carbamate insecticides. The structure of AChE has undergone some changes that resulted in different levels of transformation of differently structured AChEs. Modified forms of AChE differ among species. As a result, many different forms of cross-resistance are possible [42]). It is important to determine kinetic constants, especially Michaelis constant (Km). It is a constant that measures the enzyme's affinity toward the substrate (ACh, butyrylcholine, and ATCh). During the 1980s and 1990s, some authors [42, 55, 56, 57] indicated that altered AChE causes the resistance to carbamates and organo-phosphate compounds.

Studies on resistance to organophosphates and carbamates have shown that AChE activity of CPB is quite pronounced and easily measured. The AChE activity of the fourth instar larvae was determined by measuring the absorption, at 585µm wavelength. Total AChE activity was correlated with the determined resistance to carbamate insecticides [36].

4.2.2.2 Na+ channels – Sodium channels

In some cases where resistance cannot be explained by other causes, one can assume there has been a modification in the target-site structure. The exception is the resistance of flying insects that can lead to diminishing of the knockdown effect. In houseflies, this property is carried by the kdr gene and it can be associated with the alternation of receptors in the nerve cell membrane. Pyrethroids can predominantly affect synaptic sites, which are less sensitive in resistant housefly strains. There is some evidence that the term "change in target-site sensitivity" was coined to explain the kdr resistance factor [58, 59, 60]. The target site inactivity of motor nerves' ends to permethrin and deltamethrin is also proved. The insensitivity of binding- site of resistant strains can be a result of multiple insecticide receptors.

It is also possible that weakened binding can be a result of structural changes in proteins or changes in the structure of lipids adjacent to ion pumps in kdr-resistant strains, which can cause minor problems in the mechanism of the ion pump or diminish the process of repetitive polarization, typical for unchanged receptors. It is not clear whether the changes detected in lipid structures of neural membranes of kdr and super-kdr strains of houseflies are a factor that reduces the sensitivity of Na+ channels or these are just compensation changes, necessary for normal functioning of modified Na+ channels [56, 61], as a cause of pyrethroid resistance, which point out changes in the target-site (modification of Na+ channels), detoxification increased by oxidation, hydrolysis, and specific proteins.

4.2.2.3 GABA receptors

By using subcellular products of the neural tissue of insects, several studies have shown that cyclodienes and lindane have a neurotoxic impact by blocking the GABA receptor complex. Studies on brain tissues of cockroaches showed that resistant strains had 90% lower sensitivity of GABA receptors to cyclodienes. It was found that mutations of Rdl-genes that encode the GABAa receptor subunit caused the resistance of Hypothenemus hampei Ferrari to endosulphan [62].

4.2.2.4 ACh receptors

Acetylcholine (ACh) is a neurotransmitter that regulates a large number of vital functions. Its activity is enabled by two types of postsynaptic ACh receptors (muscarinic and nicotinic – nAChR). Nicotinic receptors mainly act as ACh activity modulators [63, 64]. The activity of insecticides is manifested through nAChR activation or blocking. The basic structure of nAChR consists of five protein subunits, mostly two identical alpha-subunits and three beta-subunits that give it a pentagonal shape. So far, scientists have detected ten different nAChR genes in insects. The number of nAChR genes implies there are much more nAChR protein subunits, whose main role is recognition when binding the receptor on one side and ACh or insecticide on the other side [65].

Every modification in the protein structure, even the smallest one, can reduce the affinity of nAChR. This mechanism can be a reason for reduced CPB sensitivity to imidacloprid when oxidative and hydrolytic enzymes are blocked [29].

4.3 Behavioral changes

This type of resistance implies the evolution of behavior, manifested in reduced exposure to toxic compounds or in the insect's ability to survive in toxic or some other kind of fatal environment. Flying insects can acquire the instinct not to dwell long on contaminated surfaces [42]. This resistance mechanism has been recorded for many insecticide classes [66]. Insects simply stop feeding or leave the treated surface.

In CPB, where management is increasingly based on growing transgenic potatoes that contain Bacillus thurigiensis δ - endotoxin, the correlation between these two resistance mechanisms is very significant. More pronounced physiological resistance was recorded [67] in the larvae that avoided transgenic potato crops [31] grown in the same field with nontransgenic potato crops.

4.4 Cross-resistance and multiresistance

Insect populations exposed to certain insecticides have an ability to simultaneously develop tolerance to other substances they have never been exposed to. Insecticide resistance is model of rapid evolution within populations and typical example of directional selection. Eradication of susceptible genotypes from field populations increases both the frequency of resistant genes, which became dominant especially in the absence of susceptible insect refugees, and the application dose of insecticide, needed to keep pest below economically damaging levels. Same physiological or biochemical mechanisms of resistance to one group of insecticides, in some cases, leads to resistance to insecticides from other group/class; such phenomenon is commonly known as cross-resistance [4].

Cross-resistance enables resistant insects to survive the exposure to insecticides with similar chemical composition to the one they are resistant to. In general, cross-resistance results in detoxifying or changing sensitivity to common biochemical and physiological damage. It also happens when one enzymatic system detoxifies more than one class of insecticides [68]. Hence, cross-resistance does not necessarily spread to all members of the same group of insects or it is limited to only closely related pesticides. Similarities in their mode of action or, sometimes the similarity of their enzymatic systems, are more important for their degradation.

On the other hand, multiresistance is resistance to insecticides from different classes. It depends on different mechanisms, so insects can develop resistance to a large number of insecticides from different classes, regardless of their chemical structure. Each new insecticide can cause one or more resistance mechanism to develop, and each developed mechanism results in resistance to similar insecticides. Multiresistance can occur when the organism develops more than one mechanisms of resistance, such as change in AChE sensitivity combined with multienzymatic detoxification, as in the case of organophosphates [68]. Rapid development of multiresistance is the most important reason why organisms quickly become resistant to new compounds introduced to replace inefficient insecticides, which is a result of persistent R-genes and their interactions manifested in several mechanisms of resistance [69].

Recorded cases of negative cross-resistance are very important. Increased resistance to one compound can lead to increased susceptibility to another [70]. Negative cross-resistance has still not been commercially exploited in field conditions, but knowledge on this mechanism can potentially be very important in practice.

Cross-resistance limits the choice of available insecticide, whereas multiresistance represents a rapid overview of insecticide selection that prevents us from reusing insecticides on resistant species for a longer period [69].

5. ROLE OF ACHE IN COLORADO POTATO BEETLE RESISTANCE TO CARBAMATES AND ORGANOPHOSPHATES

Organophosphates and carbamates are neural toxins, characterized by high toxicity and a quick action. They act inhibitory on acetylcholinesterase (AChE), during neural transmission. Organophosphates phosphorylate enzymes, whereas carbamates form an enzyme- inhibitor complex. The forming of this complex is not irreversible but the reactivation rate is 105–106 slower when compared to a similar reaction in case of ACh and AChE. The enzyme cannot disintegrate new ACh. Although organophosphates and carbamates have a similar mode of action, there are also some pronounced differences between them, mainly as a result of different binding site in the active center and due to different geometries of nucleophilic attack [71].

Acetylcholinesterase (AChE; EC 3.1.1.7) is a key enzyme in the nervous system (55), terminating nerve impulses by catalyzing the hydrolysis of the neurotransmitter acetylcholine. Carboxylesterases belong to a multifunctional carboxylesterase/cholinesterase superfamily (CCE). They are ubiquitous in most living organisms, including animals, insects, plants, and microbes. CCEs, regarding their physiological and biochemical functions could be divided in three groups: dietary/detoxification, hormone/semiochemical processing, and neuro developmental [72]. AChE is the major target for organophosphate and carbamate insecticides, which inhibit enzyme activity. Such inhibition of AChE causes excessive excitement in nerves, a blockage of neurotransmission, and the death of insects. Insensitivity of AChE to organo phosphates and carbamates is one of the important mechanisms for insecticide resistance. Changes in AChE lead to an efficient mechanism of resistance to OP and carbamate insecticides and that is site insensitivity at the target enzyme, acetylcholinesterase (AChE) [73, 74].

For testing the resistance to organophosphates and carbamates, caused by altered AChE, the determination of kinetic constants, especially the Michaelis-Menten constant (Km) is of great importance. It measures the affinity of the enzyme to its substrate [42].

The first case of AChE with decreased susceptibility to pesticides was described by Smissaert in 1961 [75]. Ioannidis et al. [76] first characterized a field population of the carbofuran-resistant CPB. The resistance was determined to be autosomal and monofactorial, leading to a decrease in AChE sensitivity to carbofuran inhibition. A study of a crude enzyme preparation [77] showed that AChE from the AZ-R strain had a 2.4-fold reduction of affinity to acetylthiocholine (ATCh) compared with AChE from a susceptible (SS) strain.

As previously recorded, altered acetylcholine esterase plays a critical role in resistance to organophosphates and carbamates [76, 78, 79]. The measurement of AChE activity is commonly used as a biomarker of exposure to different pesticides [80]. Kinetic analysis of AChE was used to explain the resistance of some insect strains and the selectivity of some organo phosphate and carbamate insecticides. A lot of different studies have described CPB resistance to OPs and CBs [46, 77, 81, 82]. Azinphosmethyl resistance has been reported in CPB; high level of resistance (136-fold) in a nearly isogenic CPB strain (AZ-R) was due to multiple resistance mechanisms, including reduced penetration, enhanced xenobiotic metabolism, and target site insensitivity [83]. Russel et al. [84] suggest that interspecific comparisons of bioassay and biochemical data suggest two major patterns of resistance to OPs/CBs resulting from an insensitive AChE: Pattern I resistance, which is generally more effective for carbamates, and Pattern II resistance, at least as effective for organophosphates as it is for carbamates and may even be specific to organophosphates in some cases.

The role of acetylcholinesterase (AChE) as the primary mechanism for removing the excitatory neurotransmitter, acetylcholine (ACh), from cholinergic synapses and its role as the target site for organophosphate and carbamate inhibitors and accumulation of ACh that results from the inhibition of AChE leads to the prolonged stimulation and, in many cases, the desensitization of the ACh receptors, eventually to severe neurological disruption, and ultimately to death [85]. Since AChE causes death, irreversible inhibitors have been developed as insecticides: organophosphates and carbamates. They have similar properties to acetylcholine but are hemisubstrates ultimately leading to irreversible inhibition of the enzyme. This inhibition leads to an accumulation of acetylcholine in the synapses (active site of the enzyme is therefore occupied and incapable of hydrolyzing its normal substrate) which in turn leaves the acetylcholine receptors permanently open, resulting in the death of the insect [75].

CPB populations resistant to azinphosmethyl contained two mutations in the AChE (S291G and R30K), which made the enzyme less sensitive to azinphosmethyl and carbofuran [73, 86]. In the strain resistant to carbofuran, the presence of two mutations (I392T and S291G) did not result in resistance, but the presence of just one (S291G) conferred high resistance to carbofuran and medium resistance to azinphosmethyl [73]. Compared to the susceptible strain, due to altered acetylcholine esterase, strain resistance to azinphosmethyl had a reduced substrate affinity for ATCh and azinphosmethyl oxon [87]. Modifications in acetylcholine esterase, resulting in resistance, may be selective. It was found that while one resistant strain was highly insensitive to arylcarbamates, another strain with the same affected enzyme was highly insensitive to organophosphates, but not arylcarbamates. Such changes in AChE made yet another resistant strain more sensitive to α-chaconine, a glycoalkaloid present in potatoes and an inhibitor of AChE. Additionally, modified AChE also had increased sensitivity to tomatine, which is also glycoalkaloid present in tomatoes [78].

Zhu and Clark [77] demonstrated that the less bulky substrates, such as ATCh, interact poorly with AChE from the AZ-R strain than AChE from the SS strain of CPB. Such structure–activity relationships may be an indication that a similar alteration in amino acid residues has taken place in the acyl pocket size in AChE from the AZ-R strain and has resulted in the altered substrate and inhibitor profile.

The target site of organophosphate (OP) and carbamate insecticides is the same; they inhibit the activity of AChE. The function of acetylcholinesterase (AChE) is degradation of acetylcholine (ACh – neurotransmitter) in the insect cholinergic synapses. Mutations in the AChE- encoding locus have been shown to confer target site insensitivity to organophosphate and carbamate insecticides, leading to modification of AChE (MACE). A range of other amino acid substitutions in AChE confer insecticide resistance, and these mutations typically reside near to or within the active site of the enzyme. Such AChE mutations, associated with insecticide resistance, mostly known as Ace in Drosophila, have also been observed in other species, including L. decemlineata. Based on bioassays and the literature, modified/insensitive AChE confers two major patterns of resistance to OPs/carbamates [84]. Pattern I resistance is characterized by significantly higher resistance ratios (RR) (much greater reduction in the sensitivity of AChE at the biochemical level) to carbamates than to organophosphate insecticides. Pattern II resistance is characterized by resistance ratios (and/or reductions in the sensitivity of AChE) that are approximately equivalent for both carbamates and OPs. There are also a few species for which an insensitive AChE has been reported and for which molecular data have been collected, but for which the resistance profiles for both OPs and carbamates have not been reported. For CPB, both patterns were registered.

In a few cases of each pattern, gene sequencing has identified the molecular nature of the alteration leading to the lowered sensitivity to inhibitors. Although it is not possible yet to relate with full confidence the mechanism by which these structural changes alter sensitivity to inhibitors. Pattern I mutations may involve changes in the active site, such as a common Gly^Ser mutation in the oxyan–ion hole, whereas Pattern II changes may result in a constriction of the cleft leading to an active site that limits the access of inhibitors and, presumably, of ACh itself. Insensitivity to inhibitors may be accompanied by a reduced ability to hydrolyze ACh. Whether this is always deleterious to the organism is unclear since it is generally considered that, as in vertebrates, AChE is present in insects at a level considerably in excess of that needed for basic neurological functions under normal physiological conditions.

Biochemical studies using an affinity-purified AChE from the SS strain established that the AChE associated with CPB possessed typical characteristics of other AChEs and consists of two different molecular forms: the major form (92%) was a hydrophilic dimer, whereas the minor form (8%) was an amphiphilic dimer. Both molecular forms had virtually identical molecular weights and isoelectric points. Amino acid analysis indicates that the mole percentages of amino acids of the AChE from CPB were highly comparable to those previously reported for AChE from Drosophila [77].

According to Zhu and Clark [77], affinity (Km) and hydrolyzing efficiency (Vmax) of AChE purified from a near-isogenic azinphosmethyl-resistant (AZ-R) strain of CPB to selected substrates, including acethylthiocholine, acetyl-(5-methyl) thiocholine, and propionythiocholine, were lower than those of AChE purified from a susceptible (SS) strain. AChE from the SS strain was significantly inhibited by higher amounts of acethylthiocholine and acetyl-(fj- methyl) thiocholine, whereas AChE from the AZ-R strain was activated by higher amounts of all four substrates examined.

Finally, it is important to notice results on toxicological tests and measuring activity of AChE of CPB populations in Serbia, resistant to OPs and carbamates [88]. The order of resistance levels for OPs and carbamates was completely opposite. Experiments showed that acetylcholinesterase (AChE) activity of CPB was very pronounced and easily measured. At a constant AChE concentration, increasing the substrate concentration will cause a positive, linear, and dependent increase in the reaction. The same applies in the reaction with constant substrate concentration and increased enzyme concentrations. AChE activity is significantly affected not only by location, but also by substrate concentration (acetylthiocholine iodide ATChI). Considering that ATChI (substrate) in increased concentrations inhibits normal AChE activity, it can be concluded that altered AChE affected the change in the population order. The total AChE activity is in correlation with the determined resistance to carbamates.

6. AN INTEGRATED APPROACH TO CPB CONTROL

The IRAC recommendation [89] is that the most effective strategy to combat insecticide resistance is to do everything possible to prevent it from occurring in the first place. It is recommended to develop and apply IRM (Integrated Resistance Management) programs as one part of a larger IPM program. Field researchers and entomologists should be focused on three basic components: pest monitoring, economic injury levels, and integration of multiple control strategies. It is essential to widely implement Economic Thresholds (ET) (use of insecticides only if pest populations are able to cause economic losses that exceed the cost of the insecticide plus application, or where there is a threat to public health). Integrated Control Strategies: Incorporate as many different control strategies as possible including the use of synthetic insecticides, biological insecticides [90, 91, 92, 93, 94], beneficial insects (predators/ parasites) [2, 95], cultural practices, transgenic plants (where allowed), crop rotation, pest- resistant crop varieties, and chemical attractants or deterrents [96, 97, 98, 99, 100].

Applications of insecticide must be timed correctly, targeting the most vulnerable life stage of the insect pest. The use of spray rates and application intervals recommended by the manufacturer and in compliance with local agricultural extension regulations is essential.

Integrated pest management (IPM), with the reduced application of synthetic insecticides, has an increasing need for alternative methods of plant protection. Together with systematic insecticide resistance monitoring [6, 12, 89], application of plant extracts with antifeedant and repellent effects could be one of the tools in efficient IPM program [101, 102, 103].

Boiteau [7] asserts that our comprehension of the potato ecosystem, along with various preventive and curative control measures, is adequate to implement a comprehensive strategy for insect pest management. The unification of concepts should encourage the integration of insect control strategies beyond individual pest management. The primary difficulty is to integrate ineffective and effective control approaches, alongside conventional and sustainable (or organic) crop protection strategies. This necessitates acquiring skills to navigate the volatility inherent in environmentally based Integrated Pest Management approaches. Active adaptive management (AAM) is a proposed method for addressing various forms of uncertainty [104]. Investigation and receptiveness to innovative concepts will be crucial for the integration of various insect management strategies and potato cultivation protection systems. [7].

REFERENCES

1. Hare, D. J. (1990): Ecology and Management of the Colorado Potato Beetle. Annu. Rev. Entomol., 35, 81- 100.
2. Hu, J. S., Gelman, D. B., Bell, R. A., Loeb, M. J. (1998): In Vitro Rearing of Edovum puttleri, an Egg Parasitoid of the Colorado Potato Beetle – Development from Egg Through the Pupal Stage. BioControl, 43, 1-16.
3. Maliszewska, J., Tgowska, E. (2012): Capsaicin as an Organophosphate Synergist against Colorado Potato Beetle (Leptinotarsa decemlineata say). J. Plant Protect. Res., 52, (1), 28-34.

4. Alyokhin, A., Y. H. Chen, M. Udalov, G. Benkovskaya, and L. Lindstrom. 2013. Evolutionary considerations in potato pest management. In: Giordanengo, P., C. Vincent, and A. Alyokhin [editors]. Insect Pests of Potato: Global Perspectives on Biology and Management. Academic Press, Oxford, UK. Pp. 543-571.

5. Nault, B., A., Kennedy, G., G. (1998): Limitations of Using Regression and Mean Separation Analyses for Describing the Response of Crop Yield to Defoliation: A Case Study of the Colorado Potato Beetle (Coleoptera: Chrysomelidae) on Potato. J. Econ. Entomol., 91 (1), 7-20.

6. Stanković, S. (2002): Monitoring Methods and Determination of Colorado Potato Beetle (Leptinotarsa decemlineata Say) Resistance Levels to Insecticides in Serbia. Faculty of Agriculture, Belgrade, 1- 96.

7. Boiteau, G. (2010): Insect Pest Control on Potato: Harmonization of Alternative and Conventional Control Methods. Am. J. Potato Res., 87 (5), 412-419.

8. Ferro, D. N., Morzuch, B. J., Margolies, D. (1983): Crop Loss Assessment of the Colorado Potato Beetle (Coleoptera: Chrysomelidae) on Potatoes in Western Massachusetts. J. Econ. Entomol., 76, 349-356.

9. Shields, E. J., Wyman, J. A. (1984): Effect of Defoliation at Specific Growth stages on Potato Yields. J. Econ. Entomol., 77, 1194-1199.

10. Zehnder, G. W., Hough-Goldstein, J. (1990): Colorado Potato Beetle (Coleoptera: Chrysomelidae) Population Development and Effect on Yield on Potatoes With and Without Straw Mulch. J. Econ. Entomol., 83, 1982-1987.

11. Alyokhin, A., Baker, M., Mota-Sanchez, D., Dively, G., Grafius E. (2008): Colorado Potato Beetle Resistance to Insecticides. Am. J. Potato Res., 85 (6), 395-413.

12. Stanković, S., Zabel, A., Kostić, M., Šestović, M. (2004): Comparative Analysis of Colorado Potato Beetle (Leptinotarsa decemlineata Say) Resistance Monitoring Methods. Pesticides, 18 (3), 159-175.

13. Stanković, S., Kostić, M., Sivčev, I., Janković, S., Kljajić, P., Todorović, G., Jevđović, R. (2012): Resistance of Colorado Potato Beetle (Coleoptera: Chrysomelidae) to Neonicotinoids, Pyrethroids and Nereistoxins in Serbia. Romanian Biotechnol. Lett, 17 (5), 7599-7609.

14. Zabel, A., Kostić, M. (1988): Testing the Efficacy and Persistence of some New Insecticides in the Control of Colorado Potato Beetle (Leptinotarsa decemlineata Say). Plant Protect., 184, 171-182.

15. Zabel, A., Rajkovic, S., Manojlovic, B., Stankovic, S., Kostic, M. (2000): Nenicotinoids in the Control of Leptinotarsa decemlineata Say. 2nd Balkan Symposium on Vegetables and Potatoes. Thessaloniki, Greece. Book of Abstracts, 541-543.

16. Zabel, A., Rajkovic, S., Manojlovic, B., Stankovic, S., Veljkovic, I. (2000): New Pesticides in Potato Protection Against the Colorado Potato Beetle (Leptinotarsa decemlineata Say.) and Late Blight (Phytophtora infestans Mont. de Bary) on Potato. 2nd Balkan Symposium on Vegetables and Potatoes. Thessaloniki, Greece. Book of Abstracts, 491-496.

17. Stanković, S., Kostić, M., Sivčev, I. (2003): Colorado Potato Beetle (Leptinotarsa decemlineata Say.) Resistance Levels to Endosulfan in Serbia. Plant Protect., 54 (1-4), 105-113.

18. Inđić, D., Vuković, S., Tanasković, S., Grahovac, M., Kereši, T., Gvozdenac, S., Savčić- Petrić, S., (2012): Screening test for detection of Leptinotarsa decemlineata (Say.) sensitivity to insecticides. Pesticides Phytomed., 27 (1), 59-67.

19. Alyokhin, A., Ferro, D. (1999): Relative Fitness of Colorado Potato Beetle (Coleoptera: Chrysomelidae) Resistant and Susceptible to the Bacillus thuringiensis Cry3A Toxin. J. Econ. Entomol., 92, 510-515.

20. Bishop, B. A., Grafius, E. (1991): An On-Farm Insecticide Resistance Test Kit for Colorado Potato Beetle (Coleoptera: Chrysomelidae). Am. Potato J., 68, 53-64.

21. Miyo, T., Keil, C. B. O., Hough-Goldstein, J. A., Oguma, Y. (1999): Inheritance of Resistance to Esfenvalerate in Colorado Potato Beetle (Coleoptera: Chrysomelidae). J. Econ. Entomol., 92, 1031-1038.

22. Jiang, W. H.; Wang, Z. T., Xiong, M. –H. (2010): Insecticide Resistance Status of Colorado Potato Beetle (Coleoptera: Chrysomelidae) Adults in Northern Xinjiang Uygur Autonomous Region. J. Econ. Entomol., 103 (4), 1365-1371.

23. Szendrei, S., Grafius, E., Byrne, A., Ziegler, A. (2012): Resistance to Neonicotinoid Insecticides in Field Populations of the Colorado Potato Beetle (Coleoptera: Chrysomelidae). Pest Manag. Sci., 68, 941–946.

24. Wegorek, P., Zamojska, J., Mrówczyński, M. (2011): Susceptibility Level of the Colorado Potato Beetle (Leptinotarsa decemlineata Say) to Chlorpyrifos and Acetamiprid in Poland and Resistance Mechanisms of the Pest to Chlorpyrifos. J. Plant Protect. Res., 51 (3), 279-284.

25. Tisler, A., M., Zehnder, G. W. (1990): Insecticide Resistance in the Colorado Potato Beetle (Coleoptera: Chrysomelidae) on the Eastern Shore of Virginia. J. Econ. Entomol., 83, 666-671.

26. Grafius, E. J., Bishop, B. A., (1996): Resistance to Imidacloprid in Colorado Potato Beetles from Michigan. Res. Pest Manag. Newslett., 7(1), 12 – 16. http://www.msstate.edu/Entomology/v8n2/art06.html.

27. Alyokhin, A., Dively, G., Patterson, M., Mahoney, M., Rogers, D., Wollam, J. (2006): Susceptibility of Imidacloprid-Resistant Colorado Potato Beetles to Non-Neonicotinoid Insecticides in the Laboratory and Field Trials. Am. J. Potato Res., 83, 485-494.

28. Alyokhin, A., Dively, G., Patterson, M., Castaldo, C., Rogers, D., Mahoney, M., Wollam, J. (2007): Resistance and Cross-Resistance to Imidacloprid and Thiamethoxam in the Colorado Potato Beetle. Pest Manag. Sci., 63, 32-41.

29. Zhao, J.Z., Bishop, B. A., Grafius, E.J., (2000): Inheritance and Synergism of Resistance to Imidacloprid in the Colorado Potato Beetle (Coleoptera: Chrysomelidae). J. Econ. Entomol., 93, 1508-1514.

30. Zhou, Z., Zhaoxu, Z., Jinhuan, P., Wenchao, G., Zhong, N., Tian, Y., Xiaa, G., Wua, J. (2012): Evaluation of the Resistance of Transgenic Potato Plants Expressing Various Levels of Cry3A against the Colorado Potato Beetle (Leptinotarsa decemlineata Say.) in the Laboratory and Field. Pest Manag. Sci. 68, 1595-1604.

31. Alyokhin, A., Ferro, D., Hoy, C., Head, G. (1999): Laboratory Assessment of Flight Activity Displayed by Colorado Potato Beetles (Coloeoptera: Chrysomelidae) Fed on Transgenic and Cry3a Toxin-Treated Potato Foliage. J. Econ. Entomol. 92, 115-120.

32. Maceljski, M. (1967): The Emergence of Resistant Potato Beetle (Leptinotarsa decemlineata [Say].) In Yugoslavia. Agriculture Tribune, 10, 891-900.

33. Šestović, M. (1972): The Resistance of Potato Beetle, Leptinotarsa decemlineata Say to Insecticides. Doctoral dissertation, Faculty of Agriculture, Belgrade.

34. Šestović, M., Peric, I. (1977): The Spectrum of Colorado Potato Beetle (Leptonotarsa

35. decemlineata Say) Resistance to Insecticides. Eighth Conference on the Application of Pesticides in Plant Protection, Hygiene and Veterinary – Porec. Proceedings, 481-495.

36. Maceljski, M., Igrc, J. (1992-1994): Studies on the Efficacy of Some Insecticides against Colorado Potato Beetle in the Years 1986-1990. Ziemniak, 1992-1994, 31-51.

37. Zabel, A. (1991): Esterase Activity and Resistance to Organophosphate Insecticides in Potato Beetle (Leptinotarsa decemlineata Say.). Faculty of Agriculture, Novi Sad.

38. Zabel, A., Kostić, M., Sivčev, I., Draganić, M., Indjić, D. (1997): Susceptibility of the Colorado Potato Beetle (Leptinotarsa decemlineata Say., Coleoptera: Chrysomelidae) to Applied Insecticides in the Republic of Serbia. Acta Horticulturae, 462, 397-405.

39. Zabel, A., Kostic, M. (1988): Testing the Efficiency and Persistence of Some New Insecticides in the Control of Colorado Potato Beetle (Leptinotarsa decemlineata Say). Plant Protect., 184, 171- 182

40. Hoy, C.W., Head, G.P., Hall, F.R. (1998): Spatial Heterogenity and Insect Adaptation to Toxins. Annu. Rev. Entomol., 43, 571-594.

41. Zlotkin, E. (1999): The Insect Voltage-Gated Sodium Channel as Target of Insecticides. Annu. Rev. Entomol., 44, 429-455.

42. Liu, N., Scott, J.G. (1998): Increased Transcription of CYP6D1 Causes Cytochrome P- 450 Mediated Insecticide Resistance in House Fly. Insect Biochem. Molecul. Biol., 28, 531-535.

43. Hassall, K. A. (1990): The biochemistry and uses of pesticides: structure, metabolism, mode of action, and uses in crop protection. 2nd ed. Weinheim: VCH.

44. Matsumura, F. (1983): Penetration, Binding and Target Insensitivity as Causes of Resistence to Chlorinated Hydrocarbon Insecticides. In: Pest Resistance to Pesticides (ed. by Georghiou G.P. and Saito T.), Plenum Press, New York, 367-387.

45. Harshman, L. G., James, A. A. (1998): Differential Gene Expression in Insects: Transcriptional Control. Annu. Rev. Entomol., 43, 671-700.

46. Terriere, L. C. (1983): Enzyme Induction, Gene Amplification and Insect Resistance to Insecticides. In: Pest Resistance to Pesticides (ed. by Georghiou G.P. and Saito T.), Plenum Press, New York, 265-298.

47. Argentine, A., Zhu, K. Y., Lee, S. H., Clark, J. M. (1994): Biochemical Mechanisms of Azinphosmethyl Resistance in Isogenic Strains of Colorado Potato Beetle. Pestic. Biochem. Physiol. 48, 63-78.

48. Wilkinson, C.F. (1983): Role of Mixed-function Oxidases in Insecticide Resistance. In: Pest Resistance to Pesticides (ed. by Georghiou G.P. and Saito T.), Plenum Press, New York, 175-207.

49. Feyereisen, R. (1999): Insect P 450 Enzymes. Annu. Rev. Entomol., 44, 507-533.

50. Hodgson, E., Kulkarni, A. (1983): Characterization of Cytochrome P-450 in Studies of Insecticide Resistance. In: Pest Resistance to Pesticides (ed. by Georghiou G.P. and Saito T.), Plenum Press, New York, 207-229.

51. Šestović, M., Peric, I. (1995): Fundamentals of the Metabolism of Insecticides: 1. Enzymes that Catalyze the Oxidation Processes of the First Phase. Pesticides, 10, 7-23.

52. Plapp, F.W., Wang, T.C. (1983): Genetic Origins of Insecticide Resistance. In: Pest Resistance to Pesticides (ed. by Georghiou G.P. and Saito T.), Plenum ress, New York, 47-71.

53. Miyata, T. (1983): Detection and Monitoring Methods for Resistance in Arthropod Based on Biochemical Characteristics. In: Pest Resistance to Pesticides (ed. by Georghiou G.P. and Saito T.), Plenum Press, New York, 99-117.

54. Anspaugh, D. D., Kennedy, G. G., Roe, R. M. (1995): Purification and Characterization of a Resistance- Associated Esterase from the Colorado Potato Beetle, Leptinotarsa decemlineata (Say.). Pestic. Biochem. Physiol. 53, 84-96.

55. Hemingway, J. (2000): The Molecular Basis of Two Contrasting Metabolic Mechanisms of Insecticide Resistance. Insect Biochem. Mol. Biol., 30, 1009-1015.

56. Hama, H. (1983): Resistance to Insecticides Due to Reduced Sensitivity of Acetylcholinesterase. In: Pest Resistance to Pesticides (ed. by Georghiou G.P. and Saito T.), Plenum Press, New York, 299-332.

57. Soderlund, D.M., Bloomquist, J.R., Wong, F., Payne, L.L., Knipple, D.C. (1989): Molecular Neurobiology: Implications for Insecticide Action and Resistance. Pestic. Sci., 26, 359-374.

58. Anthony, N.M., Brown, J.K., Feyereisen, R., ffrench-Constant, R.H. (1998): Diagnosis and Characterization of Insecticide-Insensitive Acetylcholinesterase in Tree Populations of the Sweetpotato Whitefly Bemisia tabaci. Pestic. Sci. 52, 39-46.

59. Miller, T.A., Adams, M.E. (1982): Mode of Action of Pyrethroids. In: Insecticide Mode of Action (ed. by Coats J.R.), Academic Press, New York, 3-24.

60. Miller, T.A., Salgado, V.L. (1985): The Mode of Action of Pyrethroids on Insects. In: The Pyrethroid Insecticides (ed. by Leahley J.P.). Taylor & Francis Ltd., London, 43-89.

61. Miller, T.A., Salgado, V.L., Irving, S.N. (1983): The KDR Factor in Pyrethroid Resistance. In: Pest Resistance to Pesticides (ed. by Georghiou G.P. and Saito T.), Plenum Press, New York, 353-366.

62. Nauman, K. (1998): Research into Fluorinated Pyrethroid Alcohols - An Episode in the History of Pyrethroid Discovery. Pestic. Sci., 52, 3-20.

63. Gingerich, D.P., Borsa, P., Suckling, D.M., Brun, L. (1996): Inbreeding in the Coffee Berry Borer, Hypothenemus hampei (Coleoptera: Scolytidae) Estimated from Endosulfan Resistance Phenotype Frequencies. Bull. Entomol. Res., 86, 667-674.

64. Fujii, S., Walcott, E.C., Sumikawa, K. (1999): The Action of Nicotine in the Mammalian Brain. In: Nicotinoid Insecticides and the Nicotinic Acetylcholine Receptor (ed. by Yamamoto, I., Casida, J.E.), Springer–Verlag, Tokyo, 223-237.

65. Glennon, R.A., Dukat, M. (1999): Nicotine Analogs: Structure – Affinity Relationships for Central Nicotinic Acetylcholinergic Receptor Binding. In: Nicotinoid Insecticides and the Nicotinic Acetylcholine Receptor (ed. by Yamamoto, I., Casida, J.E.), Springer– Verlag, Tokyo, 237-253.

66. Nauen, R., Ebbinghaus–Kintscher, U., Elbert, A., Jeschke, P., Tietjen, K. (2001): Acetylcholine Receptors as Sites for Developing Neonicotinoid Insecticides. In: Biochemical Sites of Insecticide Action and Resistance (ed. by Ishaaya, I.), Springer–Verlag, Berlin, 77-107.

67. Sparks, T.C., Lockwood, J.A., Byford, R.L., Graves, J.B., Leonard, B.R. (1989): The Role of Behavior in Insecticide Resistance. Pestic. Sci., 26, 383-400.

68. Head, G., Hoy, C. W. (1994): Quantitative Genetics of Behavioral and Physiological Resistance to Insecticides in Diamond Back Moth and Colorado Potato Beetle. Resist. Pest Manag. Newslett., 6, 48-49.

69. Brattsten, L.B. (1989): Insecticide Resistance: Research and Management. Pestic. Sci., 26, 329-332.

70. Metcalf, R.L. (1989): Insect Resistance to Insecticides. Pestic. Sci., 26, 333-358.

71. Nomura M, Kato Y., Miyata T (1999): Monitoring of organophosphate and carbamate resistance and identifying a combination of insecticides showing negatively correlated cross resistance in the green rice leafhopper (Nephotettix cincticeps Uhler) (Hemiptera: Deltocephalidae). Appl Entomol Zool. 34, 525-30.

72. Corbett, J.R., Wright, K., Baillie, A.C. (1984): The Biochemical Mode of Action of Pesticides. Second edition, Academic Press, London.

73. Lu, F-G., Fu, K-Y., Li, Q., Guo, W-C., Tursun, A., Li, G-Q. (2014): Identification of Carboxylesterase Genes and their Expression Profiles in the Colorado Potato Beetle Leptinotarsa decemlineata Treated with Fipronil and Cyhalothrin. Pesti. Biochem. Physiol., doi: 10.1016/j.pestbp.2014.12.015.

74. Kim, H.J., Dunn, J.B., Yoon, K.S., Clark, J.M. (2006): Target Site Insensitivity and Mutational Analysis of Acetylcholinesterase from a Carbofuran-Resistant Population of Colorado Potato Beetle, Leptinotarsa decemlineata (Say). Pesti. Biochem. Physiol., 84, 165-179.

75. Yu-Xin Chai, Guo-Ying Liu, Jin-Jun Wang (2007): Toxicological and Biochemical Characterizations of AChE in Liposcelis Bostrychophila Badonnel (Psocoptera: Liposcelididae). Pesti. Biochem. Physiol., 88, 197-202.

76. Fournier, D. (2005): Mutations of Acetylcholinesterase which Confer Insecticide Resistance in Insect Populations. Chemico-Biologic. Interact., 157-158, 257-261.

77. Ioannidis, P.M., Grafius, E.J., Wierenga, J.M., Whalon, M.E., Hollingworth, R.M. (1992): Selection, Inheritance and Characterization of Carbofuran Resistance in the Colorado Potato Beetle (Coleoptera: Chrysomelidae). Pestic. Sci. 35, 215-222.

78. Zhu, K.Y., Clark, J.M. (1995): Comparison of Kinetic Properties of Acetylcholinesterase Purified from Azinphosmethyl-Susceptible and Resistant Strains of Colorado Potato Beetle. Pestic. Biochem. Physiol. 51, 57-67.

79. Wierenga J.M., Hollingworth R.M. (1993), Inhibition of Altered Acetylcholinesterases from Insecticide-Resistant Colorado Potato Beetles (Coleoptera: Chrysomelidae). J. Econ. Entomol. 86, 673-679.

80. Stanković, S., Zabel, A., Kostić, M., Manojlović, B., Rajković, S. (2004): Colorado Potato Beetle (Leptinotarsa decemlineata Say) Resistance to Organophosphates and Carbamates in Serbia. J. Pest Sci., 77, 11-15.

81. Mohamadi, M.M., Mossadegh, M.S., Hejazi, M.J., Goodarzi, M.T., Khanjani, M., Galehdari, H. (2010): Synergism of Resistance to Phosalone and Comparison of Kinetic Properties of Acetylcholinesterase from Four Field Populations and a Susceptible Strain of Colorado Potato Beetle. Pesti. Biochem. Physiol., 98, 254-262.

82. Argentine, J.A., Lee, S.H., Sos, M.A., Barry, S.R., Clark, J.M. (1995): Permethrin Resistance in a Near Isogenic Strain of Colorado Potato Beetle. Pestic. Biochem. Physiol. 53, 97-115.

83. Zhang, A., Dunn, J.B., Clark, J.M. (1999): An Efficient Strategy for Validation of a Point Mutation Associated with Acetylcholinesterase Sensitivity to Azinphosmethyl in Colorado Potato Beetle, Leptinotarsa decemlineata (Say). Pestic. Biochem. Physiol. 65, 25-35.

84. Malekmohammadi, M., Hejazi, M.J., Mossadegh, M.S., Galehdari, H., Khanjani, M., Goodarzi, M.T. (2012): Molecular Diagnostic for Detecting the Acetylcholinesterase Mutations in Insecticide-Resistant Populations of Colorado Potato Beetle, Leptinotarsa decemlineata (Say.). Pesti. Biochem. Physiol., 104, 150-156.

85. Russell, R., Claudianos, C., Campbell, P.M., Horne, I., Sutherland, T.D., Oakeshott,

86. J.G. (2004): Two Major Classes of Target Site Insensitivity Mutations Confer Resistance to Organophosphate and Carbamate Insecticides. Pesti. Biochem. Physiol., 79, 84-93.

87. Hollingworth RM, Dong K (2008) The biochemical and molecular genetic basis of resistance to pesticides in arthropods. In: Global pesticide resistance

in arthropods, Whalon M.E., Mota-Sanchez D., Hollingworth R.M., Eds., CAB International, Oxfordshire, U.K., p. 40.

88. Kim H.J., Clark, J.M. (2002): Evaluation of Resistant Point Mutations in Recombinant Acetylcholinesterase of Colorado Potato Beetle. Abstr. Papers Am. Chem. Soc., 224:U103-U10.

89. Clark J.M. (1997): Insecticides as Tools in Probing Vital Receptors and Enzymes in Excitable Membranes. Pesti. Biochem. Physiol., 57, 235-254.

90. Zabel, A., Stankovic, S., Kostic, M., Sivcev, I., Kuzevski, J., Kostic, I., Krnjajic, S (2015): Acetylcholinesterase [AChE] Activity of Colorado Potato Beetle Populations in Serbia Resistant to Carbamates and Organophosphates. Romanian Biotechnological Letters (in press: under approval). http://www.irac-online.org/about/resistance/management/Zabel, A., Manojlović, B., Rajković, S., Stanković, S., Kostić, M. (2002): Effect of Neem Extract on Lymantria dispar L. (Lepidoptera: Lymantridae) and Leptinotarsa decemlineata Say. (Coleoptera: Chrysomelidae). J. Pest Sci., 75, 19-26.

91. Ascher, K.R.S., Meisner, J., Klein, M., Rejesus, R.S., Obra, J.B., Zabel, A., Kostic, M., Manojlovic, B., Stankovic, S., Tomasi, O., Sekulovic, D., Beimain, S.R., Golob, P., Andan, H. F., Atarigiya, J., Chare, F.A., Can, P., Gupta, P., Siddiqui, M.R., Joseph, S., Shivpuri, A., Gupta, R.B.L (2000): Abstracts of Presentations on Selected Topics at The XlVth International Plant Protection Congress (IPPC) 2. NEEM (Azadirachta indica), Phytoparasitica: 28 (1), 87-90.

92. Trdan, S., Cipar, A., Bergant, K., Andjus, Lj., Kac, M., Vidrih, M., Rozman, L. (2007): Effect of Temperature on Efficacy of Three Natural Substances to Colorado Potato Beetle, Leptinotarsa decemlineata (Coleoptera: Chrysomelidae), Acta Agriculturae Scandinavica Section B-Soil and Plant Science, 57 (4), 293-296.

93. Gökçe, A., Whalon, M.E., Çam, M.E., Yanar, Y., Demirta, İ., Gören, N. (2006): Plant Extract Contact Toxicities to Various Developmental Stages of Colorado Potato Beetles (Coleoptera: Chrysomelidae), Ann. Appl. Biol., 1-6.

94. Safaei Khorram S., M., Taher Nasabi, N., Jafarnia, S., Khosroshahi, S. (2011): The Toxicity of Selected Monoterpene Hydrocarbons as Single Compounds and Mixtures against Different Developmental Stages of Colorado Potato Beetle, Leptinotarsa decemlineata Say. (Coleoptera: Chrysomelidae). J. Entomol., 8 (5), 404-416.

95. Trdan, S., Vidrih, M., Andjus, L., Laznik, Z. (2009): Activity of Four Entomopathogenic Nematode Species against Different Developmental Stages of Colorado Potato Beetle, Leptinotarsa decemlineata, (Coleoptera, Chrysomelidae), Helminthologia, 46 (1), 14-20.

96. Kostić, M., Zabel, A., Popov, V., Ristić, M., Stanković, S., Manojlović, B., Rajković, S. (2001): Effects of Essential Oils of Some Species of the Genus Tanacetum on Potato Leaf Attractiveness to Larvae of the Colorado Potato Beetle. Food in the 21st century, Book of Abstracts, 136-137.

97. Kostić, M., Kostić, I., Marković, T., Jevđović, R., Stanković, S., Todorović, G., Nedić, N. (2012): Disruption of Attractant Properties of Potato Foliage on Leptinotarsa decemlineata Say by the use of Salvia officinalis L. Essential Oil. Proceedings of the 7th Conference on Medicinal and Aromatic Plants of Southeast European Countries, Serbia, pp. 351-356.

98. Zabel, A., Manojlovic, B., Stanković, S., Thomasi, O., Kostić, M. (2000): The Efficacy of Essential Oil of Myristica fragrans Hout to Leptinotarsa decemlineata Say. and Lymantria dispar L. Plant Protect., 233-234, 211- 220.

99. Stević, T., Kostić, M., Stanković, S., Ristić, M., Soković, M. (2003): Effect of Thujone to Bioagents. Maced. Pharmaceut. Bull., Book of Abstracts, 49 (1,2), 165.

100. Kostic, M., Dražić, S., Zabel, A., Manojlovic, B., Rajkovic, S., Stankovic, S. (2002): Disturbance of Potato Attraction for Colorado Potato Beetle Larvae by Essential Oils of Some Species of the Genus Tanacetum. Arch. Agric. Sci., 221-222, 87-99.

101. Kostic, M., Stanković, S., Ristic, M., Jevdjovic, R., Rajkovic, S. (2003): Effect of Essential Oils of Genus Tanacetum Plants to the Attractiveness of Potato Foliage for Adults of Colorado Potato Beetle. Medic. Raw Mater., 23, 69-82.

102. Kostic, M., Stevic, T., Stankovic, S., Soković, M., Ristic, M. (2003): Effect of Camphor on Biological Agenses, VI Conference of Plant Protection in Serbia, Abstracts, 112-114.

103. Kostić, M., Dražić, S., Popović, Z., Stanković, S., Sivčev, I., Živanović, T. (2007): Developmental and Feeding Alternations in Leptinotarsa decemlineata Say. (Coleoptera: Chrysomelidae) caused by Salvia officinalis L. (Lamiaceae) Essential Oil. Biotechnol. Biotechnologic. Equip., 21, 4.

104. Shea, K., Possingham, H.P., Murdoch, W.W., Roush, R. (2002): Active Adaptive Management in Insect Pest and Weed Control: Intervention with a Plan for Learning. Ecologic. Applic., 12, 927-936.

3 Chapter

SPIROTETRAMAT – A SUBSTITUTE FOR THE MANAGEMENT OF PARASITIC SUCKING INSECTS AND ITS ENVIRONMENTAL PERSISTENCE

1. INTRODUCTION: AN OVERVIEW

Globally, farmers' crops are under assault from numerous insect pests, leading to heightened production costs and potential total crop loss. Consequently, there is an ongoing pursuit of chemicals or formulations to manage emerging pests or those that have acquired resistance. Environmentally friendly pesticides that demonstrate efficacy are highly esteemed; these multifunctional products can be incorporated into optimal agricultural practices and biological control, facilitating an integrated pest management program without the necessity for compounds detrimental to non-target organisms. In 2008, Bayer synthesized a molecule from tetramic acid, known as spirotetramat, which had previously exhibited insecticidal characteristics, in response to the demand for innovative and superior goods. [1].

Spirotetramat functions as an inhibitor of lipid biosynthesis and offers a novel solution for managing problematic insects, including the apple wholly whitefly and whitefly biotype Q, which inflict significant damage on agricultural crops and have developed resistance to conventional pesticides employed for their control [1, 2]. This research was conducted to review the genesis and chemical features of spirotetramat, examining potential applications, its environmental fate (in soil, air, and water), its metabolic processes in plants, and the likelihood of resistance development.

2. THE ORIGIN OF SPIROTETRAMAT

Evidence exists indicating that compounds derived from the structural unit of tetramic acid have biological activities across a wide spectrum; antibiotic tirandamicin A and the phytotoxin of tenuazonic acid are examples of this type of compounds found in nature [3]. Furthermore, there are synthetic compounds that are utilized as herbicides and insecticides as in the case of spirotetramat (commercial name Movento®) developed by Bayer CropScience as an insecticide [3, 4]. The discovery of this compound came about through research into improving myticide and herbicide activity by Bayer with compound derived from tetramic acid. As a first step in this research, compounds were synthesized with the Bucherer-Bergs reaction of tetramic acid (compound 1a in Figure 1] and its acetyl derivative (compound 1b in Figure 1), splitting the 1-amino-4-methyl-carboxylic acid methyl ester cyclohexane. According to the research, they observed a significant improvement in herbicidal activity in comparison with analogous compounds where spirocyclic were not substituted. In another research, t they also reported that the compound 1b demonstrated excellent myticide activity and was highly effective against the aphid Myzus persicae [5].

Previous evaluations in the investigation led to derivatives of spirocyclic tetramic acid alkoxy- substituted. In this case, the mechanism is the synthesis of Strecker and the splitting of the 4- metoxy-1-amino-ciclohexane-carbonitrile (compound 2 in Figure 2) and they were able to obtain and isolate isomer mixes; the least present isomer had good control of Myzus persicae, which at that time was close to the efficacy of the best aphicide, imidacloprid. However, the results demonstrated a disadvantage in that they saw an increase in the herbicidal activity of the sample [5].

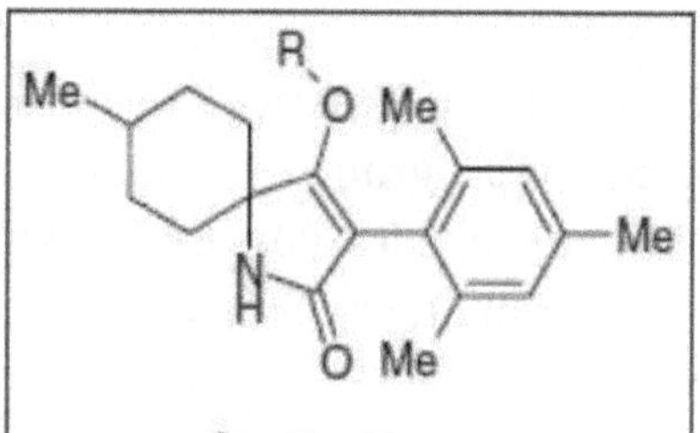

1a R = H
1b R = CO-Me

Figure 1. Compounds derived from tetramic acid with improved herbicidal activity.

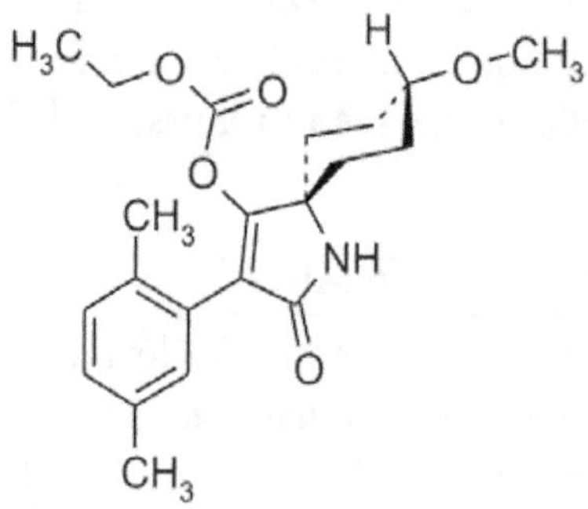

Figure 2. Molecular structure of spirotetramat.

3. PEST CONTROL

Spirotetramat is an insecticide that targets sucking insects in their juvenile stage such as aphids (Aphis spp., Myzus spp., Dysaphis spp., Toxoptera spp., Phorodon humuli), rice aphids (Phylloxera spp.), psyllids (Psylla spp., Paratrioza cockerelli), mealybugs (Pseudococcus spp., Planococcus spp.), and whiteflies (Bemisia spp., Trialeurodes vaporariorum). Table 1 lists the studies into the relationship of the effectiveness of spirotetramat in different insects [1, 6].

As mentioned earlier, the aphicide activity of spirotetramat is effective in the immature stages where incomplete ecdysis can be observed, the insect cannot completely shed its exoskeleton, thus impeding its growth; however, what has been observed in nymphs is that they appear to be immobile and they dry up quickly. The efficacy of spirotetramat on adult insects is reduced due to their mobility; they tend to produce nymphs that die within 24 hours or the nymphs are non-fertile, thus reducing the procreation and fertility of the future generations [7]. Spirotetramat has demonstrated excellent efficacy on peach, cotton, and plum aphids that are 3-4 days old. It has also been observed that on female adult whiteflies (B tabacco) treated with spirotetramat (40 and 200 ppm), the number of eggs produced is a function of the applied doses (major reduction 90% and 60%), including a concentration of 8 ppm, 80% of the eggs do not hatch. It was also observed that the way of contact of the insecticide influences its effectiveness on the control of the insects; it has a major effect if spirotetramat is ingested orally than if it is by direct contact with the insect [1, 8].

Another laboratory study suggested that spirotetramat can be utilized in a safe integrated pest management program for the control of the cabbage aphid, as there is less mortality in comparison of other insecticides of the marmalade hoverfly Episyrphus balteatus, which is a natural aphid predator; furthermore, the fertility of the treated adult syrphids is not affected [9]. In another study to determine the collateral damage of spirotetramat on the wasp Anagyrus, a grapevine mealybug parasite, it was found that there was no detectable mortality on the parasite after 24 hours of application; there were no adverse effects on the development of the parasite in the pupa stage inside the mummified mealybug, nor were there any effects on the emergence of the new Anagyrus [10]. It must be pointed out that in an integrated pest management program where the arthropod Galendromus occidentalis is used for biological control, the use of spirotetramat is not recommended given that at concentrations of 0.228 g a.i.L-1, there was a mortality rate of 90% for the eggs, and 100% for the larvae [11]. This was also similar for the toxicity results for Tamarixia radiata, a parasitoid of citrus Asian psyllid (Diaphorina citri Kuwayama), that with an application dose of 0.8 mL L-1 with water did not present favorable conditions for its development and it was highly toxic [12].

Table 1. Organisms controlled with spirotetramat under different conditions.

Reference	Dose applied	Organism controlled	Pest location
Moens et al., 2011 [9]	75 g a.i.ha^{-1}	Brevicoryne brassicae L.	Cabbage
Jamieson et al., 2010 [13]	3.36 g a.i.100 L^{-1}	*Orchamoplatus citri*).	L. C.
Mansour et al., 2011 [10]	120 mL h L^{-1}	Planococcus ficus	L. C.
Page-Weir et al., 2011 [14]	40 mL a.i.100 L^{-1}	Bactericera cockerelli	Tomato & potato
Smiley et al., 2011 [15]	88-110 g a.i.ha^{-1}	Heterodera avenae	Wheat roots
Duvaresch et al., 2008 [16]	120 g a.i.ha^{-1}	Aphis gossypii	Cotton
Kay & Herron, 2010 [17]	144 g a.i.ha^{-1}	Frankliniella occidentalis	Peppers
Fu & Del Real, 2009 [18]	60-120 g a.i.ha^{-1}	Planococcus ficus	Vine
Marcic et al., 2012 [19]	200, 60, 18 mg a.i.L^{-1}	Tetranychus urticae	L. C.
Frank & Lebude, 2011 [20]	1.7 oz 100 gal^{-1}	Adelges tsugae	Fir
Elizondo & Murguido, 2010 [21]	0.5 and 0.6 L ha-1	Myzus persicae Sulzer, Bemisia tabaci Gennadius, Thrips palmi Karny	Potato (Solanum tuberosum L)

*a.i. Active ingredient
*L. C. Laboratory conditions

4. FATE OF SPIROTETRAMAT IN THE PLANT AND ENVIRONMENT

It is important to note that after applying insecticides on crops or on the soil, it is possible that the active ingredient is not absorbed permanently by the soil or that it can mobilize to bodies of water. There exist a lot of physical and chemical and microbiological factors that can determine the fate of the products used in plant protection, some are: hydrolytic degradation and photochemical, biological transformation and mineralization, absorption and movement of the active ingredient, as well as the degraded products in the soil. It is important to note that the above mentioned processes depend on the chemical structure and the physical properties of the compound used, as well as the soil, the vegetation, and the climatic conditions [22].

4.1 Factors affecting the fate of spirotetramat

Considering that spirotetramat has no acidic properties or alkaline in aqueous solutions, it also stands out that soil pH and that of the aqueous systems have no influence on the physico-chemical properties of the spirotetramat. The solubility and lipophilicity of water are important because they provide us with information on the mobility and solubility of spirotetramat in water; if they are low (0.0299 g L-1), this indicates good soil absorption, resulting to very low risk of infiltrating into aquifers. With a base vapor pressure of 5.6 x 10-9 Pa and Henry's constant

6.99 x 10-8 Pa, it can be concluded that there is no possibility of spirotetramat volatizing in any significant form [22, 23]. It is important to consider the properties of the metabolite BYI08330- enol (referred as "enol" from now on); the enol form possesses properties slightly acidic (pKa=5.2). Furthermore, due to its high solubility in water (2.7 g L-1 at pH 7), it presents a risk of possible leaching into subterranean waters; same as with spirotetramat, the volatility of the enol form possesses no significant role [22].

4.2 Plant metabolism of spirotetramat

Before the creation of spirotetramat, there were only systemic insecticides that were only capable of moving in one way, those that enter the plant then move to different locations within the plant; however, this travel was only one way going up the xylem. The advantage of spirotetramat is that once it penetrates under the leaf of a plant, it is transformed by a hydrolytic split to spirotetramat-enol that due to its physiochemical properties is capable of moving up and down through the phloem, which allows it to reach and access pests that are difficult to reach, such as the grape mealy bug (Planococcus ficus) [18]. In contrast with systemic insecticides that travel only one way, such as the case with imidacoprid, spirotetramat-enol being a systemic metabolite with double lanes can protect new leaves generated after the application and it can even protects the roots [1, 18].

Reference [24] determined the metabolism of spirotetramat in apple, cotton, lettuce, and potato (Table 2). In the cultivars analyzed, the main residues found were the father (BYI08330) and three dominant metabolites, BYI08330 enol, BYI08330 enol-glucoside, and BYI08330 cetohydroxy, which is in accordance with those reported by [25] (Figure 3). However, in apples, they detected a fourth metabolite, BYI08330 monohydroxy, with a considerable percentage (around 15.6%); meanwhile in the potato tuber, the main metabolite was BYI08330-enol, along with the absence of the father compound. To conduct those studies, a foliar application of spirotetramat OD-100 was applied, where the dose administered was 167 g a.i.ha-1 for lettuce, which was equivalent to the maximum recommended by the manufacturer; the same was administered for apple, potato, and cotton, with rates of 576, 308, and 264 g a.i.ha-1, respectively, which were equivalent to 2.5, 1.1 at 1.8 and 0.85 times the dose recommended per each season.

It is important to consider that even using higher doses than the recommended, none of the residual concentrations found will surpass the maximum residual limit (MRL) established by the Environmental Protection Agency (EPA) and the Codex Alimentarius of the FAO/OMS [25]. It was observed that the residual concentration of insecticide on apple leaves, potato leaves, and lettuce was superior to the apple fruit, potato (tuber), and the cotton seed. In the three leaves analyzed (lettuce, apple, and potato leaves) the father compound was found to be above 49% of the total residues, which may indicate that the major part of the compound recovered as residue remains in the leaves without being metabolized.

Table 2. Proportions and principal metabolites of spirotetramat in apple, cotton, lettuce, and potato.

Compound (mg kg^{-1})	Apple	Apple (leaves)	Lettuce	Cotton (seed)	Potato (tuber)	Potato (leaves)
Spirotetramat (BYI08330)	0.32	26.37	1.75	<0.001	-	5.455
BYI08330-enol	0.01	4.26	0.56	0.047	0.168	0.870
BYI08330-enol glc	0.03	-	0.36	0.004	0.006	0.395
BYI08330-cetohydroxy	0.05	1.09	0.20	0.011	0.018	2.745
BYI08330-monohydroxy	0.10	-	-	-	-	-
Total residues	0.61	36.63	3.13	0.119	0.225	11.057
MRL[a]	0.7	-	8	0.3	1.6	-
MRL[b]	0.7	-	7	-	0.8	-

[a]MRL, [26]
[b]MRL, [25]
Source: [24], modified

4.3 Fate of spirotetramat in the soil

It is necessary to investigate the degradation of the active compounds in the soil since it is possible that part of the insecticide will reach the soil directly or indirectly after being applied to a crop. The most important process to consider in the soil is the degradation by micro-organisms under aerobic conditions. However, there are other factors that could contribute, such as the abiotic chemical degradation expressed as photolysis on the soil surface and also hydrolysis; other physical processes involved such as leaching, a translocation that can make it more profound in the soil; volatility; and the evaporation from the plant or from the soil surface [22]. However, these same researchers [22] observed that spirotetramat under aerobic soil conditions will degrade rapidly after 1-2 days, dissipating more than 90%. At the same time, during the testing period the two major metabolites generated were BYI08330-enol (maximum 24.3%) and BYI08330-cetohydroxy (maximum 16.3%), two of the dimers enol BYI08330-MA-amida (maximum 6.4%), and lastly two minor metabolites, BYI08330-desmethyl-enol (maximum 3.7%) and BYI08330-oxo-enol (maximum 1.2%) [23, 27].

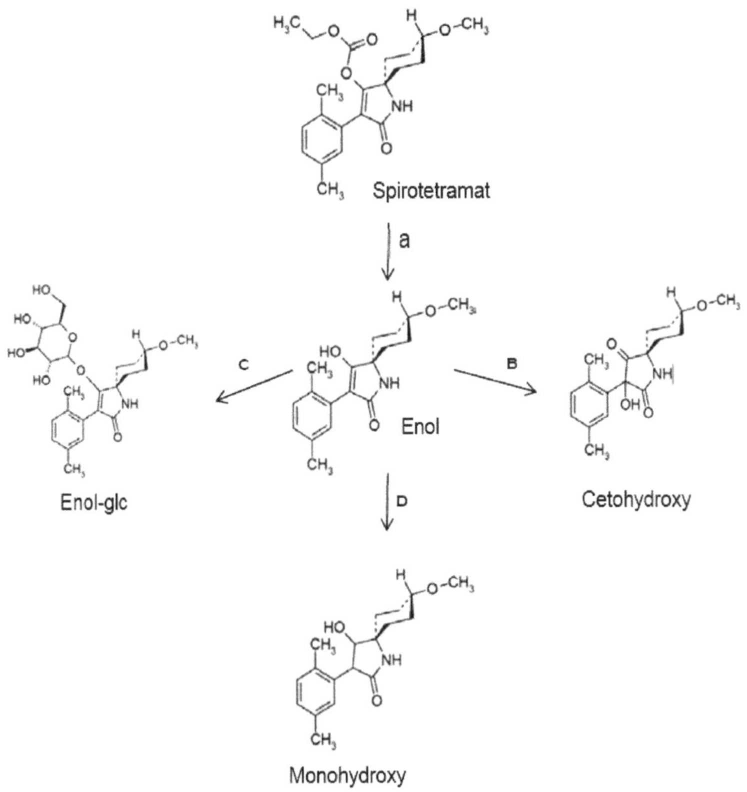

Figure 3. Principal reactions and metabolite of spirotetramat in plants: a) Hydrolytic split, b) Oxidation of the Pyrrole group, c) Conjugation of the hydroxyl group BYI08330-enol with glucose, and d) Reduction.

In the study designed for 127 days, under aerobic conditions, spirotetramat was degraded rapidly; a day after the application, only 53.6% and 72.2% of the substance was detected. There were two principal metabolites identified, BYI08330-cetohydroxy (maximum 25.3%) and BYI08330-enol (maximum 7.8%); there were also three minor metabolites detected, these were confirmed using the previous method established in laboratory studies. It was also observed that for the aerobic soil metabolism, under acidic extraction, the metabolite BYI08330-enol was partially unstable, and that like spirotetramat it dissipated using a two-phase kinetic [22].

As was mentioned earlier, the velocity of degradation of spirotetramat in the soil under aerobic conditions was very rapid. Under laboratory conditions, the degradation time (DT50), was from 0.14 days (geometric average); for the majority of the scenarios it was 0.21 days. In situations with trails under outside climatic conditions, spirotetramat also degraded rapidly, with a DT50 average of approximately 2 days. The velocity of degradation for BYI08330-enol in the soil under aerobic conditions was 0.08 days (DT50); this information allows us to conclude that this metabolite is the one that will degrade rapidly [22].

The soil degradation studies under field conditions with spirotetramat demonstrate that the dissipation velocity DT50 was between 0.3 and 1.0 days; the dissipation of 90% (DT90) was between 1.1 and 3.5 days. In the case of the combined residues of spirotetramat (BYI08330- enol, BYI08330-cetohydroxy), the DT50 was between 5.0 and 23.4 days, the DT90 had a range of 16.7 to 77.8 days. The residues of spirotetramat were not found to be below the shallow layer (0-15 cm), due to the possibility of the presence of leaching in subterranean waters was not probable. Considering that within 14 days after the application of spirotetramat it degraded to concentrations below 0.5 μg kg-1, the possibility of the accumulation of residues in the soil one year later after the first application is low [22, 27].

The photo-transformation of spirotetramat on the soil surface does not represent a process of degradation relevant to conditions of solar radiation. The trials undertaken to evaluate photo- transformation on the soil surface reveal that there are no different products derived from this effect after the application of spirotetramat [22, 23].

On the other hand, the anaerobic degradation in the soil follow almost the same route as under aerobic conditions, that is to say that no different metabolites are formed than those observed under aerobic conditions and it is concluded that it degrades rapidly [22, 23, 27].

Based on the literature discussed previously, the main route of spirotetramat dissipation in soil is the degradation to enol-BYI08330 and BYI08330-cetohydroxy; these followed by a degradation to non-extractable residues and mineralization to CO2. Concerning the mobility of the spirotetramat, the results showed that this pesticide can be classified as low mobility in soil. In the case of the BYI08330-enol, the strongly retained portion is considered stationary, while the weak form, as well as the BYI08330-cetohydroxy bound fraction possesses an intermediate leaching potential through the soil [22, 27].

4.4 Fate of spirotetramat in the aquatic environment

The research trials conducted demonstrate that spirotetramat is susceptible to degradation under biotic and abiotic processes in darkness as well as solar light. With reference to the abiotic degradation, the hydrolytic degradation becomes a relevant mechanism for the degradation of spirotetramat in the environment, especially under neutral and alkaline conditions. The half- life under hydrolytic conditions (20°C) at pH 7 is from 13 days, and at pH 9 it is less than half a day. On the other hand, hydrolysis does not represent a relevant degradation mechanism with regards to BYI08330-enol in the environment, the half-life at the pH range of 4 to 9, at 25°C is expected to be about a year [22].

The results of the photo-transformation in water demonstrate that this mechanism contributes in a significant way to the elimination of spirotetramat in natural water.

In systems with water/ sediment, spirotetramat is degraded rapidly through the metabolites BYI08330-enol and BYI08330-cetohydroxy. In the same system under anaerobic conditions, spirotetramat degrades rapidly, mainly into the metabolite BYI08330-enol. From the previous information and the evaluation of drinking water exposure, the use of spirotetramat does not represent a risk to human health [22, 23].

According to the results of toxicological studies isolated in Ceriodaphnia dubia, it was observed that mixing spirotetramat with an agricultural adjuvant (Destiny) caused more damage together than each one separately; this does not indicate synergy, but that each compound causes a certain level of mortality, and together the effect of the mixture is additive. This suggests that no further study is needed to determine which mixes of insecticides and adjuvants are causing damage to aquatic organisms [28].

4.5 Interaction of spirotetramat with the air

With a base vapor pressure of 5.6 x 10-9 Pa for spirotetramat and 1.2 x 10-10 Pa for l BYI08330- enol, it is expected that none of the two compounds will volatize when applied to the leaves or to the soil surface. Furthermore, considering the estimated life of these compounds in the air (maximum 3 hours); they are not expected to be able to travel in a gaseous state over large distances and as a result they cannot accumulate in the air [22, 27].

5. FIELD STUDIES OF SPIROTETRAMAT IN GRAPEVINE

It is essential to understand the course of action and toxicity of pesticides through the development of methods and procedures of bioassays because some chemicals (such as organo-phosphates, carbamates, and pyrethroids) can express their toxicity in a maximum time frame of 48 hours; other reduced risk pesticides such as spirotetramat, spinetoram, novaluron, chlorantraniliprole, and flubendiamide express their toxic effects several days after treatment [11]. For the majority of systemic insecticides, the primary route of entry is through the xylem, normally through the roots after directed applications. A foliar treatment of spirotetramat can be translocated acropetally and basipetally; it can also be ambimobil (movement across the xylem and phloem) and by this manner it can supply systemically to the top and bottom of the plant with a great potential for control of grape phylloxera [29].

The profile of the insecticide residuals in grape is influenced not only by its penetration and properties of translocation, but also by the active growth pattern of the vine that will produce an effect of dilution of the residues. If the compound is applied on the mature leaves through a foliar application, the drop in the residues will depend on the environmental degradation of the compound, given that the life size is constant during this stage [30]. Spirotetramat is an insecticide compatible with an integrated pest management approach; it has provided a new mode of action against sucking insects such as whiteflies, psyllids, and aphids [22]; it is effective in reducing the stages of A aurantii (citrus pest) and allows for the survival of the primary parasite A melinus [31].

In reference [29], they concluded that the goal of obtaining laboratory data would be useful for the implementation of a strategy to implement when using pesticides; it is essential to understand the attributes of the pesticide in question, the target organism (pest or beneficial), and the ecosystem that these organisms are present in.

6.THEPOSSIBLEGENERATIONOFRESISTANCETOSPIROTETRAMAT

Resistance to insecticides and myticides is one of the serious obstacles in the effective management of pests, and is a clear example of evolution and natural selection. True resistance is produced when there is a structural genetic change that could be hereditary. In contrast, tolerance is the natural ability of a population of arthropods to tolerate the toxic effects of a specific insecticide. This can occur through a physiological adaption in just one generation but by the same toxin, it can lose the effect if the insects are not exposed again to this toxin. Actually, the insecticide and myticide resistance in grape in North America is not a problem due to the existence of management programs [29].

Two types of resistance are recognized—behavioral and physiological. Behavioral resistance is defined as the capacity of the arthropod to avoid toxic doses that ordinarily would be lethal. On the other hand, physiological resistance is a question of hypersensitivity of the arthropod exposed to the compound, which depends on three factors: reduction in the penetration of the toxin, a better way to detoxify, and desensitizing the target destination [29].

The cases of documented insecticide resistance in aphids within the group of ketoenols is for spirodiclofen, where there have been strains observed in the laboratory and field populations of mites Tetranychus urticae [32-34], Panonychus citri [35], and Panonychus ulmi [36]. This information indicates a possible risk of resistance in aphids to spirotetramat. Recent studies by Pan and collaborators [37] report a strain of cotton aphid that develops spirotetramat resistance of 11.97 times by adults and 441.26 times by adult nymphs, in comparison with the susceptible strain. However, these lack the cross resistance to existing insecticides and for this reason it is considered a new tool in the management of insecticidal resistance for cotton aphid.

A proteomics study based on identification and analysis of proteins associated with the mechanism for tolerance to spirotetramat in Aphis gossypii Glover detected approximately 493 associated protein points that possibly may confer resistance to spirotetramat for the cotton aphid [38]. Knowledge generation involving proteomic resources are expected to contribute to a better understanding of the development of resistance to spirotetramat.

7. CONCISE SUMMARY

Spirotetramat functions as a lipid biosynthesis inhibitor and demonstrates significant efficacy against key aphid species, including whitefly. The activity of spirotetramat is predominant in the immature developmental stages of aphids may occasionally result in partial ecdysis, leading to diminished productivity and fertility. The findings indicate that spirotetramat may be utilized within a secure integrated pest management framework that incorporates biological control; nevertheless, there exists conflicting evidence in this regard, necessitating independent evaluation for each specific organism employed in biological control.

Upon detection within the plant, spirotetramat is converted into its enol form and the metabolite cetohydroxy, which are the two primary breakdown products. The likelihood of spirotetramat leaking into groundwater is minimal, and the probability of soil accumulation is similarly exceedingly low. Conversely, the findings in aquatic environments indicate that, under specific situations, accumulation may occur; hence, appropriate precautions should be observed in applications near these systems. Moreover, no evidence was found suggesting the potential accumulation of spirotetramat in the atmosphere. Finally, to date, proteomic investigations have indicated specific kinds of resistance to spirotetramat in cotton aphids.

REFERENCES

1. Nauen R, Reckmann U, Thomzik J, Thielert W. Biological profile of Spirotetramat (Movento®)—a new two-way systemic (ambimobile) insecticide against sucking pest species. Bayer CropScience Journal. 2008;61(2):245-277.
2. Brück E, Elbert A, Fischer R, Krueger S, Kühnhold J, Klueken AM, Nauen R, Niebes JF, Reckmann U, Schnorbach HJ, Steffens R, van Waetermeulen X. Movento, an innovative ambimobile insecticide for sucking insect pest control in agriculture: Biological profile and field performance. Crop Protection. 2009;28:838-844.
3. Mallinger A, Nadal B, Chopin N, Le Gall T. One-Pot synthesis of 3-aryltetramic acids. European Journal of Organic Chemistry. 2010;6:1142-1148. DOI: 10.1002/ejoc. 200901092.
4. Zhao JH, Ji MH, Xu XH, Cheng JL, Zhu GN. Synthetic derivatives of spiromesifen and their bioactivity research. Chinese Chemical Letters. 2009;20:1307-1310.
5. Fischer R, Weiß HC. Spirotetramat (Movento®)—discovery, synthesis and physico chemical Properties. Bayer CropScience Journal. 2008;61(2):127-140.

6. Babcock JM, Gerwick CB, Huang JX, Loso MR, Nakamura G, Nolting SP, Rogers RB, Sparks TC, Thomas J, Watsona GB, Zhua Y. Biological characterization of sulfoxaflor, a novel insecticide. Pest Management Science. 2010;67:328-334.

7. Kühnhold J, Klueken AM, de Maeyer L, van Waetermeulen X, Brück E, Elbert A. Movento®, an innovative solution for sucking insect pest control in agriculture: Field performance in fruits and vegetables. Bayer CropScience Journal. 2008;61(2):279-306.

8. Vermeer R, Baur P. Movento® product development: Custom-made formulations for an exceptional active ingredient. Bayer CropScience Journal. 2008;61(2):141-157.

9. Moens J, De Clercq P, Tirry L. Side effects of pesticides on the larvae of the hoverfly Episyrphus balteatus in the laboratory. Phytoparasitica. 2011;39:1-9. DOI 10.1007/ s12600-010-0127-3.

10. Mansour R, Suma P, Mazzeo G, Grissa Lebdi K, Russo A. Evaluating side effects of newer insecticides on the vine mealybug parasitoid Anagyrus sp. near pseudococci, with implications for integrated pest management in vineyards. Phytoparasitica. 2011;39:369-376. DOI: 10.1007/s12600-011-0170-8.

11. Lefebvre M, Bostanian NJ, Thistlewood HMA, Mauffette Y, Racette G. A laboratory assessment of toxic attributes of six 'reduced risk insecticides' on Galendromus occidentalis (Acari: Phytoseiidae). Chemosphere. 2011;84:25-30.

12. Hall DG, Nguyen R. Toxicity of pesticides to Tamarixia radiata, a parasitoid of the Asian citrus psyllid. BioControl. 2010;55:601-611.

13. Jamieson LE, Page-Weir NEM, Chhagan A, Curtis C. The efficacy of insecticides against Australian citrus whitefly (Orchamoplatus citri). New Zealand Plant Protec tion. 2010;63:254-261.

14. Page-Weir NEM, Jamieson LE, Chhagan A, Connolly PG, Curtis C. Efficacy of insecticides against the tomato/potato psyllid (Bactericera cockerelli). New Zealand Plant Protection.

15. Smiley RW, Marshall JM, Yan GP. Effect of foliarly applied spirotetramat on reproduction of Heterodera avenae on wheat roots. Plant Disease. 2011;95(8):983-989. DOI: 10.1094/ PDIS-01-11-00172011;64:276-281.

16. Duvaresch D, Papa G, Rotundo M, Furlan RS, Takao W. Effect of new insecticide (spirotetramat) in mixture with neonicotinoid on the control of aphids, Aphis gossypii (hemiptera:aphididae), in cotton. Beltwide Cotton Conferences, Nashville, Tennessee, January 8-11, 2008:1381-1384.

17. Kay IR, Herron GA. Evaluation of existing and new insecticides including spirotetramat and pyridalyl to control Frankliniella occidentalis (Pergande)

(Thysanoptera: Thripidae) on peppers in Queenslandaen. Australian Journal of Entomology. 2010;49:175-181.

18. Fu AA, Del Real A. Estudio de evaluación de la efectividad biológica del insecticida spirotetramat en el control del piojo harinoso de la vid Planococcus ficus (hemiptera:pseudococcidae) en la costa de Hermosillo, Sonora. XII Congreso Internacional en Ciencias Agrícolas. ICA-UABC 2009;29:295-299.

19. Marcic D, Petronijevic S, Drobnjakovic T, Prijovic M, Peric P, Milenkovic S. The effects of spirotetramat on life history traits and population growth of Tetranychus urticae (Acari: Tetranychidae). Experimental and Applied Acarology. 2012;56:113-122. DOI 10.1007/s10493-011-9500-2.

20. Frank SD, Lebude A. Season-long insecticide efficacy for hemlock woolly adelgid, Adelges tsugae (hemiptera: adelgidae), management in nurseries. Florida Entomologist. 2011;94(2):290-295.

21. Elizondo AI, Murguido CA. Spirotetramat, nuevo insecticida para el control de insectos chupadores en el cultivo de la papa. Fitosanidad. 2010;14(4):229-234.

22. Babczinski P, Hellpointner E. Environmental fate of Spirotetramat (Movento). Bayer CropScience Journal. 2008;61(2):181-201.

23. FAO (2011). SPIROTETRAMAT, First draft prepared by Mr. Stephen Funk, US Environmental Protection Agency, USA. Available from http://www.fao.org/fileadmin/templates/agphome/documents/Pests_Pesticides/JMPR/Evaluation08/Spirotetramat.pdf.

24. Sur R. Metabolism of spirotetramat (Movento®) in plants. Bayer CropScience Journal. 2008;61(2):227- 243. Codex alimentarius. Available from http://www.codexalimentarius.net/pestres/data/commodities/index.html?id=CM347&expand=TP2.

25. Maus C. Ecotoxicological profile of the insecticide spirotetramat. Bayer CropScience Journal 2008;61(2):159-180.

26. EPA (2011). Available from http://www.gpo.gov/fdsys/pkg/FR-2011-05-18/html/ 2011-11937.htm. Australian Pesticides and Veterinary Medicines Authority (2009). Available from http://www.apvma.gov.au/registration/assessment/docs/prs_Spirotetramat.pdf.

27. Chen XD, Stark JD. Individual- and population-level toxicity of the insecticide, spirotetramat and the agricultural adjuvant, Destiny to the Cladoceran, Ceriodaphnia dubia. Ecotoxicology. 2010;19:1124-1129.

28. Bostanian NJ, Wise JC, Isaacs R. Pesticides for Arthropod Control in Vineyards. In: Bostanian NJ, Vincent Ch, Isaacs R, editors. Arthropod Management in Vineyards:

29. Pests, Approaches, and Future Directions. Springer Science+Business Media BV; 2012. p. 53-90. DOI: 10.1007/978-94-007-4032-7_4.

30. Timmeren S van, Wise JC, Vandervoort C, Isaacs R. Comparison of foliar and soil formulations of neonicotinoid insecticides for control of potato leafhopper, Empoasca fabae (Homoptera: Cicadellidae), in wine grapes. Pest Management Science. 2011;67:560-567.

31. Garcerá C, Ouyang Y, Scott SJ, Moltó E, Grafton-Cardwell EE. Effects os Spirotetramat on Aoinidiella aurantii (Homoptera: Diaspididae) and its parasitoid, Aphytis melinus (Hyminoptera: Aphelinidae). Horticultural Entomology. 2013;106(5):2126-2134.

32. Van Pottelberge S, Van Leeuwen T, Khajehali J, Tirry L, Genetic and biochemical analysis of a laboratory-selected spirodiclofen-resistant strain of Tetranychus urticae Koch (Acari: Tetranychidae). Pest Management Science. 2009;65:358-366.

33. Khajehali J, van Nieuwenhuyse P, Demaeght P, Tirry L, van Leeuwen T. Acaricide resistance and resistance mechanisms in Tetranychus urticae populations from rose greenhouses in the Netherlands, Pest Management Science. 2011;67:1424-1433.

34. Ilias A, Roditakis E, Grispou M, Nauen R, Vontas J, Tsagkarakou A. Efficacy of ketoenols on insecticide resistant field populations of two-spotted spider mite Tetranychus urticae and sweet potato whitefly Bemisia tabaci from Greece. Crop Protection. 2012;42:305-311.

35. Hu J, Wang C, Wang J, You Y, Chen F. Monitoring of resistance to spirodiclofen and five other acaricides in Panonychus citri collected from Chinese citrus orchards. Pest Management Science. 2010;66:1025-1030.

36. Kramer T, Nauen R. Monitoring of spirodiclofen susceptibility in field populations of European red mites, Panonychus ulmi (Koch) (Acari: Tetranychidae), and the cross-resistance pattern of a laboratory-selected strain. Pest Management Science. 2011;67:1285-1293.

37. Pan Y, Yang C, Gao X, Peng T, Bi R, Xi J, Xin X, Zhu E, Wu Y, Shang Q. Spirotetramat resistance adaption analysis of Aphis gossypii Glover by transcriptomic survey, Pesticide Biochemistry and Physiology. 2015. DOI: 10.1016/j.pestbp.2015.04.007.

38. Xi J, Pan Y, Wei Z, Yang C, Gao X, Peng T, Bi R, Liu Y, Xin X, Shang Q. Proteomics based identification and analysis proteins associated with spirotetramat tolerance in Aphis gossypii Glover. Pesticide Biochemistry and Physiology. 2015;119:74-80.

Chapter 4 MANAGEMENT STRATEGIES FOR INSECT RESISTANCE IN BT MAIZE

1. INTRODUCTION: AN OVERVIEW

The history of integrated pest management in Brazilian soybeans is associated with the evolving paradigm of pest control that emerged in the 1960s, a time when global awareness of the hazards of excessive pesticide application was heightened [1, 2]. This fact led to government regulations aimed at diminishing the usage of these poisons by providing incentives for the implementation of integrated pest management programs across various crops. At that point, the notion of integrated management gained prominence and is now regarded as a significant technological advancement. The Integrated Pest Management (IPM) seeks to amalgamate several management strategies rather than depending solely on insecticide application [3]. The approach fundamentally entails a decision-making process that employs the coordinated use of several techniques to optimize the management of all pest categories in a sustainable and economically viable manner [4]. This management paradigm has proliferated globally and reached Brazil, where it has been swiftly integrated for pest control, particularly for maize cultivation.

Brazilian agriculture has evolved in recent years, with significant yield gains in many economically important crops, including maize. Besides the use of agricultural inputs with quality and cutting-edge technologies, the climate in general has contributed to the increase in production. Despite these favorable factors, phytophagous insects also continue to be a cause for concern in agribusiness because of its great ability to adapt to changes seen in production systems. The solution adopted to reduce the losses arising from the injury caused by pests has been, in most cases, the application of chemical insecticides. It is relatively easy to understand the reasons that have led the farmer to choose chemical control as a control tactic of "recognized" insects to cause economic losses.

The general understanding is that chemical insecticides have indisputable advantages: low cost, acts quickly, little demand in knowledge, and can be used to control various pest species. Other causes for probably using chemical control almost in a predominant way is the lack of knowledge about other control tactics or even the lack of conviction about the effectiveness of these alternative methods such as integrated pest management.

The efficiency of chemical control is often not the expected by the simple fact that it depends on several technical factors, which most often are not considered when applying the product. Sprayer type, application type, nozzles for spraying, droplet size needed for good plant cover, solution volume, application speed, climatic factors (such as wind speed, rainfall, temperature, and humidity), phenological stage of the plant and of the pest target, attack site and economic level damage of the pest, and even the applicator qualification are some factors that can compromise the action of the applied product. Therefore, when these factors are not considered, the probability of not achieving the expected results is high.

Upon encountering problems in managing a certain pest, the farmer must collaborate with an expert to assess the reasons for the ineffectiveness of the strategies implemented thus far. Failure to conduct this analysis invariably increases the danger of erroneous decision-making, potentially exacerbating an already identified condition. For instance, developing new applications or altering active ingredients without acknowledging that the causes may not be associated with the utilized product. Excessive applications incur increased costs, with no evidence supporting their efficacy in adequately controlling the insect.

2. INTEGRATED PEST MANAGEMENT (IPM)

The failure to reduce the population of a phytophagous species recognized as a key pest in a given situation usually occurs by not using the principles of integrated pest management (IPM). By its name, it is already implied that the solution of phytosanitary issues should not be thought of in a single tactic, even though apparently it may be "very practical" and convenient for the farmer. Because of lack of implementation of IPM, it is easy to understand statements and questions, such as the following: there are cases of pest resistance to insecticides; it is necessary to increase the dose of products; there are chemical residues in the soil, water, and harvested products; there are pesticides effects on flora and fauna; cost of control is prohibitive.

Brazil is already becoming a major agrochemical consumer in the world. Information of this nature in the media needs to be changed. What is expected is to reach the point of having an IPM program where there is satisfaction of both the farmer and the consumer, including the protection of the environment as a whole [5].

2.1 Fall armyworm (Spodoptera frugiperda) and Helicoverpa armigera

In maize (Zea mays L.), historically, fall armyworm has been the main pest. However, it is interesting to note that even with the advance in science in providing new technologies for their control, the pest is still present in the agroecosystem causing losses to agribusiness, even with use of means for their control. The lack of IPM should again be highlighted. As a main feature, the pest has as hosts several cultivated plant species or native species, available year- round in Brazil. Furthermore, general climatic conditions in the country are not limiting for their development.

As a result of this, the moth can be detected all year round in traps containing pheromone as attractive. That is, there is the real possibility of having the presence of caterpillars in their hosts just after the emergence of the plant and also throughout the plant cycle.

The moth of S. frugiperda lays its eggs in clusters, and each cluster contains up to 300 eggs. At hatch, the aggregate larvae begin to feed, in a short period of time, in the plant where the eggs were placed, by ragged feeding the leaf, leaving a characteristic symptom. Subsequently, they migrate into adjacent plants. In the migration process, the caterpillar produces a thread that is adhered to the leaf and is projected into the air by the wind and is easily carried to other plants. Often, in the new plant, the caterpillar goes toward the interior of the leaves still rolled ("whorl") without causing the symptom of "ragged feeding."

This symptom has been used as an indicator to control the pest. However, for the reasons mentioned, it may underestimate the level of infestation. The caterpillar phase lasts around 21 days, influenced mainly by temperature. When the initial infestation coincides with the "whorl" phase, the caterpillar remains housed in this location. When fully developed, by reaching the size between 4 and 5 cm, it leaves the whorl and heads to the ground, where it becomes known as the pupa stage. Eleven days after this stage, the adult emerges, which will restart another cycle. The period of time between laying the eggs and the appearance of a young adult insect varies between 35 and 45 days. Therefore, from the same oviposition, there could be at least three discrete generations during a maize cycle. However, because the flow of moths is constant, it normally occurs in overlapping generations, and therefore caterpillars and postures of different ages can be found in the same plant at the same time. This fact is usually a complicating factor for applying insecticides via spraying. Caterpillars of different ages require different doses of the applied insecticide, either chemical or biological. Therefore, the management of the pest is essential to know not only the pest population density but also the distribution by age-group of insects. It is not an easy task when making the decision on the need for control after sampling only based on symptoms of damage such as the scraping of leaves. A decision on the need to control based on captured moths in pheromone trap, when placed in the area soon after planting, is a more efficient procedure than those based on the injury symptom of the pest.

Because the continuous flow of S. frugiperda moths is not uncommon, the presence of postures and caterpillars in more developed maize plants.

However, such insect stages are not easily observed in the plant when there is no more whorl. Often the insect can be found feeding in the tassel or ear, in silks or directly in the grains, causing direct damage to yield, because any method of control via spraying is difficult to apply in these places. In addition to machines handling difficulties in the target area, there is also the difficulty of reaching the pest, protected by the leaves or the ear husks. The presence of larvae of S. frugiperda in the ear can be as common as corn earworm itself, the Helicoverpa zea. Insect infestations in the ear can cause severe damage to the farmer, as it jeopardizes the expected yield. Therefore, alternative methods should be prioritized to such pests in maize. Obviously, to reduce the population density of fall armyworm in the ear, there must be a proper pest management in the previous stages of insect development.

Morphologically, the new species is very similar to Brazilian corn earworm (H. zea) and also presents a very similar life cycle. However, its potential for destruction to the preferred hosts is undeniable. Due to these similarities, the problem that initially occurred in Bahia was ascribed to H. zea.

The oviposition is usually performed on the style stigma of maize. At hatch, the larvae consume grains in development. Secondary bacterial infections are common in the ear. The larvae can also feed on the new leaves of the whorl, from the most developed leaves and from the tassel of the plant. Mobility, polyphagia, and high reproductive rate are attributes that differentiate H. armigera from H. zea. The caterpillars are quite aggressive, occasionally carnivorous, and can be cannibals when the opportunity arises. If disturbed, they drop from the plant and wound up on the ground. Caterpillars turn into pupae within a cocoon silk, some centimeters below the soil surface. Under favorable conditions, the development cycle can be completed in just over a month. Therefore, several generations per season are possible, especially in warmer areas.

In the tropics, reproduction continues throughout the year. H. armigera, also called the "old world caterpillar," is usually found in parts of Europe, Asia, Africa, and Australia, while H. zea, the caterpillar of the "new world," is common in the Americas. The pest is more abundant in maize during the phase of "silking," when the adult female lays the egg individually on the style stigma. Adults feed on nectar or on other exudates from different plant species. The young larvae tend to feed initially on the style stigma but soon start to feed on the grain in formation. There are six larval stages, and the fully developed larva measures about 40 mm long. Third, instar caterpillars (8–13 mm long) and so on account for 90% of all food consumed (and thus its damage). Large caterpillars (above 24 mm) are the most harmful ones once they consume approximately 50% of their diet, when they are between the fifth and sixth instars. Therefore, control measures should be directed when the caterpillars are still small (less than 10 mm).

The pupa is dark brown, measuring between 14 mm and 18 mm in length, with a smooth surface, rounded, with two parallel spines on the posterior end. The moth has a wingspan between 30 and 45 mm. Females lay over a thousand eggs in her lifetime. The ease with which the pest acquires resistance to insecticides has been considered a hallmark of the species in areas where the pest usually occurs. At these locations, the development of resistance has been most extensively documented for synthetic pyrethroids, but already there is record of resistance to other groups of insecticides as carbamates and organophosphates.

The migration movements of the species could explain the resistance propagation. In regions of origin, research has shown that maize is among the preferred hosts of the pest, followed by soybeans and cotton. In Brazil, the simultaneous presence of these three crops in the same region is common, as occurred in western Bahia, the starting point of an outbreak of the pest. However, H. armigera can survive in more than 300 taxa of plants.

2.1.1 Control strategies

The occurrence of insects in the ear, in general, makes the management more complex. Besides the difficulty of monitoring, there is also the difficulty of reaching the pest, protected by the leaves or the ear husks. Therefore, alternative methods should be prioritized for such pests. In the specific case of the fall armyworm, one should make a proper management also during the vegetative stage of corn.

2.1.2 Biological control

The production system for maize is the pest combat, including species that attacks the ear. A first reason is the less frequent use of chemical insecticides. This fact can be explained by the use of Bacillus thuringiensis Berliner (Bt) plants, whose consequence was a significant reduction in foliar sprayers to control fall armyworm during the growing phase of the plant. Additionally, in the case of pests that attack the ears, because the caterpillars stay housed under the straw, which reduces its exposition to chemical spraying, there is greater difficulty in controlling by other methods.

The egg phase has been considered critical in the life cycle of many species of insects belonging to the order Lepidoptera. For example, to H. armigera always occurs a high rate and natural mortality, reaching values above 88%, mainly in the first 3 days of oviposition. Such index can reach 95%, considering the mortality of eggs and the first larval stages. Significant indexes have also been verified for H. zea.

Species of Trichogramma mainly (Hymenoptera: Trichogrammatidae) and, in a lower degree, Telenomus (Hymenoptera: Scelionidae) are common egg parasitoids. Among larvae parasitoids, the most common include Cotesia spp. and Microplitis croceipes (Cresson) (Hymenoptera:

Braconidae); Campoletis spp. (Hymenoptera: Ichneumonidae); and Eucelatoria spp. and Archytas marmoratus (Townsend) (Diptera: Tachinidae).

Considering the commercial existence and experience in releasing Trichogramma in Brazil and in other countries, this biological control agent is recommended for both conventional and Bt maize planting. The inundative release of parasitoid should be associated with the monitoring of moths in the target area.

This monitoring is carried out with traps containing synthetic sexual pheromone, specific to each type of target pest. The release of the parasitoid can be made by the distribution of card plants containing parasitized eggs near the emergence of the adult parasitoid or the direct adult release. As the parasitoid has an objective to target the pest, it can also be used in soybeans, cotton, and other crops where pests cause economic damage, regardless of the size of the cultivated area.

Obviously, one should consider that chemical insecticides required for other targets must not be applied at the same time of the release of the biological control agent.

Reduced use of chemical insecticides, through the use of applied biological control, leads to the gradual return of other biological control agents. In maize, over 100 insect species have already been described as predators of phytophagous species that feeds on both eggs to larvae. Some species prey in both the immature stage, as in the adult stage. Among the most common predators are lady beetle Hippodamia convergens Guerin-Meneville and Coleomegilla maculata DeGeer (Coleoptera: Coccinellidae), lacewings such as Chrysoperla spp. (Neuroptera: Chryso□ pidae), minute pirate bugs such as Orius spp. (Hemiptera: Anthocoridae) and Geocoris spp. (Hemiptera: Lygaeidae), and earwigs such as Doru luteipes (Dermaptera: Forficulidae) and Euborelia spp. (Dermaptera: Carcinophoridae).

2.1.3 Microbial control

Viruses, bacteria, and fungi have also been used against pests of maize. Especially for H. armigera control, in the literature, although they mention the use of baculovirus, they highlight the increase use of Bt. The control efficiency with microorganism depends largely on the period and the application technique because caterpillars cannot be protected within the ear.

In Brazil, there is a great experience in using baculovirus to control fall armyworm. In other countries, a commercial product based in nuclear polyhedrosis virus (NPV) already exists, such as, for example, in the USA, to control H. zea and H. armigera. To achieve success in the control ear pests with virus, the product must be applied in order to hit the target, both for the attack location as compared to the caterpillar development stage, which cannot be greater than 10 mm in length.

Similarly to the virus, maize ear pests can be controlled by the Bt-based products. However, the efficiency of the products depends essentially on adjustment of the solution volume (liters of spray solution/ unit area) that can be evaluated by the use of sensitive papers which should obtain a minimum number of 30 drops cm-2.

2.1.4 Chemical control

The same cares from microorganism applications are valid for the use of chemical insecticides. In addition to the restrictions, the possibility of a negative action of product on populations of natural enemies is considered. This fact is critical when it comes to reaching a target pest that is generally protected against the action of the applied products.

On the other hand, the exposure period of the pest to the action of the chemical is very small, and therefore, the application of pesticides must follow a strict pest monitoring system and thus avoid adverse effects on nontarget insects.

Eggs and larvae are often not sampled in corn because eggs are difficult to detect among the silks and caterpillars are generally being within the ear, making it a costly and low-precision process.

The moths, however, can be monitored by light traps and pheromone traps. Both genders are caught in light traps and only males are attracted by the pheromone. Both types of traps give an estimate of when the moths invade or emerge in a given area. However, pheromone traps are easier to use because they are selective. The pheromone is usually used in conjunction with a suitable trap, the inverted cone type, or the Delta type. Moreover, the presence of three to five moths per night is sufficient to indicate that pest control measures must be taken.

Therefore, improvements in cultural practices to maintain and enhance the impact of natural enemies represent an excellent strategy to improve the perspective for the natural biological control. Growing plants around the main crop and that attract natural enemies, such as sunflower, should be encouraged. The "trap crop" is often suggested for several species of pest, including the ear pest complex. It should, however, consider the high degree of preference of moths to lay eggs on maize in early stage of development of silks. Planting small plots of maize before the main crop can be interesting because the farmer can thus eliminate the initial infestation of the pest before their population grows enough to cause damage to the main crop.

In areas where pest populations initially develop into weeds and then disperse for major crops, the elimination of these plants by mowing or using herbicides or even applying insecticides can significantly reduce damage to neighboring crops.

2.2 Sugarcane borer (Diatraea saccharalis)

Indirect losses caused by this pest are more important economically because of the galleries built inside the stalk, thus the plants become more susceptible to tipping, tassel infertility, and reduced productivity and still favor the entry of opportunistic pathogens. According to EMBRAPA [6], by attacking the interior of the stalk of the plant, the larvae cause damage that can result in losses between 10% and 50% on yield. The highest losses are results from attacks in the internodes that are closer to the ear because it results in interference in the movement of nutrients produced by the plant, which are carried to a higher production of leaves instead of grain production.

The adult, with nocturnal habits, has the aspect of moth, with the forewings of a straw-yellow color, some brownish drawings, and whitish hind wings and a 25-mm wingspan. The caterpillars measure approximately 22–25 mm long, with brown head and whitish/yellowish body with numerous dark spots.

Regarding the cycle, oviposition is made in maize leaf after mating, generally on the dorsal side. The number of eggs in each oviposition is from 5 to 50, with an imbricated posture, resembling snake leather or fish scale. Immediately after hatching, and upon reaching the second instar, they enter the stem. Its attack can be identified by the inlet and outlet holes, as well as the longitudinal opening of the maize stalk, where the presence of the caterpillar or the passageway left by it is observed.

In high infestations, the attack of this insect can cause losses up to 21% in production. It can attack 65 plant species such as sugarcane, maize, millet, sweet sorghum, wheat, grain sorghum, and rice, besides many other grasses (Poaceae) and spontaneous weeds such as Sorghum halepense, Paspalum sp., Panicum spp., and Holcu ssp. Moreover, Andropogon ssp. The larvae damage maize in various ways: in small plants, by attacking the whorl, causing holes in the leaf blade to the death of the meristem. In more developed plants, they open galleries, feeding on the stem. These galleries are usually longitudinal but may present circular aspect, making the plant very susceptible to falling. Damages can also occur in the ear, allowing the cross infestation with weevils Sitophilus spp.

2.2.1 Methods of control

2.2.1.1 Chemical control

Depending on the behavior of this pest, chemical control usually does not present satisfactory result, unless the attack begins very early. In this case, seed treatment with systemic insecticides or pyrethroid sprays directed toward the base of the plant gives good results.

2.2.1.2 Biological control:

In the past 60 years, the biological control of this pest in sugarcane crop has been successful with the caterpillar parasitoid Cotesia flavipes and, more recently, with the egg parasitoid Trichogramma galloi and may be extended to the control methodology for the maize crop.

2.2.1.3 Mechanical control

Elimination of crop residues and host plants, especially grasses (Poaceae), help reduce the infestation for the next crop season.

2.3 Black cutworm (Agrotis ipsilon)

Black cutworm, from the genus Agrotis, constitute an important group of insect pests, mainly due to damages to the large number of cultivated plants and their wide geographic distribution. Agrotis ipsilon is the main species of black cutworm referred to in Brazil and is a polyphagous insect, which attacks mainly horticultural crops [7]. It can also attack other species of different plant families, in crops such as maize, soybeans, beans, and cotton [8].

The adults of this pest are moths with a 35-mm wingspan, whose anterior wings are brown with some black spots, and posterior are hyaline white, with a gray lateral edge [9]. Eggs are deposited on the shoot of the plant, stalks, stems, or on the ground near host plants; they are whitish and may be found individually or in groups. Each female can lay over a thousand eggs in a lifetime [8].

After the first instar, the caterpillars are directed to the ground, where they remain protected during the day. They measure up to 5 cm in length, are robust, smooth, and in a variable coloration, with a predominance of dark gray and brown with black spots. They have nocturnal habits and are housed in the soil under debris during the day [11].

Regarding the cycle, after 4 days of the oviposture on the leaves, the caterpillars emerge. After approximately 30 days, they become pupae and remain in the soil for a period of 10 to 20 days until they become adults. The process varies 34–64 days (egg: 4; caterpillar: 20–40; pupae: 10– 20). A female can lay up to 1,260 eggs, with a preoviposition period of 3 days [8].

The caterpillars attack at night, and to find them during the day, you need to revolve the soil at the base of the host plant. The main damage occurs on the establishment period of the crop when the caterpillars cut the young plants—seedlings of up to 20 cm—tumbling them and may cause high reduction of the stand. However, attacks in older plants can occur, which in this case will demonstrate the presence of cut leaves or galleries open at the stem base (they can cause the symptom of "dead heart") or more shallow roots.

When the death of plant is not observed, the attack causes tillering. It is not common to see small caterpillars exerting plant cutting activity; they often destroy the leaf blade and the petiole [8, 9].

2.3.1 Methods of control

To have an effective system of control for this pest, we recommend the use of various tactics of control, individually or harmoniously, creating a management strategy based on cost- benefit analyzes and with a reduction on the impact on the farmers, the society, and the environment adopting IPM.

2.3.1.1 Cultural control

Early desiccation is a practice that can reduce the infestation of Agrotis spp. since the moths prefer to lay eggs on plants or crop residues still green. The highest incidence of attack occurs in areas of not cleaned and heavy soil. In this way, the correct postcultivation management is indispensable to keep the pest below the economic injury level [9].

2.3.1.2 Insecticide application technology

Due to the nocturnal habit of this pest, another management tactic that is important is the quality of pesticide application technology. This must be done directing the jet spray to the base of the plant, preferably in the early evening and with a high solution volume [8].

2.3.1.3 Seed treatment

Due to the nocturnal habit of the pest and the difficulty of being hit directly by pesticides, seed treatment with systemic insecticides can be very effective to control this pest. This practice has shown to be even more efficient in areas with history of high occurrence and recurrence.

2.3.1.4 Chemical control

As an emergency control, chlorpyrifos can be used in spraying, preferably in the early evening.

2.3.1.5 Biotechnology

The use of genetically modified seeds with insecticidal proteins can be a tool to control this pest but is more effective to control small caterpillars [8].

2.4 Cornstalk borer (Elasmopalpus lignosellus)

It is very difficult to manage cornstalk borer in sandy soils (well drained) and under cerrado vegetation (savannah) (especially in the first year of cultivation) in dry periods with high temperatures, in particular in the first 30 days after emergence. Just as the black cutworm, the cornstalk borer causes damage also known as "dead heart" and causes significant losses in the stand.

The moth of nocturnal habits has a 1.5- to 2.5-cm wingspan and has gray-yellowish wings. It lays eggs preferably in the base of plants or in the soil, which are initially clear, but with the approach of the hatching become dark red. The caterpillar has blue-green color, with brown, purple, or dark brown transverse stripes, and measures about 1.5 cm [8].

It is a sporadic pest, however, polyphagous; it feeds from diverse crops (such as soybeans, maize, and cotton), with great capacity for destruction in a short period of time, especially between VE and V3 stages. After hatching, the caterpillar scrapes the plant leaves and starts its penetration in the stem remaining in this location during the day. It builds a shelter with web and dirt, which is attached to the gallery's opening also made by it, where droppings are being accumulated. Its damages are associated with drought after plant emergence, and the greatest damages are observed in conventionally tilled fields, with light, well-drained soil, and lower damages in sites with tillage and irrigation.

In maize, it feeds inside the stem and goes upward toward the growing point of the plant (apical bud), eventually damaging it, causing reduction in size or even death of the youngest leaves, a symptom known as "dead heart." In certain situations, the attack symptoms of cornstalk borer do not necessarily cause the dead heart but shoots at the base of the plant and present symptoms very similar to the attack of green belly stink bug (Dichelops spp.).

In soybeans and cotton, cornstalk borer feeds on the stem and branches of seedlings, causing wilting, drying, tipping, and even death. In larger plants, the pest opens galleries inside the stem. The damage is greater when the attack occurs early in the development of culture, when the young plants are eaten and have less ability to recover. During the larval stage, the insects are highly mobile and can migrate from dead plants to live ones and can cause major damage and even failure in planting lines. They also cause drying and death of plants, necessitating replanting [8].

2.4.1 Methods of control

2.4.1.1 Chemical control

Can be accomplished by seed treatment with systemic insecticides. Insecticides applied soon after the appearing of the pest have not shown satisfactory results, making the best option the preventive control.

2.4.1.2 Cultural control

In regions with high incidence of pest, increased seed density per area may be an alternative. Maintaining humidity also contributes to the decrease of the attack of this pest [8].

2.5 Corn earworm (H. zea) and (H. armigera)

Due to the moth habit of depositing eggs on the plant stigma and the caterpillar developing inside the ear, H. zea is called corn earworm.

It has pronounced larval movement in different crops and is aggressive when touched, adopting a defensive posture. The pupal development occurs in the soil and can occur optional diapause depending on weather conditions.

H. armigera has a higher attack spectrum than H. zea. In addition to maize, cotton, soybean, and tomato crops, the preferred targets of H. zea, it also attacks beans and sorghum, which causes damages to vegetative and reproductive structures.

Caterpillars of Helicoverpa spp. perform predation of other species of caterpillars and also on the same species (cannibalism) [8].

They have a high fertility rate and can occur up to 11 generations of the pest, with night oviposition preferably and capacity of laying 2,200–3,000 eggs on host plants, but with no predilection for specific parts of the plant [8].

For this reason, it feeds inordinately of all plant structures at an early stage, with preference for the reproductive structures in final stages of development [8].

2.5.1 Methods of control

One of the key points for success in controlling H. armigera and H. zea is to correctly identify the pest in the field, mainly due to its similarity to Heliothis virescens, the tobacco budworm.

It presents different behavior in relation to this pest, with aggression and resistance to insecticides based on synthetic pyrethroid characteristics [11], the joint use of agricultural practices and the integrated management of pests in a correct manner are essential.

2.5.2 Integrated pest management

The use of integrated agronomic systems, combining knowledge of the target pest, the constant monitoring of the crops that are in the system, and the adoption of practices that aimed cultural control and biological maintenance, combined with the use of biotechnologies to fight pest, are suitable forms of maintenance and control of Helicoverpa spp.

2.5.3 Chemical control

The use of insecticides from the chemical group diamides has shown satisfactory control in the fight against the pest.

2.5.4 Adoption of Bt maize is occurring rapidly

With only 6 years of the release of its cultivation by CTNBio, over 70% of the Brazilian maize crops were coming from transgenic crops, and it is projected to increase to 81%, which represents the cultivation area with intensive use of technology [12].

2.6 Corn rootworm (Diabrotica speciosa)

Among the six species of Diabrotica occurring in the tropics, Diabrotica speciosa is distinguished by economic importance to maize crops. This species is a polyphagous pest widely distributed in Brazilian states and in some countries in South America. The adults damage the shoots of various crops such as horticultural crops (solanaceous, cucurbits, crucifers plants), beans, soybeans, sunflower, and maize, causing defoliation and in some cases are vectors of pathogens. When adults feed, it transmits numerous viruses to plants. The viruses are easily transmitted mechanically and produce highly antigenic responses. The transmission of the virus from one insect to another is associated with the contact to the regurgitated material, defecated or through contaminated hemolymph. In the order Coleoptera, species of Cerotoma and Diabrotica genres are the most important vectors of viruses in the Americas. The larva has been considered one of the most important underground pests of crops such as maize, wheat, other cereals, and potato. The economical loss caused by the larva for these crops has been significant in the southern states and in some areas of the Southeast and Midwest. In the South, areas where soils are usually rich in organic matter and retain higher humidity favors the biology of larvae. In irrigated areas of the Southeast and Midwest, where several host crops are grown in succession, the damage has been representative. The larvae feed on the roots, reducing the plant's ability to absorb water and nutrients, making them less productive and subject to lodging, causing losses when harvesting is performed mechanically. For the maize crop, losses have been reported in the yield varying between 10% and 13% due to the attack, when high infestation of this pest occurs [13].

The adults are greenish color presenting three yellow spots on each shard, black tibia and tarsus and brown head, being called "patriot." They measure about 6 mm in length. Males are smaller than females. Adult longevity, the pace of oviposition and fertility depend on the type of food they feed on in the larval and adult stages.

The longevity may vary from 41.8 to 55.5 days for the males and from 51.6 to 58.5 days for the females. The oviposition is held in the soil around the plants. The eggs are yellow and measure 0.5 mm in diameter. The incubation period ranges from 6 to 8 days. The larva phase goes through three instars, and the larvae reaches 10 mm long, with whitish coloring, brown head, and a chitinized dark plate in the last abdominal segment. The average larval period is 18 days. The prepupa average period is 5 days and pupal period is 7 days. The life cycle varies from 24 to 40 days.

The temperature is a climate factor that affects the rate of development of the immature stages as well as the longevity of adults and reproduction [13].

2.6.1 Methods of control

Chemical control has been the most widely used method for controlling various species of Diabrotica. In Brazil, research works about the control of D. speciosa larvae attacking maize crop are scarce, complicating the recommendation of insecticides and the application method to control this pest, while in other countries, information about the control of other species of the genus is abundant.

The persistence of insecticides has been considered an important factor in the control of Diabrotica larvae. Ideally, the pesticide persists in the soil for 6 to 10 weeks, providing protection to the plant in the most susceptible period to pest [14]. As a result, treatment of seeds with insecticides has shown problems in the control of the larvae. The use of granular insecticides or spraying in the planting groove is effective alternatives to control the larvae [13].

Biological control is a promising tactic for managing this pest. Several natural enemies are described attacking adults and larvae of D. speciosa. The ones with most frequent occurrence are Celatoria bosqi (Dip., Tachinidae), Centistes gasseni (Hym., Braconidae), fungi Beauveria bassiana, Metarhizium anisopliae, and Paecilomyces lilacinus. The control of larvae, especially with fungi, has great potential to be implemented in field conditions. As a strategy for the use of cultural control, it is important to consider that soil moisture and preparation method can affect the population of larvae. Adults have a clear preference for oviposition in darker soils with higher organic matter levels and moisture [13].

3. INSECT RESISTANCE MANAGEMENT TO THE BT TECHNOLOGY

The control of insect pests in agriculture has been done mainly through the application of chemical insecticides. However, chemical insecticides have lost effectiveness due to the selection of populations of resistant insects and cause adverse environmental effects.

In this context, the biological insecticide B. thuringiensis (Bt) has emerged as an alternative for the control of insect pests of agriculture. The Cry proteins produced by Bt have demonstrated a high specificity, and there is no evidence that directly affect natural enemies [15] as well as vertebrates [16]. These features have made the development of transgenic plants producing Cry proteins in its solubilized form possible, which give the property of resistance to insect pests. In the sequence, we will discourse about these proteins, as they are the mechanisms of action in the target insect, and their most important applications.

Bt is a gram-positive bacteria, strictly aerobic, which during its life cycle has two main stages: vegetative growth, which bacteria replicate by splitting, and sporulation, which is differentiating bacteria in the spore. Bt is considered a ubiquitous bacteria since it has been isolated from around the world in many different systems, such as soil, water, plant leaves, and dead insects, among others. In the sporulation phase, Bt bacterium is characterized by producing a parasporal body known as "crystal," which is a protein nature and has insecticidal properties. The crystal protein is formed by proteins called δ-endotoxins, also known as Cry or Cyt proteins. δ-Endotoxin proteins have been found active against insects of Lepidoptera, Coleoptera, Diptera, Hymenoptera (ants), and also against other invertebrates such as nematodes, flatworms, and protozoans.

As mentioned, there are two types of δ-endotoxins: Cry and Cyt proteins. So far, more than 733 Cry genes and 38 different Cyt genes have been cloned and sequenced [17]. This is certainly a valuable arsenal for insect pest control. The nomenclature of δ-endotoxin is based solely on the similarity of the primary sequence. By definition, any parasporal protein that presents any toxic effect on body verified by bioassay or any protein that presents similarities with the Cry proteins are considered a Cry protein. Currently, Cry proteins have been found in other species of bacteria such as Clostridium bifermentans (classified as Cry16A and Cry17A) with activity to mosquitoes. The Cyt are Bt parasporal proteins that exhibit hemolytic activity.

Cry proteins are sorted and divided into 73 groups and several subgroups, and Cyt proteins into two different groups and subgroups, based on the similarity of the amino acid sequence. The Arabic numeral designates an identity of 45% (for example, Cry1, Cry2, etc.), the capital letter corresponds to 45–78% identity (cry1A, cry1B, etc.), the lowercase letter corresponds to the identities of 78–95% (Cry1Aa, Cry1Ab, Cry1Ac, etc.), and the Arabic numeral at the end of the nomenclature indicates more than 95% identity (Cry1Aa1, Cry1Aa2, etc.).

The symptoms observed in susceptible insect larvae when Bt crystals and spores are ingested are as follows: cessation of intake, intestinal paralysis, diarrhea, complete paralysis, and eventually death. In general, it is accepted that the Cry proteins are forming pores, which cause an osmotic imbalance in epithelial cells since proteins bind to receptors of the cell surface digestive system.

The Cry proteins are produced as a protoxin that needs to be proteolytically processed by proteases present in the gut of susceptible insects. This proteolytic processing releases toxic fragments to the insect (protein in the solubilized form), with a mass between 55 and 65 kDa, which interact with receptor proteins present in the microvilli of intestinal cells of the target insect. Subsequently, the proteins bind to the intestinal membrane forming a lytic pore.

Despite low similarity of Cry proteins, in some cases less than 25%, these have a similar structure composed of three domains. The domain I, composed of seven

α and amphipathic antiparallel helices, where six of them surrounds the helix α5. This is the domain that forms the ion pore. Domain II consists of three folded β-sheet and three handles, where the most structural difference is observed. This is the domain less conserved among Cry proteins. However, its sequence and tertiary structure play an important role in the specificity of the protein since the handles interact with the receiver located in the microvilli of the midgut epithelial cells. Domain III consists of two antiparallel β folded sheets forming a sandwich and is also involved in the interaction with receptors.

Table 1. General summary of maize plants genetically modified approved for marketing in Brazil.

Commercial name	Events	Protein	Applicant	Year of approval
YieldGard*	MON810	Cry1Ab	Monsanto	2007
TL**	Bt	Cry1Ab PAT	Syngenta	2007
Herculex**	TC1507	Cry1F PAT	DuPont and Dow AgroSciences	2008
YR YieldGard/ RR2**	NK603 and MON810	CP4-EPSPS Cry1Ab	Monsanto	2009
TL/TG**	Bt11 and GA21	Cry1Ab PAT mEPSPS	Syngenta	2009
Agrisure Viptera*	MIR162	VIP 3Aa20	Syngenta	2009
HR Herculex/ RR2**	TC1507 and NK603	Cry1F PAT CP-4EPSPS	DuPont	2009
VTPRO*	MON89034	Cry1A.105 Cry2Ab2	Monsanto	2009
TL TG Viptera**	Bt11, MIR162, and GA21	Cry1Ab VIP3Aa20 mEPSPS	Syngenta	2010
VTPRO2**	MON89034 7 NK603	Cry1A.105 Cry2Ab2 CP4-EPSPS	Monsanto	2010
Optimum Intrasect RR2**	MON810, TC1507, and NK603	Cry1A.105 Cry2Ab2 Cry1F PAT CP4-EPSPS	DuPont	2010
Optimum Intrasect**	TC1507 and MON810	Cry1F Cry1Ab PAT	DuPont	2011
VTPRO3**	MON89034 and MON88017	Cry1A.105 Cry2Ab2 Cry3Bb1 CP4-EPSPS	Monsanto	2011
Herculex XTRA maize	TC1507 x DAS-59122-7	Cry1F PAT Cry34Ab1 Cry35Ab1	DuPont and Dow AgroSciences	2013

*Insect resistant.

**Insect resistant and herbicide tolerant.

Source: CTNBio [20].

The aminopeptidase N (APN) is a protein from the family of cadherins (BtR) and have been proposed as potential recipient of Cry1A proteins in Lepidoptera. The APN is a protein with an apparent mass of 120 kDa, which is anchored to the membrane via a glycosylphosphatidyl group inositol (GPI). There is evidence that the interaction of the protein with the cadherin receptor promotes an additional cut in the extreme amino terminus of the Cry protein by facilitating the formation of an oligomer or "pre-poro" formed by four monomers, which is responsible for membrane insertion and pore formation. For the "pre-poro" to be inserted in the membrane, it is necessary to interact with the APN receptor. The proteins anchored in the membrane by GPI are preferably distributed in specific regions of the membrane, known as lipid rafts, which have specific characteristics due to the high content of cholesterol and glycolipids. The interaction of the Cry protein of the "pre-pore" with the APN facilitates the insertion of oligomer in the lipid rafts on the membrane, resulting in pore formation [18].

The Bt technology relies on the transfer and expression of resistance genes to insect pest in maize, isolated from the bacteria B. thuringiensis Berlinger (Bt) [19]. The preservation of susceptibility to Bt toxins in pest populations depends on resistance management programs (IRM). Table 1 presents a summary of the most important technologies for maize crop.

3.1 Considerations about the refuge area

The main IRM strategy is the use of "high dose/refuge," which involves the use of high dose of Bt protein in plants, promoting high mortality of heterozygotes associated with the planting of refuge, i.e., a proportion of the crop in which it must be planted a non-Bt variety, allowing the survival of susceptible individuals to mate with possible resistant ones [21]. A protein may have high dose activity for a pest species and moderate or low dose to others, which does not impair the IRM because it is expected a simultaneous action of other mortality factors, such as natural enemies [22]. In this scenario, the adoption of the refuge area is also key to the IRM.

The explanation for cases of resistance to Bt crops appears to be related to the nonuse of high dose/refuge [23] strategy, particularly the nonadoption of refuge [24, 25].

The configuration of refuge areas may vary, but basic criteria of size and proximity to the Bt crops based on the target pest bioecology should be followed [22] so that these areas produce consistent proportions of adults for mating and maintaining susceptibility. In Figure 1, specified examples of refuge areas settings are shown.

Figure 1. Examples of refuge areas settings.

A good example of an alternative method for pest control, especially in Brazil, against fall armyworm, was the development and release of genetically modified plants such as Bt maize, a technology adopted with incredible rapidity in Brazil. Unfortunately, used without proper care, there are already complaints from different parts of the country in a few years of use about the presence of caterpillars and their damage above expectations. In fact, the expectation of farmers is that there would be no injuries from this pest in the crop. For fear of having economic losses, the chemical control so far left as low priority is back to be used in some areas of higher incidence even in Bt maize. Therefore, the importance of the technology should always be emphasized, however, pointing out that it alone will not solve the numerous phytosanitary problems in maize or other crops. For various reasons, since the commercial release of Bt maize, there was already a concern for the proper management of the technology to prevent breakdown of resistance by target pests. All good agricultural practices generally conveyed along with the acquisition of the seed must be strictly followed. Such practices include adopting refuge areas.

Until 2007, the scenario of maize crop in Brazil was of growing losses by caterpillar's attacks. Problems with fall armyworm, black cutworm, corn earworm, cornstalk borer, and sugarcane borer increasingly frightened the farmer, who had little efficiency in the control of these pests using insecticides.

Quickly, the Bt technology in maize significantly reduced the problems with chewing insects, causing the erroneous impression that the technology was "bulletproof," meaning that nothing needed to be done and that all IPM practices could be left aside. However, with passing time and the intensive use of technology, the problems with insect resistance began to appear.

Resistance can be defined as a biological and evolutionary phenomenon that occurs in response to selection pressure exerted by the different control agents.

The evolution of resistance consists in the selection of resistant individuals that are naturally present in nature, leading to increased frequency of these individuals or their genes in the pest population, leading eventually to restrictions control agent efficiency. Unlike foliar insecticides, the Bt crops carry a much higher selection pressure on populations of insect pests that are target to control due to continued expression of insecticidal toxins over the crop growth period. This causes a higher risk of pest developing resistance to Bt technology.

The continuous expression of insecticidal proteins throughout the cycle of Bt plants and this rapid adoption represent threats to its durability, the strong selection pressure on the pest insects [23, 26]. Indeed, cases of resistance to Bt toxins have been reported for maize pests such as S. frugiperda [27–29] and Diabrotica virgifera [24].

According to Kumar et al. [30], the use of refuge areas, represented by planting susceptible varieties surrounding soybean crops sown with Bt varieties, is the main strategy to prevent the development of resistance.

Although avoiding the phenomenon of resistance of pests to insecticides (chemical or biological) should be a constant concern in the case of Bt crops, the recommended strategy involves actions that require time and use of machines, which may result in hatred of farmers to compliance, resulting in lower lifetime varieties with this feature.

The aggravating factor is that due to the characteristic (or genetic) of resistance when we start to see damage in the field in a technology with medium-high dose, the frequency of alleles is now probably around 10%, with the chance that, with continuous exposure to technology, the population will be at a much higher proportion of resistant individuals in a few generations.

Since the launch of the first Bts, several companies warned that the IPM practices should not be set aside and, especially, the refuge area should be established in all farms. The refuge, which is the planting of at least 10% of the area with a non-Bt hybrid maize, allows the survival of insects susceptible to Bt technology. The preservation of these susceptible insects allows the crossing with possible resistant insects, resulting in a progeny of susceptible insects.

However, few farmers planted the refuge area, and when they did, spraying in such areas were constant in order to obtain the productivity in the area. The result was that even when present, in many cases, the refuge areas did not work effectively in the maintenance of susceptible insects that would mate with any resistant insect coming from the Bt areas. In the absence of susceptible insects from the refuge area, any surviving insects resistant to exposure to Bt mated with each other, allowing relatively rapid increase of the resistance alleles and increased amounts of resistant individuals in the field.

Now that the resistance break of fall armyworm to Cry1F technologies is a reality, the question is, Is the refuge still necessary and beneficial for this technology?

The answer is certainly yes, because there are other pests that are controlled by the Cry1F protein as the sugarcane borer; other insects are likely to also develop resistance in case the best management practices are not applied, and in case there is no maintenance of susceptible individuals by adopting the structured refuge. The refuge is essential to maintain the efficiency of this control. In addition, all the technologies in the market today will have their increased durability and benefit from the adoption of best management practices and refuge areas for planting.

Knowing that the refuge areas are part of the IPM and the insect resistance management, how should we proceed to make the correct use?

As previously mentioned, poor adherence of refuge use or the many insecticide applications in the refuge, eliminating susceptible individuals, resulted in an ineffective resistance management system, which favored a faster resistance evolution rate.

It is known that only the adoption of refuge is not enough to maintain the effectiveness of the technology and should also be considered to manage the use of insecticides in agriculture. The refuge should be as a donor area of susceptible insects so that they can mate with any resistant insects and the result is susceptible individuals in larger quantities. Therefore, it is necessary to maintain differential applications between the refuge and Bt crop so that the application rate of insecticide in the refuge should be lower than in the fields. Basically, we have to think of resistance management in Bt area and management of economic damage in the refuge area.

3.2 Early desiccation followed by insecticide

The previous crops as well as weeds and volunteer plants in the environment can host the main pests that attack maize in the initial phase, influencing the predominant species and the initial pressure of pests. Thus, in the no-tillage system, pest pressure in the early stage of the crop can be greater when compared to conventional tillage.

In the case of the presence of pests in the area, it is recommended that the application of insecticide be followed by preplanting desiccation, aiming the reduction of the initial population of pests, which are the most challenging for seed treatment; the control of resident caterpillars in later instars, which can cause early damage even in Bt maize crops; and the maintenance of the initial stand of the crop.

Regarding the early cover crop desiccation, it aims to provide dry straw on the ground, facilitating the operation of planting and promoting the protection of the soil. The optimal timing of herbicide applications may vary according to weather conditions and the cropping system used.

It is recommended to make two herbicide application; in the first period of approximately 30 days before planting, thus avoiding the presence of green mass at the time of sowing, and in the second desiccation shortly before planting in order to control the first flow of weeds after the first desiccation.

We highlight some benefits of desiccation performed at the right moment: more efficient use of insecticide in the second desiccation, as the green cover reduces its intensity with the first desiccation (eliminating the umbrella effect for insecticide); better plantability: easier cut of the straw by planter; availability of dry straw in the crop germination period: protection of soil moisture; reduction of possible allelopathic effects of the previous crop as the main crop; and ease in weed control in the postemergency phase of the crop, if necessary.

3.3 Weed control

Some weeds may host insect pests of succeeding crops, allowing a significant amount to survive in the areas of cultivation in the off-season period. In addition, weeds can be sources of caterpillars in later instars, which presents major difficulty to control by the Bt technology. Some practices may contribute to a better control of weeds, as well as prevent resistance to herbicides:

- Do not leave fallow areas: use integrated practices of weed management during the year, focusing on the handling of the seed bank (crop rotation and covers).
- Start growing in clean area: make an effective control early in the preplanting and, if necessary, use a preemergent in high pressure areas of weed.
- Use the dose and the correct moment of the application of products in good management system, in compliance with the best application conditions.
- Use the postharvest management: use the association of herbicides with different modes of action.
- Monitor the results of the implemented management strategy, preventing the establishment of remnant populations of weed in the crop.
- Use the best agronomic practices to maximize crop competitiveness with weeds, also avoiding seed dispersal by agricultural implements.

Regarding the management of volunteer plants after the maize crop, it is common the occurrence of germinação of remaining grains from previous crop spontaneously;

The amount and timing of germination of these maize kernels, producing crop residues (also known as "tigueras"), depends on many factors, being the quality of the previous harvest one of the most important; herbicides called graminicides are the main management tool of these plants. Volunteer plants are controlled until the V3/V4 stage to obtain consistent and quick controls. Weed competition is prevented with subsequent soybean crop, making the early management of volunteer plants.

3.4 Seed treatment

Seed treatment (ST) is a practice that seeks control of underground and initial culture pests, a period of great susceptibility to pests. The damage caused by these pests results in crop failures due to the attack on the seeds after planting, damage to roots after germination, and shoots of newly emerged plants. The correct choice of chemical is essential to the success of this operation. We recommend using products from broad spectrum to provide efficient control of the initial pests of the crop complex, which will bring results as the protection of plants in the initial development phase, broad-spectrum pest control, and maintenance of the initial stand of the crop.

3.5 Crop rotation

Crop rotation consists of alternating the planting of different species of crops in the same agricultural area. The choice of species for crop rotation should take into account economic factors, pests, diseases, and fertilization, among others.

To obtain maximum efficiency, improving productivity capacity of the soil, the planning of crop rotation must consider, preferably commercial plants and, whenever possible, involving species that produce large amounts of biomass and rapid development, cultivated singly or intercropped with commercial crops.

Among the benefits of crop rotation in pest management in Bt maize, the highlights are as follows: improved physical and chemical properties of the soil, reduction of disease inoculum source for subsequent crops, reduction of the initial population of some insect pests of the crop, aid in weed management, and ability to switch herbicides for the control and increase in the system productivity.

4. NEW SOURCES OF RESISTANCE

The interaction of plant-herbivore insects occurs in various combinations of genotypes and environments, which makes its coevolution process broad and diverse. For this reason, plants and insects can provide a wide range of mechanisms, which make them resistant to attack or able to circumvent the acquired resistance. Thus, when considering the coevolution as a dynamic process, we must be sure that the natural resistance or artificially acquired by an organism may be short-lived or long-lasting, but difficultly can occur permanently. On the other hand, the duration of plant resistance will be greater as lower the speed on the evolution of resistance in the insect target, in other words, we must focus on strategies to reduce the selection pressure on the target. It is precisely in this aspect that the search for new genes that may confer resistance to insects fits. For example, using more than one resistance gene in a genetically modified plant, it is possible to prolong the emergence of resistant individuals, especially if these genes relate to different sources of resistance as a toxin and a compound that attracts a natural enemy target.

The prospect of important genes in plant–insect interaction has the fundamental objective of assisting in the preparation of new alternatives, both with the identification of genes that make plants resistant or susceptible to insect attack, as with the identification of genes that are associated with the insect's ability on circumvent the defenses of their hosts. Knowledge of the physiology of insects resistant to Bt toxins, for example, is important to the discovery of new targets (genes or genetic polymorphisms).

Otherwise, other Bt toxin proteins or other natural enemies of herbivorous insects may also represent new alternatives resistance.

In this sense, studies aiming at prospecting for new important genes in plant-herbivore insect interactions can concentrate on the plant by identifying mRNA expressed (transcriptome) [31], proteins (proteomics) [32], or metabolites (metabolomics) synthesized in specific tissues and moments of the interaction, or they may focus on the insect by the use of the same tools applied to tissues or moments fundamental to the success of interaction, such as the study of the digestive proteins secreted in the midgut and that enable herbivores [31] or the study of regulatory elements of metamorphosis [34]. Alternatively, prospecting studies can focus on the interaction of model organisms for which there is already high amount of generated knowledge (genomic knowledge and tools to produce genetic alterations), such as Arabidopsis–Scaptomyza flava interaction (Drosophila) [35], or may focus on a single study or specific response mechanism by, for example, the application of a compound that is known to cause a direct defense response in plants [36].

Different strategies can be useful for gene prospecting, including comparative analyzes of transcriptoma, proteomics, metabolomics, and the functional study of genes by mutagenesis, overexpression, and gene silencing. Indeed, comparative analyzes can be exploited as ideal strategies for global exploration of important genes in plant–insect interaction. Such analyzes can be conducted in order to compare important genes in plant–insect interaction in different environmental conditions [37] in resistant and susceptible plants [38] in injured plants by different insects [39] and others.

Global prospection strategies achieved particular prominence with the use of new technologies of DNA sequencing to characterize transcriptoma (RNA-seq).

With RNA-seq strategies, it is possible to generate billion bases of information in single runs (at a much lower cost than Sanger sequencing), which allows access to regulatory genes, represented by one or a few mRNAs [40] and covering full-length cDNAs [41].

Although the global strategies of gene prospecting are potentially unlimited, the success of identifying real candidates depends on the development of an efficient experimental design. On the work of Li et al. [37], the defense mechanisms of two soybean varieties, that is, resistant and susceptible to an aphid, were studied using microarrangements of cDNA, and the collection period after an infection determined by the time necessary to the insect reaches the xylem vessel elements in the plant, about 8 hours in the resistant cultivar and 3.5 hours in the susceptible cultivar.

The large-scale study of metabolites produced by plants in the presence of insect pests also consists in an innovative possibility of seeking alternatives for its control and the identification of genes or important metabolic pathways. In soybean leaves [42], it was observed that constitutively leave extracts of PI 227687 contain the isoflavonoid genistein and seven flavonol glycosides, including rutin [43], by studying the leaf extract resistant to insect genotypes PI 274454, "IAC-100," and PI 229358, which identified and quantified the flavonol rutin and the isoflavonoid genistein.

Their identification and their role in the interactions of insects with soybean plants can guide geneticists in order to keep them in descendant generations as part of the defense armory of plants against herbivores. To study if the insect resistance of genotypes PI 227687, PI 274454, and "IAC 100" is due to chemicals present in their constitution, they used extracts of these genotypes mixed to artificial diet. By the results obtained, Piubelli et al. [43, 44] found that those strata negatively affect the biology of Anticarsia gemmatalis. Additionally, studies have shown that the flavonol rutin causes antibiosis in Trichoplusia ni (Hübner) [45, 46].

In general, although the Bt strategy to control lepidopteran still is the world's most important in controlling pests, new sources of resistance may operate independently or may also be added to the Bt strategy so as to promote their own maintenance of Bt resistance in commercialized transgenic plants.

Molecular biology tools have supplemented the information generated by morphological and behavior studies, contributing to the elucidation of issues in the fields of taxonomy, ecology, pests population genetics, parasitoids, predators, and entomopathogenic bacteria. Its resolving power has allowed increased knowledge on the occurrence of cryptic species, differentiation of insect races, and separation of microorganisms species indistinguishable by morphological characters. These tools also have wide application in genetic studies of resistance to insecticides and toxins and in the determination of genes associated with these phenomena. On the other hand, they have facilitated the breeding works to plant resistance to insects, as well as the transformation of the beneficial organisms to increase pest control potential. Considering its potential and reducing reagent costs and simplifying processes, we expect a growing application in basic and applied fields of entomology and its related areas.

5. CONCISE SUMMARY

The initial challenge is to devise innovative methodologies for the application of integrated pest management principles that align with the evolving realities of the field, particularly in tropical regions where soybean is cultivated, its incorporation into more intricate production systems, the significant overlap of common pests across various crops within the same system, and the extensive area of cultivation. The framing of this issue has become increasingly significant over the past decade, causing considerable concern among farmers on the scarcity of available workers for fieldwork.

Creating biologically-based control technologies that are both technically viable and economically competitive will pose a challenge, considering the variety of pests in production systems and the influence of alternative control methods, particularly insecticides and fungicides, on biological control agents.

The introduction of Bt crops represents a significant advancement in pest management strategies for several crops, including maize, canola, cotton, and, soon, soybeans. The utilization of Bt crops not only regulates significant Lepidoptera species but also yields a beneficial externality by safeguarding insects that serve as predators or parasitoids of pests, owing to the reduced application of insecticides that indiscriminately affect these biological control agents. Nonetheless, there exists a persistent risk of the establishment of Lepidoptera populations that are resistant to the toxin produced by Bt crops, attributable to farmers' noncompliance with refuge strategies. The occurrences are analogous across many crops with Bt cultivars or types, and certain pests infest several crops that possess Bt events, heightening the possibility of the evolution of resistant populations.

REFERENCES

1. Carson, R.L. Silent spring. Boston: Houghton Mifftlin Co., 1962. 368p.
2. Van Den Bosch, R. The pesticide conspiracy. New York: Doubleday & Co. Inc., 1978. 212p.
3. Kogan, M. Integrated pest management: historical perspectives and contemporary developments. Annual Review of Entomology, v. 43, p. 243–270, 1998.

4. Prokopy, R.J.; Kogan, M. Integrated pest management. In: Resh, V. H.; Cardé, R.T. (Ed.). Encyclopedia of Insects. New York, Academic Press, 2003, p. 4–9.

5. Cruz, I. Circular encartado na revista Cultivar Grandes Culturas. Caderno Técnico, Ed. 168, 12p. 2013.

6. EMBRAPA—Empresa Brasileira de Pesquisa Agropecuária. A Broca da Cana-de- Açúcar, Diatraea saccharalis, em Milho, no Brasil. Sete lagoas: Embrapa milho e sorgo, 2007. (Circular Técnica 90).

7. Bento, F. M. M.; Magro, S. R.; Fortes, P.; Zério, N. G.; Parra, J. R. P. Biologia e tabela de vida de fertilidade de Agrotis ipsilon em dieta artificial. Pesquisa Agropecuária Brasileira, v. 42, n. 10, p.1369–1372, 2007.

8. Schneider, et al. Manual de pragas do milho, soja e algodão.Sementes Agroceres, 2013.148p.

9. Gallo, D.; Nakano, O.; Silveira Neto, S.; Carvalho, R. P. L.; Baptista, G. C.; Berti Filho, E.; Parra, J. R. P.; Zucchi, R. A.; Alves, S. B.; Vendramin, J. D.; Marchini, L. C.; Lopes, J. R. S.; Omoto, C. Entomologia agrícola. Piracicaba: FEALQ, 2002. 920p.

10. Santos, H. R.; Nakano, O. Dados biológicos sobre a lagarta-rosca Agrotis ipsilon (Hufnagel, 1767) (Lepidoptera, Noctuidae). Anais da Sociedade Brasileira Entomológica do Brasil, v. 11, p. 33–48, 1982.

11. Ávila, C. J.; Vivan, L. M.; Tomquelski, G. V. Ocorrência, aspectos biológicos, danos e estratégias de manejo de Helicoverpa armigera (Hübner) (Lepidoptera: Noctuidae) nos sistemas de produção agrícolas. Circular Técnica 23, EMBRAPA, Dourados—MS, 2013.

12. Galvão, A.; Attie, J.; Menezes, L.; Cunha, J.; Bisinotto, F. Relatório biotecnologia. Uberlândia: Céleres, 2012.

13. Viana, P. A. Manejo de Diabrotica speciosa na cultura do milho. Circular Técnica 141.SeteLagoas, MG Setembro, 2010. 6p.

14. Levine, E.; Oloumi-Sadeghi, H. Management of diabroticite rootworms in corn. Annual Review of Entomology, Palo Alto, v. 36, p. 229–255, 1991.

15. Bravo, A.; Gill S. S.; Soberón M. Bacillus thuringiensis mechanisms and use. Comprehensive Molecular Insect Science Elsevier BV, p. 175–206, 2005.

16. Bravo, A.; Gill, S. S.; Soberón, M. Mode of action of Bacillus thuringiensis Cry and Cyt toxins and their potential for insect control. Toxicon. v. 49, n. 4 p.423–435, 2007.

17. Crickmore, N. et al. Bacillus thuringiensis toxin nomenclature. Disponível em: <http:// www.lifesci.sussex.ac.uk/Home/Neil_Crickmore/Bt/>. Acessoem: 27 jan. 2015.

18. Bravo, A.; Gómez, I.; Conde, J.; Muñoz-Garay, C.; Sánchez, J.; Miranda, R.; Zhuang, M.; Gill, S. S.; Soberón, M. Oligomerization triggers binding of a Bacillus thuringiensis Cry1Ab pore-forming toxin to aminopeptidase N receptor leading to insertion into membrane microdomains. Biochimica et Biophysica Acta, v. 1, p. 38–46, 2004.

19. Carneiro, A. A.; Guimarães, C. T.; Valicente, F. H.; Waquil, J. M.; Vasconcelos, M. J. V.; Carneiro, N. P.; Mendes, S. M. Milho Bt: teoria e prática da produção de plantas transgênicas resistentes a insetos-praga. Sete Lagoas: Embrapa Milho e Sorgo, 2009. 25 p. (Embrapa Milho e Sorgo. Circular técnica, 135).

20. CTNBio. Comissão Técnica Nacional de Biossegurança. Ministério da Ciência, Tecnologia e Inovação. 2015. Acesso em: 05/jun. Disponível: <http://www.ctnbio.gov.br/>.

21. Bernardi, O.; Albernaz, K. C.; Valicente, F. H.; Omoto, C. Resistência de insetos-praga a plantas geneticamente modificadas. In: Borém, A.; Almeida, G. D. de. Plantas geneticamente modificadas: desafios e oportunidades para regiões tropicais. Visconde de Rio Branco: Suprema, 2011. p. 179–204.

22. Martinelli, S.; Omoto, C. Resistência de insetos a plantas geneticamente modificadas. Biotecnologia,Ciência e Desenvolvimento, Brasília, DF, v. 34, p. 67–77, 2005.

23. Huang, F.; Andow, D. A.; Buschman, L. L. Success of the high-dose/refuge resistance management strategy after 15 years of Bt crop use in North America. Entomologia Experimentalis et Applicata, Amsterdam, v. 140, n. 1, p. 1–16, 2011.

24. Gassmann, A. J.; Petzold-Maxwell, J. L.; Keweshan, R. S.; Dunbar, M. W. Field- evolved resistance to Bt maize by Western Corn Rootworm. PLoS One, Cambridge, v. 6, n. 7, 2011.

25. Kruger, M.; Van Rensburg, J. B. J.; Van Den Berg, J. Transgenic Bt maize: farmers' perceptions, refuge compliance and reports of stem borer resistance in South Africa. Journal of Applied Ecology, London, v. 136, n. 1–2, p. 38–50. 2012.

26. Tabashnik, B. E.; Gassmann, A. J.; Crowder, D. W.; Carrière, Y. Insect resistance to Bt crops: evidence versus theory. Nature Biotechnology, London, v. 26, n. 2, p. 199–202, 2008.

27. Matten, S. R.; Head, G. P.; Quemada, H. D. How governmental regulation can help or hinder the integration of Btcrops into IPM programs. In: Romeis, J.; Shelton, A. M.; Kennedy, G. G. Integration of insect resistant genetically modified crops within IPM program. New York: Springer, 2008. v. 5, p. 27–39.

28. Storer, N. P.; Babcock, J. M.; Schlenz, M.; Meade, T.; Thompson, G. D.; Bing, J. W.; Huckaba, R. M. Discovery and characterization of field resistance to Bt maize: Spodoptera frugiperda (Lepidoptera: Noctuidae) in Puerto Rico. Journal of Economic Entomology, Lanham, v. 103, n. 4, p. 1031–1038, 2010.

29. Villela, F. M. F.; Waquil, J. M.; Vilela, E. F.; Siegfried, B. D.; Foster, J. E. Selection of the fall armyworm, Spodoptera frugiperda (SMITH) (Lepidoptera: Noctuidae) for survival on Cry 1A(b) Bt toxin. Revista Brasileira de Milho e Sorgo, v. 1, n. 3, p. 12–17, 2002.

30. Kumar, S.; Chandra, A.; Pande, K.C. Bacillus thuringiensis (Bt) transgenic crop: an environment friendly insect-pest management strategy. Journal of Environmental Biology, v. 29, p. 641–653, 2008.

31. Smith, C. M.; Liu, X.; Wang, L. J.; Liu, X.; Chen, M.; Starkey, S.; Bai, J. Aphid feeding activates expression of a transcriptome of oxylipin-based defense signals in wheat involved in resistance to herbivory. Journal of Chemical Ecology, v. 36, p. 260–276, 2010.

32. Wei, Z.; Hu, W.; Lin, Q.; Cheng, X.; Tong, M.; Zh U, L.; Chen, R.; He, G. Understanding rice plant resistance to the brown planthopper (Nilaparvata lugens): a proteomic approach. Proteomics, v. 9, p. 2798–2808, 2009.

33. Chi, Y. H.; Sal, R. A. Z.; Balfe, S.; Ahn, J. E.; Sun, W.; Moon, J.; Yun, D. J.; Lee, S. Y.; Higg, T. J. V. I; Pittendr, B. I.; Murd, L. L. O.; Zh U-Sal Zman, K. Cowpea bruchid midgut transcriptome response to a soybean cystatin—costs and benefits of counter- defence. Insect Molecular Biology, v. 18, p. 97–110, 2009.

34. Fu, Q.; Liu, P. C.; Wang, J. X.; Song, Q.S.; Zhao, X.F. Proteomic identification of differentially expressed and phosphorylated proteins in epidermis involved in larval-pupal metamorphosis of Helicoverpa armigera. BMC Genomics, v. 10, p. 600, 2009.

35. Whiteman, N. K.; Groen, S. C.; Che Vasc O, D.; Bear, A.; Bec Kwith, N.; Greg Ory, T. R.; Den Oux, C.; Mammarella, N.; Ausubel, F. M.; Pierce, N. E. Mining the plant–herbivore interface with a leaf mining Drosophila of Arabidopsis. Molecular Ecology, v. 20, p. 995–1014, 2011.

36. Matthes, M. C.; Bruce, T. J. A.; Ton, J.; Verrier, P. J.; Pickett, J. A.; Nap Ier, J. A. The transcriptome of cis-jasmone-induced resistance in Arabidopsis thaliana and its role in indirect defence. Planta, v. 232, p. 1163–1180, 2010.

37. Broekgaarden, C.; Poelman, E. H.; Steenhui S, G.; Voorrips, R. E.; Dicke, M.; Vosman, B. Responses of Brassica oleracea cultivars to infestation by the aphid Brevicoryne brassicae: an ecological and molecular approach. Plant Cell and Environment, v. 31, p. 1592–1605, 2008.

38. Li, Y.; Zou, J.; Li, M.; Bilgin, D. D.; Vodkin, L. O.; Hartman, G. L.; Clough, S. J. Soybean defense responses to the soybean aphid. New Phytologist, v. 179, p. 185–195, 2008.

39. Ehlting, J.; Chowrira, S. G.; Mattheus, N.; Aeschliman, D. S.; Arimura, G. I.; Bohlmann, J. Comparative transcriptome analysis of Arabidopsis thaliana infested by diamond back moth (Plutella xylostella) larvae reveals signatures of stress response, secondary metabolism, and signalling.BMC Genomics, v. 9, p. 20, 2008.

40. Gilardoni, P. A.; Schuck, S.; Jüngling, R.; Rotter, B.; Baldwin, I. T.; Bonaventure, G. Super SAGE analysis of the Nicotiana attenuate transcriptome after fatty acid-amino acid elicitation (FAC): identification of early mediators of insect responses.BMC Plant Biology, v. 10, p. 66, 2010.

41. Pauchet, Y.; Wilkinson, P.; Vogel, H.; Nelson, D. R.; Reynolds, S. E.; Heckel, D. G.; ffrench-Constant, R. H. Pyrosequencing the Manduca sexta larval midgut transcriptome: messages for digestion, detoxification and defence. Insect Molecular Biology, v. 19, p. 61–75, 2010.

42. Hoffmann-Campo, C.B. Role of the flavonoids in the natural resistance of soyabean to Heliothis virescens (F.) and Trichoplusia ni (Hübner). 1995. 165 f. PhD dissertation, The University of Reading, Reading, 1995.

43. Piubelli, G. C.; Hoffmann-Campo, C. B.; Moscardi, F.; Miyakubo, S. H.; Oliveira, M.C.N. Are chemical compounds important for soybean resistance to Anticarsia gemmatalis.Journal of Chemical Ecology, v. 31, p. 1515–1531, 2005.

44. Piubelli, G.C. Bioatividade de genótipos de soja resistentes a A. gemmatalis Hübner (Lepidoptera: Noctuidae) e interações de suas substâncias químicascom inimigos naturais. 2004. 126f. Tese (Doutorado)—Universidade Federaldo Paraná, Curitiba, 2004.

45. Hoffmann-Campo, C.B.; Harborne, J.B.; Miccafferry, A.R. Préingestiveand post-ingestive effects of soya bean extracts and rutin on Trichoplusia ni growth. Entomologia Experimentalis et Applicata, v. 98, p. 181–194, 2001.

46. Bellés, X. Beyond Drosophila: RNAi in vivo and functional genomics in insects. Annual Review of Entomology, v. 55, p. 111–28, 2010.

5 Chapter — CHALLENGES IN REGULATING INSECT OVA

1. INTRODUCTION: AN OVERVIEW

The primary reproductive method of most insects is oviparity, characterized by egg laying. The literature on the control of insect eggs is sparse, despite eggs being regarded as the most challenging life stage to exterminate. Instances of this are observable in both agricultural and urban pest species. The management technique frequently involves disregarding the eggs, awaiting their hatching, and thereafter targeting the nymphs that emerge, as they are more susceptible to extermination. Throughout numerous treatments, the eggs remain unaccounted for as they do not represent a nuisance stage (they neither bite nor feed) and are less conspicuous than other stages. Nevertheless, the eggs that survived the treatment will soon hatch, resulting in a reinfestation. The infestation may require time to escalate into a significant issue, at which point reevaluation or reinitiation of the therapy may be warranted.

Insect eggs possess adaptive methods to improve their survival, such as encasing the embryo within an eggshell. The eggshell, known as the chorion, serves as the primary barrier for the growing embryo from external stimuli. A primary environmental stressor confronting pest bug species is human application of pesticides. The distinctive composition of the eggshell renders the majority of insecticidal agents ineffective in penetrating it.

Insecticides that penetrate the eggshell must reach the target spot within the embryo to be efficacious. Several insects have demonstrated enzymatic resistance to pesticides during the egg stage. Eggs subjected to repeated pesticide treatment exhibited elevated enzyme activity for the degradation of insecticides. The amalgamation of diminished penetration via the eggshell and pesticide resistance renders eggs an exceptionally challenging life stage to exterminate.

2. INSECT EGGSHELL MORPHOLOGY

Insect eggs are remarkably structured to provide the developing insect embryo with protection, simultaneously providing a barrier against insecticide penetration. This is evident from the array of structural designs observed on eggshells across insect species, which have specifically evolved to provide the developing insect ultimate protection in the environment. There are many variations in the respiratory structures, shape, size, coloration, and the chorionic structure of eggs, all seemingly evolved to withstand environmental stress. The diversity of eggshells and the structural complexity of the eggshell (Figure 1) are evident in three butterfly eggs in the families Nymphalidae and Saturniidae. The amazing diversity of insect eggs has received little attention from the scientific community. H. E. Hinton's three- volume work on the biology of insect eggs [1] represents one of the most significant contributions to understanding insect eggs.

Various studies have attempted to characterize the insect eggshell, or chorion, and its layers. Eggshell morphological descriptions have revealed that many insect eggs have detailed sculpturing on the outside of the eggshell. The sculpturing is usually comprised of multisided geometrical shapes, arranged on the eggshell in an aesthetically pleasing pattern.

Not all insect eggshells are perfectly symmetrical. For instance, the outer surfaces of the chorion in true bugs oftentimes are geometrical but the shapes are irregular in shape and size. Furthermore, some insect eggs lack the geometrical shapes extending from the eggshell and are completely smooth. The formation and shape of the eggshell is highly dependent on the outline shape of the mother's follicle cells that synthesize the eggshell.

In the family Pentatomidae (stink bugs), the chorion is characterized by the surface structure, termed either "spinose" or "coarse" [2]. "Spinose" refers to insect eggs that have projections arranged in patterns that extend outwardly from the surface. The term "coarse" refers to indented pit structures on the outer eggshell surface [2].

The chorion is produced within the female's ovariole by the follicular epithelium. In the simplest form, the chorion is typically comprised of three layers (exochorion, endochorion [inner and outer], and vitelline membrane) [2, 3]. The vitelline membrane is the innermost layer that surrounds the embryo. A few insect studies have further subdivided the eggshell layers into waxy layers and crystal chorionic layers, which most probably serve as the main barriers in the eggshell against water loss. However, these structures and layers differ between insect families and species, depending on their habitat and individual respiratory and water requirements.

Gravid female insects have accessory glands that secrete gluelike substances, sometimes referred to as "cement" to adhere their oviposited eggs to a substrate. These glue substances have to be able to withstand environmental stress for the duration of embryonic development before the larvae is ready to emerge from the eggshell. Much like the eggshell, these glue substances are primarily comprised of proteins [4]. When considering the penetration of insecticides, the glue sheath adds an extra layer of protection for the developing embryo. The head louse glue sheath is especially problematic from the perspective of lice control.

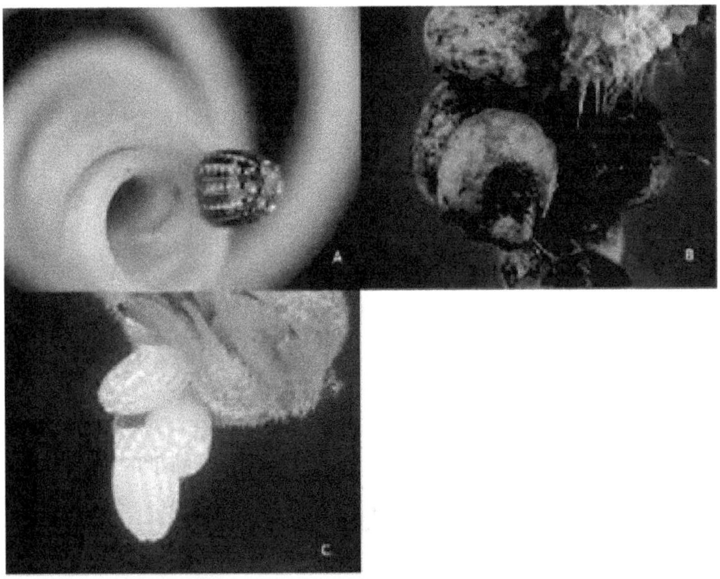

Figure 1. Automontage photographs of three butterfly eggs. (A) An egg from the gulf fritillary, Agraulis vanillae; family: Nymphalidae. (B) Eggs from the luna moth, Actias luna; family: Saturniidae. Note the exuvia from the recent molt of the larvae. (C) Eggs from the zebra longwing, Heliconius charithonia; family: Nymphalidae.

Anyone who has experienced head lice is well aware of the glue sheath that surrounds a head louse egg, called a nit. The glue sheath laid down by the mother from her collateral glands adheres the nit to the hair shaft. This makes nits incredibly difficult to remove from hair. The nit comb, which is a widely used management technique for head lice, has very fine teeth designed to brush nits from the hair shaft. The sheath covers the entire egg, except at the operculum where respiration occurs [5]. Understanding the components of the sheath could allow researchers to develop novel ways to denature the protein components of the sheath or to coat the sheath and prevent the embryo from breathing [5].

The embryo inside of the eggshell hatches, or encloses, through the egg cap, called the operculum. The operculum is usually located on the anterior pole of the egg. The border of the operculum is comprised of multiple, small, uniform-shaped holes along the circumference of an egg, usually aligned side by side. The appearance of these holes is similar to a loose-leaf notebook with a perforated edge on each page that allows pages to easily be torn out. The perforations on the egg make the operculum easier to break open, thus allowing the first instar larvae to push through the operculum during eclosion. This process is taxing to the small larvae, so many larvae have a specialized spine, or egg burster, on their heads to assist with hatching from the operculum [1]. In addition to an egg burster, the larvae will grow in size by engulfing air and amniotic fluid, creating pressure inside of the egg until they expand enough to break free from the eggshell.

In addition to the eggshell layers, there are structures present on the eggshell for respiration (aeropyles) and fertilization (micropyles) and also inner eggshell structures for the movement of oxygen (pillars, sometimes also referred to as struts or columns). The pillar, or column structures, can be observed easily in the scanning electron micrograph of a hatched bed bug egg (Figure 2). These structures that open into the eggshell are potentially sites that would allow insecticides to enter the insect egg. The insect eggshell must maximize embryo respiration while preventing water loss. The eggshell is designed not only to limit water loss from the egg but also to limit excessive water from entering the eggshell. Consequently, water-based insecticidal preparations do not easily penetrate insect eggshells.

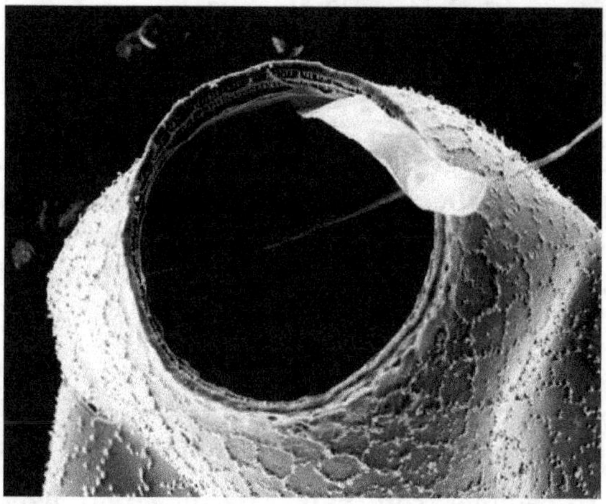

Figure 2. Scanning electron micrograph of a hatched bed bug egg. The egg has been cut in half, and the operculum has been removed for visualization purposes. The inner layers, including the respiratory struts and columns, can be observed.

Insect eggs are very small; therefore, they have an increased surface area-to-volume ratio [6]. Having a large proportion of the insect egg's surface exposed to the environment in relation to the internal volume makes the egg even more prone to desiccation. Multiple studies have been conducted to quantify the balance between respiration and water conservation of insect eggs [6–8]

3. RESPIRATION AND WATER CONSERVATION OF INSECT EGGS

The respiratory systems in insect eggs are very different from systems in other insect life stages. The principle idea is still the same; there must be openings on the eggshell to allow the exchange of gases with the atmosphere and mechanisms to move gases throughout the eggshell. Typically, insect embryos respire through tiny pores called aeropyles located on the outer eggshell, which allows for efficient gas exchange and reduces water loss during respiratory activities (Figure 3). To reduce water loss, many terrestrial insect eggs have a reduced numbers of aeropyles [6]. The aeropyles connect to an inner air-filled space that is referred to as the pillar system, which moves oxygen from the outer eggshell layers to the inner embryo. Insect eggs may have their aeropyles located on the outer rim of the operculum, as seen on many hemipteran eggs, or, if there is no defined rim, they will be located on the border of the operculum. Alternatively, aeropyles can also be scattered across the egg surface.

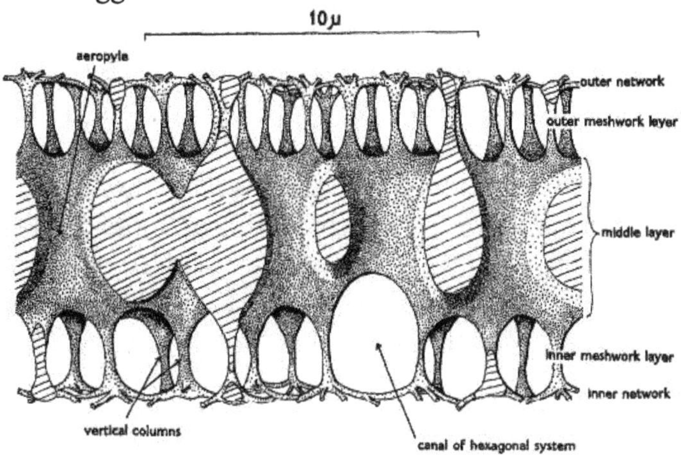

Figure 3. Respiratory system of the common house fly. Aeropyles can be seen extending through the eggshell and the various columnar sections and meshwork that compose the entire respiratory structure (Hinton 1967 [7]).

Considering there are an estimated 1 million insects on earth, there are numerous modifications and exceptions to the respiratory systems of insect eggs dependent upon the environment and respiratory requirements of individual species. While terrestrial insects have an inner pillar system, aquatic insect eggs primarily respire with a plastron. The plastron is a gas filled air layer below the outer chorion of the egg shell. The plastron acts as a physical gill that allows eggs to respire under water [6]

Some terrestrial insect eggs that are oviposited in environments flooded with water may also have a plastron. In eggs with a large plastron membrane, only small parts of the plastron are permeable, limiting water loss. Terrestrial eggs typically do not have a plastron but do have a gas layer directly under the outer surface of the chorion and connected to the aeropyles. Aeropyles of terrestrial insects occasionally flooded by rainwater are located on the ends of respiratory horns. These horns extend above the water surface to allow for more efficient uptake of oxygen when the egg is surrounded by a layer of water [6]. These respiratory horns are essential to allow the insect to breathe during periods of heavy rain when an insect egg lacks the plastron respiratory system.

The eggshell layers and elaborate respiratory systems of insect eggs are modified to conserve water for the insect egg. Terrestrial insect eggs are provided from their mother all of the water necessary for survival and development at the time of oviposition. Therefore, the embryos must conserve this limited amount of water during their development while also respiring. Water loss occurs across the chorion and is correlated to oxygen consumption requirements of the embryo. The more gas exchange that occurs, the more vulnerable the insect embryo is to water loss.

The waxy layer in the insect eggshell most probably is the main barrier in the eggshell that prevents water loss. There was a significant increase in water loss in Manduca sexta eggs when the waxy layer was dissolved by organic solvents [8]. In addition to the waxy layer, the serosal membrane that envelopes the embryo has been shown to also protect insect eggs against desiccation [9]. RNAi technology has been used to prevent the development of the serosa in the beetle Tribolium castaneum [9]. Following RNAi treatment, the serosa-less eggs were subjected to a range of different humidities. Eggs exposed to the lowest humidities could not retain water and fewer eggs hatched compared to eggs that still had their serosal membrane intact.

Environmental factors and embryonic development can exacerbate water loss in insect eggs. Elevated environmental temperatures and low relative humidity can increase water loss greatly. Insect eggs have adapted to these conditions by having an altered density or number of eggshell chorionic layers. The developing embryo also has different metabolic requirements as it progresses in development. As insect embryos develop into larvae, the metabolic rates and water loss rates increase [8]. These metabolic requirements peak shortly before the embryo hatches from the eggshell.

4. INSECT EGG MANAGEMENT

Eggs are undoubtedly the hardest insect life stage to kill with insecticides. Regardless of the application method, the tough eggshell that covers the egg prevents the entry of many insecticides, including water-based, oil-based, fumigants, and even some mechanical control methods.

An ovicide is a term used for an insecticide that specifically targets the egg stage. Smith and Salkeld [10] proposed three requirements for an ovicide to work: (1) the egg has to be in a location where it will be exposed so lethal concentrations of toxicant can reach it, (2) the egg has to be susceptible to the toxicant, and (3) a large enough proportion of eggs have to die from the toxicant in order for the treatment to be justified. The first requirement, exposure, is very important when considering insect eggs.

Agricultural pests, especially those that feed on plants, will commonly hide their eggs under leaves or may insert their eggs inside of the plant. Consideration of where the insect lays its egg in relation to an insecticide application is needed to guarantee an efficient treatment. The microclimate that the leaves create for the insect eggs is essential for their survival. Larger leaves, which absorb more sunlight, get much hotter than smaller leaves and can potentially kill eggs [11]. To limit sun exposure, insect eggs are often hidden underneath leaves and thus escape direct spray during an insecticidal treatment, resulting in an insect outbreak once those eggs hatch.

Systemic insecticides are often used for control of agricultural pests, but the use of these formulations presents a large problem for controlling the egg stage as far as exposure is concerned. These products are applied to the soil and then are taken up into the plant's xylem. Systemic products work well against immature and adult stages of sap-sucking pest insects, but these products do not work on eggs because the egg stage does not feed on the plant. Thus, the egg stage will survive and be a potential source for reinfestation.

If the eggs are not well hidden and the toxicant reaches the egg, the chemical still has to penetrate the eggshell and ultimately reach the embryo in order to be effective. Thus, the egg has to be susceptible to the toxicant. Variations in egg susceptibility could be due to the eggshell and differences in the chemical composition between eggs of different species [12] or could be due to enhanced resistance mechanisms of the embryo. Lastly, the toxicant must work well enough to kill a majority of the eggs in the population to be considered a viable option for control.

4.1 Insecticides for insect egg control

Normally, oil-based insecticides penetrate the insect eggshell more readily than water-based insecticides. The eggshell is comprised of a waxy component that allows the passage of oil- based products rather than water-based formulations. Early research has suggested that petroleum oils may also act by covering the aeropyles and causing egg mortality by limiting oxygen supply to the embryo [10]. Although most oils can be expected to penetrate the insect eggshell more easily, essential oils have been found to have difficulty penetrating eggs of the confused flour beetle and the Mediterranean flour moth compared to penetration through the cuticle of other life stages of the same insects, thus resulting in lower toxicity to eggs of these organisms [13]. Although oil-based products normally work better than water-based formulations in killing insect eggs, water has been shown to penetrate some insect eggshells. For instance, eggshell permeability to water has been demonstrated in the migratory locust, Locusta migratoria migratoriodes [14]. Empty, hatched migratory locust eggs were filled with water then placed into an osmotic solution, and water was observed both entering and leaving the eggshells. However, there is limited information regarding insect eggshells and their permeability to various chemicals. Consequently, the permeability of the insect eggshell to water, oil, or other chemical constituents of insecticides is not well established in the scientific literature.

Determining the mode of action of ovicidal insecticides is difficult. Formamidine insecticides, which have been evaluated against tobacco budworm (Lepidoptera: Noctuidae) eggs, were found to increase the levels of octopamine titers in eggs after treatment [15]. Formamidine insecticides have been shown to have a novel mode of action on insects by mimicking the actions of octopamine [16], which regulates insect behavior and energy metabolism. These results suggested that an increase in octopamine during embryogenesis could be playing a role in increased mortality of eggs treated with formamidine insecticides.

Paraoxon actively inhibits cholinesterase in Pieris eggs and was shown to prevent 100% of eggs from hatching [17]. Similarly, when house cricket eggs were exposed to carbamate insecticides, there was also a decrease in cholinesterase activity in eggs following exposure, but the insecticides did not prevent eggs from hatching [18].

Triatoma infestans eggs may be capable of detoxifying the organophosphate parathion with acetylcholinesterase enzymes [19]. The eggs produced elevated levels of acetylcholinesterase after being treated with parathion. The embryos were fully developed within the egg following treatment with parathion but never hatched from the eggshell. Therefore, the authors suggested that the embryos developed their nervous system during a later developmental stage and the parathion did not have an effect until the nervous system was fully developed.

All of these studies provide examples of insecticides permeating the eggshell and reaching the embryo, resulting in a physiological response to the insecticide. Unfortunately, most studies have only evaluated whether or not a particular insecticide has ovicidal action, thus lacking information on mode of action and penetration of the ovicide.

4.2 Fumigation of insect eggs

Fumigation has been found to be highly effective against eggs of several stored product pests. Eggs of four different species of common stored product pests, the almond moth (Cadra cautella), the Indian mealmoth (Plodia interpunctella), the lesser grain borer (Rhyzopertha dominica), and the red flour beetle (T. castaneum), have been evaluated to determine time, temperature, and pressures that were required for mortality by fumigation [20]. As temperatures increased and pressure decreased, time to mortality was reached in a shorter amount of time. Pressures above 100 mmHg and temperatures below 22.5°C were not practical to reach mortality because exposure times had to be increased drastically for the fumigant to cause egg mortality.

As is the case with liquid insecticides, studies have also shown that the egg stage is the most problematic life stage to kill with fumigants [12]. The main cause for low mortality is the impermeability of the eggshell. When eggs of Schistocerca gregaria were treated with sulfuryl fluoride, the gas was retained primarily in the proteinaceous portion of the eggshell instead of penetrating into the embryo [12]. In addition, egg age influences the efficacy of sulfuryl fluoride [20]. Eggs of the Mediterranean flour moth aged 1–2 days were the most tolerant to sulfuryl fluoride compared to younger and older eggs. Lower doses of sulfuryl fluoride were required to kill the Mediterranean flour moth as temperatures were increased from 15°C to 25°C.

Fumigation of stored product pests has been limited by EPA registration and regulations [20]. This concern has instigated investigations for alternative nonchemical control methods for treating stored-product insects. An alternative control method to fumigation for stored- product pests that has been evaluated is the use of a vacuum system in storage bins to limit oxygen availability. However, eggs were also the most difficult life stage to kill using this method. This is not too surprising because an insect egg is very small and thus requires minute amounts of oxygen compared to the immature and adult stages of insects [20]. Therefore, extremely low levels of oxygen are required to create a hypoxic environment that will kill eggs.

4.3 Transovarial transport of insecticides

The transovarial transport of insect growth regulators from mother to offspring has been shown to cause a considerable reduction in egg hatch and viability. Insect growth regulators affect the development and occasionally the reproduction of insects. There are two main types of insect growth regulators: juvenile hormone analogs and chitin synthesis inhibitors. Juvenile hormone analogs (JHAs) mimic the natural juvenile hormone present in insects and can cause multiple physiological and morphological problems. Chitin synthesis inhibitors (CSIs), as the name suggests, inhibit the proper formation of chitin between insect molts. CSIs can cause insects to have malformed, thinner cuticles and ruptured intestines. Many of the symptoms will lead to eventual death of the insect if they do not die at the time of molt.

The transovarial transport of insect growth regulators from adult female insects to eggs have been evaluated in several insect species. Different insect species have varying tolerances and responses to pyriproxifen. For example, none of the eggs of gravid whitefly, Bemisia tabaci, females treated with the juvenile hormone analog pyriproxifen hatched [21]. Alternatively, when pyriproxifen was applied to female adults of the common green lacewing, Chrysoperla carnea, it had little effect on preventing their eggs from hatching [22]. Transovarial transport has also been documented with diflubenzuron, a chitin synthesis inhibitor. Treatment of adult female C. carnea with diflubenzuron resulted in 100% mortality of eggs at the highest tested dose.

The penetration of insecticides into the female's ovaries and elsewhere in the female body has been evaluated by using [14C]-labeled isotopes [22]. Most of the pyriproxifen in the female adult C. carnea was excreted within a couple of days, so it was probably not present before oviposition [22]. The use of the radio-labeled isotopes showed that diflubenzuron, unlike pyriproxifen, was absorbed more slowly and was retained within the female's body, which explains the high levels of toxicity to eggs with this insecticide.

4.4 Fungi as biological control agents for eggs

Entomopathogenic fungi have been used on a variety of insect species as a biological control agent, but the egg stage has not been found to be highly susceptible to fungal pathogens in several cases. Eggs of the greenhouse whitefly, Trialeurodes vaporariorum, were found to be nonsusceptible to the fungus Aschersonia aleyrodis [23]. No fungal spores or discoloration of the eggs was observed, and there were no differences in mortality between eggs that were treated with the fungus compared to eggs that had not been treated. However, the fungal spores were persistent for several days, so when the first instar larvae emerged from the egg, they became infected with the fungus.

Five different fungi, (Beauveria bassiana, Metarhizium anisopliae, Metarhizium flavoviride, Paecilomyces farinosus, and Paecilomyces fumosoroseus) were tested against eggs of the curculionidae beetles, Otiorhynchus sulcatus and Sitona lineatus [24]. S. lineatus eggs were much more tolerant of the fungal pathogens compared to O. sulcatus eggs. O. sulcatus eggs were found to be susceptible to all fungi except for B. bassiana. Only one fungus, M. flavoviride, was found to be moderately effective against S. lineatus eggs, causing 32% egg mortality. No other fungal treatments resulted in egg mortality to S. lineatus eggs.

5. INSECTICIDE RESISTANCE IN INSECT EGGS

Insect eggs, like other insect stages, vary in their susceptibility to insecticides. Susceptibility differences between eggs from different insect species may be due to variability in chorionic adaptations that facilitate the uptake of oxygen [10]. Fewer aeropyles or smaller aeropyles may reduce the amount of insecticide that can enter the egg. In addition to respiratory structures, insecticides may also enter the chorion through the micropyles that allow fertilization of the egg [25]. Furthermore, modifications of the chorionic structures of the insect eggshell may enhance or reduce the penetration of insecticides.

Another important consideration when determining susceptibility of eggs to insecticides is the age of the egg. A freshly laid egg that has not fully developed is usually more susceptible to insecticides compared to an egg that has aged several days. Insect egg susceptibility to insecticides also changes during embryonic development [9]. The eggshell itself may harden during development, and the embryo may produce enzymes that break down insecticides.

Few studies have focused on insecticide resistance in eggs. Insecticide resistance in eggs from different insect species has been demonstrated when resistance was already quantified in the adult or larval stages [26–30]. Eggs collected from insect strains with resistance in the adult stage have been shown to have similar resistance. These studies determined that insecticide resistance had developed in eggs but did not determine the type or mechanisms of resistance.

Bed bugs, Cimex lectularius, have been extensively studied as adults with regards to insecticide resistance. Molecular technology has revealed that adult bed bugs have three types of resistance to pyrethroid insecticides: enhanced enzyme detoxification, KDR resistance, and target-site insensitivity. Recent research has revealed that bed bug eggs are also highly resistant to pyrethroids (Figure 4). Eggs collected from a strain that is considered pyrethroid-susceptible (Harlan) were much easier to kill with the insecticide deltamethrin, compared to eggs collected from strains (Richmond and Epic Center) where pyrethroid resistance was previously determined in adults.

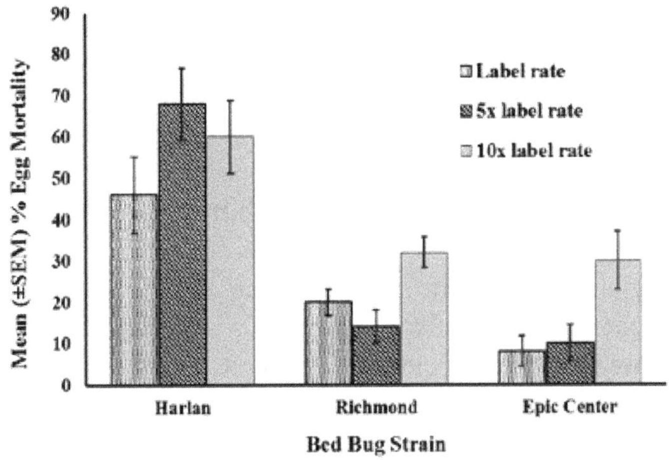

Figure 4. Mean percent mortality of eggs from a susceptible strain of bed bug eggs (Harlan) and two pyrethroid-resistant strains (Richmond and Epic Center) that were treated with the insecticide deltamethrin (0.05%). Bed bug eggs from the resistant strains had much lower mortality values compared to the susceptible strain eggs at each tested rate, showing that resistant eggs were much more difficult to kill. This study determined that resistance had already developed in the egg stage of bed bugs (Taken from Campbell et al. 2015 [30]).

Head lice, Pediculus humanus capitis (Phthiraptera: Pediculidae), have been shown to be highly resistant to pyrethroid insecticides [26]. Eggs, nymphs, and adults were evaluated from three different resistant head louse populations. Eggs were found to be highly resistant to permethrin in populations that had already demonstrated a high resistance to pyrethroid insecticides in adults and nymphs. This study suggests that there are similar resistance mechanisms within head louse eggs and adults from the same population.

Resistance patterns between eggs and first instars of Reduviid bugs, T. infestans, have been evaluated [31]. Insecticide resistance varied between eggs from different populations of T. infestans that were collected throughout Argentina and Bolivia. Eggs from a resistant strain that were aged several days were found to be as resistant to deltamethrin as the first instars (Toloza et al. 2008). Also, eggs from the resistant strain were found to be resistant to lambda- cyhalothrin but susceptible to fipronil and fenitrothion. First instar nymphs from a resistant strain had similar patterns of resistance as the eggs.

Eggs have also been found to be more tolerant than adult stages of the lesser grain borer, R. dominica, to the fumigant phosphine. Not only were eggs more tolerant to phosphine, eggs that were collected from adults that were previously screened for resistance were harder to kill with phosphine compared to eggs collected from strains that were previously determined to be susceptible to phosphine [27]. Screening for resistance in the adult stage against the fumigant phosphine was a reliable indicator for determining resistance in the egg stage of the lesser grain borer, with 9 of the 10 strains of adults that were phosphine resistant also exhibiting resistance in the egg stage. Eggs of Liposcelis bostrychophila, a stored product pest, were found to have a delay in embryonic development when fumigated with phosphine. This delay in embryonic development seems to be a method of resistance, causing control failures with phosphine treatments because the eggs were able to survive treatment this creates a problem for grain storage facilities because they may later find a reinfestation of the pest after the eggs hatch. Much remains to be understood regarding insect eggs and their resilience to various management strategies. The deficiency in understanding egg biology, physiology, and regulation, in contrast to other developmental stages, is regrettable given the significance of this period for prospective management techniques. Most research assessing the efficacy of pest control measures have predominantly overlooked the egg stage, concentrating instead on adult or immature stages. Neglecting the egg stage during treatment implementation allows the eggs to hatch, potentially resulting in reinfestation. Consequently, additional research on the effectiveness of control treatments for eggs is required, particularly in instances where the eggs are readily available and amenable to treatment. Insect eggs must not be overlooked in pest management strategies only due to their diminutive size or lack of biting or feeding behavior. Control measures aimed at insect eggs are beneficial since they eradicate pests before to inflicting any damage.

REFERENCES

1. Hinton H. Biology of Insect Eggs, Volume I, II, III. Oxford, England: Pergamon Press Inc.; 1981.
2. Wolf K, Reid W, Rider D. Eggs of the stink bug Acrosternum (Chinavia) marginatum (Hemiptera: Pentatomidae): a scanning electron microscopy study. J Submicrosc Cytol Pathol. 2002;34:143–150.
3. Wolf, K, Reid W. Egg morphology and hatching in Mormidea pictiventris (Hemiptera: Pentatomidae). Can J Zool. 2001;79:726–736.
4. Li D, Huson M, Graham L. Proteinaceous adhesive secretions from insects, and in particular the egg attachment glue of Opodiphthera sp. moths. Arch Insect Biochem Physiol. 2008;69:85–105.
5. Burkhart, C. G., Burkhart, C. N. Head lice: scientific assessment of the nit sheath with clinical ramifications and therapeutic options. J Am Acad Dermatol. 2005; 53:129–133.
6. Hinton H. Respiratory systems of insect egg shells. Annu Rev Entomol. 1969;14:343- 368.
7. Hinton, H. The respiratory system of the egg-shell of the common housefly. J Insect Physiol. 1967;13:647–651.
8. Woods, H. Water loss and gas exchange by eggs of Manduca sexta: trading off costs and benefits. J Insect Physiol. 2010;56:480–487.
9. Jacobs C, Rezende G, Lamers G, van der Zee M. The extraembryonic serosa protects the insect egg against desiccation. Proc R Soc Lond B Biol Sci. 2013;280;1–8. DOI: 10.1098/rspb.2013.1082.
10. Smith E, Salkeld E. The use and action of ovicides. Annu Rev Entomol. 1966;11:331- 368.
11. Potter K, Davidowitz G, Woods H. Insect eggs protected from high temperatures by limited homeothermy of plant leaves. J Exp Biol. 2009;212:3448–3454.
12. Outram I. Factors affecting the resistance of insect eggs to sulphuryl fluoride—II: the distribution of sulphuryl-35 S fluoride in insect eggs after fumigation. J Stored Prod Res. 1967;3:353–358.
13. Tunc I, Berger B, Erler F, Dagli F. Ovicidal activity of essential oils from five plants against two stored-product insects. J Stored Prod Res. 2000;36:161–168.

14. Browning T. Permeability to water of the shell of the egg of Locusta migratoria migratorioides, with observations on the egg of Teleogryllus commodus. J Exp Biol 1969;51:99- 105.

15. Sparks T, Kirst H, Herlein M, Larson L, Worden T, Thibault S. Biological activity of the spinosyns, new fermentation derived insect control agents, on tobacco budworm (Lepidoptera: Noctuidae) larvae. J Econ Entomol. 1998;91:1277–1283.

16. Evans P, Gee J. Action of formamidine pesticides on octopamine receptors. Nature. 1980;287:60–62.

17. David W. The systemic insecticidal action of paraoxon on the eggs of Pieris brassicae (L.). J Insect Physiol. 1959;3:14–27.

18. Hartman, M. Insecticidal Action of Carbamates on the Eggs of the House Cricket. J Econ Entomol. 1972;65:638–640.

19. de Villar M, Wood E, Zerba E, De Licastro S, Casabe N. Cholinesterases and eserine- resistant esterases in the developing embryo of Triatoma infestans and its role as targets for inhibition in the ovicide action of parathion. Comp Biochem Physiol C. 1980;67:55–59.

20. Mbata G, Phillips T, Payton M. Mortality of eggs of stored-product insects held under vacuum: effects of pressure, temperature, and exposure time. J Econ Entomol. 2004; 97:695–702.

21. Ishaaya I, Horowitz R. Pyriproxyfen, a novel insect growth regulator for controlling whiteflies: mechanisms and resistance management. Pesticide Sci. 1995;43:227–232.

22. Medina P, Smagghe G, Budia F, del Estal P, Tirry L, Vinuela E. Significance of penetration, excretion, and transovarial uptake to toxicity of three insect growth regulators in predatory lacewing adults. Arch Insect Biochem Physiol. 2002;51:91–101.

23. Fransen J, Winkelman K, van Lenteren J. The differential mortality at various life stages of the greenhouse whitefly, Trialeurodes vaporariorum (Homoptera: Aleyrodidae), by infection with the fungus Aschersonia aleyrodis (Deuteromycotina: Coelomycetes). J Invertebr Pathol. 1987;50:158–165.

24. Poprawski T, Marchal M, Robert P. Comparative susceptibility of Otiorhynchus sulcatus and Sitona lineatus (Coleoptera: Curculionidae) early stages to five entomopathogenic Hyphomycetes. Environ Entomol. 1985;14:247–253.

25. Beament J. The role of cuticle and eggshell membranes in the penetration of insecticides. Ann Appl Biol. 1952;39:142–143.

26. Cueto G, Zerba E, Picollo M. Evidence of pyrethroid resistance in eggs of Pediculus humanus capitis (Phthiraptera: Pediculidae) from Argentina. J Med Entomol. 2008;45:693–697.
27. Bell C, Hole B, Evans P. The occurrence of resistance to phosphine in adult and egg stages of strains of Rhyzopertha dominica (F.) (Coleoptera: Bostrichidae). J Stored Prod Res. 1977;13:91–94.
28. Leonard B, Graves J, Sparks T. Ovicidal effects of selected insecticides against pyrethroid-susceptible and-resistant tobacco budworm. Southwest Entomol. 1991.
29. Ho S, Goh P. Deltamethrin as a potential ovicidal pyrethroid against Plutella xylostella L. Toxicol Lett. 1984;22:161–164.
30. Campbell B, Miller D. Insecticide resistance in eggs and first instars of the bed bug, Cimex lectularius (Hemiptera: Cimicidae). Insects. 2015;6:122–132.
31. Toloza A, Germano M, Cueto G, Vassena C, Zerba E, Picollo M. Differential patterns of insecticide resistance in eggs and first instars of Triatoma infestans (Hemiptera: Reduviidae) from Argentina and Bolivia. J Med Entomol. 2008;45:421–426.

6 Chapter

CURRENT STATUS OF INSECTICIDE RESISTANCE IN AEDES AEGYPTI (DIPTERA: CULICIDAE) IN MEXICO

1. INTRODUCTION: AN OVERVIEW

Aedes aegypti serves as the principal urban vector for the viruses responsible for dengue, Chikungunya, and yellow fever [1–4]. The females are predominantly endophagic and endophilic diurnal vectors that preferentially feed on people. They consume several blood meals prior to laying an egg batch, so enabling a single infectious female to transmit the virus to multiple individuals. The females deposit their eggs in receptacles located in the peridomestic environment, where the immature larvae and pupae subsequently develop [10,11]. Ae. aegypti is prevalent in densely populated regions of Mexico at elevations up to approximately 1,500 meters above sea level [12]. Aedes albopictus, the principal vector of dengue and Chikungunya viruses, is prevalent in rural environments [13–15]. This human biter was imported to Texas in the 1980s and has proliferated extensively throughout northern and central Mexico, as well as the southernmost region of the country. [16–18].

Urban surroundings have facilitated the prevalence and proliferation of Ae. aegypti in 30 out of 32 states in Mexico, excluding Tlaxcala and the Federal District, hence resulting in the endemic transmission of dengue and, more recently, Chikungunya since 2014. [22].

2. STUDY AREA

Mexico is in the southern part of North America, between 14° 32 and 32° 43 North and 86° 42 and 118° 27 West. The country is divided administratively into 31 states and one federal district (Mexico City). Mexico has a land area of 1,964,375 km^2. It is surrounded by the Gulf of Mexico and Caribbean Sea to the east, the United States of America to the north, the Pacific Ocean to the south and west, and Belize and Guatemala to the southeast. The main features of the physiography of Mexico are Northern and Southern Plateau. Two mountain chains, the Sierra Madre Oriental on the east and The Sierra Madre Occidental on the west, leave plains along the shores of the Gulf of Mexico and the Pacific Ocean. The Sierra Madre Oriental obstructs the circulation of air from the Gulf of Mexico toward Northern and

Central Mexico. This characteristic on physiography allows a variety of climates, and the altitude performs a dominant effect on temperature. The prevalent climate conditions are dry to arid in the country. The North territory (47.7%) presents arid and semiarid conditions (23.5%), subhumid with 7 months of long dry season prevailing in Central Mexico, 16.3% presents dry tropical mainly the shores, 12.4% of the territory located in Southern Mexico presents humid tropical climate, and in both mountain chains small areas with humid temperate climate are found [23].

3. THE SITUATION OF DENGUE AND CHIKUNGUNYA FEVER IN MEXICO

A recent study estimated that up to 390 million dengue virus (DENV) infections, including close to 100 million cases of dengue disease manifestations, occur annually across the world [24]. Ae. aegypti and DENV were widely distributed in the Americas in the early 1900s, but a campaign against yellow fever initiated in 1947 and continued to the early 1970s resulted in both the mosquito and its associated viruses being eliminated from most of Central and South America and from Mexico [25]. Success then bred failure as resources were diverted to other health problems. From the 1980s, Ae. aegypti has reemerged in the Americas, facilitated by uncontrolled urbanization providing ample opportunities for mosquito breeding and population growth [1,25]. The mosquito now has regained the full extent of its range from a century ago. This reemergence of the vector combined with global trafficking of DENV in infected humans through increased air travel have led to the Americas becoming hyperendemic, with cocirculation of all four DENV serotypes in many areas [26,27]. In Mexico, dengue reemerged with a DENV-1 outbreak in 1979 followed by outbreaks of the other serotypes in the next two decades [25,28,29]. In 2009, a major dengue epidemic with nearly 250,000 reported clinical cases occurred in Mexico (55,363 confirmed) and this epidemic has continued through 2010 with ~58,000 clinical cases (22,352 confirmed) , 2011 with ~68,000 cases (15,578 confirmed) , 2012 with ~165,000 cases (84,612 confirmed) , 2013 with ~232,000 cases (62,330 confirmed) , 2014 with 125,000 cases (32,100 confirmed), and 2015 (up to September 25, 2015 ~125,000 clinical cases with 13,454 confirmed) [30].

The autochthonous transmission of Chikungunya in the Region of the Americas was first detected on December 2013. By July 2014, an imported case was reported in Mexico, two more imported cases appeared on September 5. By the end of 2014, a total of 131 autochthonous confirmed cases were reported as well as a total of 13 imported cases; and ~7,500 confirmed cases by October 2, 2015 [30].

4. VECTOR CONTROL IN MEXICO

Since 1950, operational vector control programs in Mexico have used a series of insecticides to control Ae. aegypti [31]. The organochlorine insecticide DDT was used extensively for indoor house spraying from 1950 to 1960 and was used in some locations as recently as 1998. In recent decades, the chemical control of mosquito larvae has relied on the use of organophosphate insecticides with temephos as the active ingredient. The adulticide malathion was used for ultra-low volume (ULV) space spraying from 1981 to 1989. An oil-based formulation of chlorpyrifos was registered for use in some locations in Mexico to control the adult stage of the mosquito from 1996 to 1999. The organophosphates as adulticides were replaced by pyrethroids according to the Norma Official Mexicana NOM-032-SSA2-2002 [32]. The pyrethroid permethrin was applied as a sole adulticide in Mexico for more than a decade.

On June 1, 2011, a new policy was published in NOM-032-SSA2-2010 [33] that established the characteristics of the insecticides to be used for vector control in Mexico. The selection of the insecticides should be based on vector resistance, effectiveness, and safety related to exposure. The list of insecticides has since been updated each year [34]. A new policy published on April 16, 2015 (NOM-032-SSA2-2014) [35], maintained the same requirements practically as the regulation published in 2011.

5. INSECTICIDE RESISTANCE IN AE. AEGYPTI – A THREAT TO ITS CONTROL

The extensive use of DDT to control Ae. aegypti in Mexico and other parts of the Americas during the 1950s and 1960s resulted in the development of resistance [36]. This action was unfortunate because both DDT and pyrethroids target voltage-gated sodium channels in the insect nerve sheath where structure-related interactions occur in specific regions of the sodium channels that prolong their opening and produce paralysis. Indeed, the similar mode of action probably produced cross-resistance to pyrethroids in DDT-resistant Ae. aegypti [37–41].

Pyrethroid resistance is clearly increasing despite the initial optimism over their rapid action and novelty [42]. Evidence of resistance to permethrin insecticide used in Mexico for more than 10 years in Ae. aegypti populations in Mexico due to enzymatic mechanisms such as α- aand β-esterases was reported in Baja California North and South [43], in Quintana Roo, south f Mexico [31], and some states of northeast Mexico [44]. More recently, Aponte et al. [45] found increased levels of esterases and glutathione S-transferase related with resistance to DDT, permethrin, and deltamethrin in Ae. aegypti populations from the state of Guerrero located on the west coast of Mexico.

The presence of a kdr mutation V1016I in the voltage-gated sodium channel gene is also associated with resistance to pyrethroids. This mutation was originally found in a permethrin resistant strain from Isla Mujeres, off the coast of Cancun [46,47].

High frequencies of this resistance allele were subsequently found in collections of Ae. aegypti from 78 sites in Mexico with some of the highest frequencies detected in collections from Veracruz state [48,49].

Flores et al. [50] reported an extensive monitoring of the frequency of kdr Ile1,016 in Ae. aegypti populations from Merida, Yucatan, south of Mexico, as part of the "Casa Segura" project. Ae. aegypti collections were characterized by both molecular kdr and biochemical resistance to pyrethroid insecticides such as permethrin and deltamethrin. Ile1,016 allele frequencies varied among collection sites ranging from 0.14 to 0.98. Within Merida City, fifteen collection sites had medium to high homozygote frequencies. The lowest Ile/Ile homozygote frequencies corresponded to small towns nearby Merida City.

A second mutation F1534C on the IIIS6 domain of the same gene was also detected in Ae. aegypti populations from Guerrero state located on the west coast of Mexico [45] and the Yucatan Peninsula [51] conferring resistance to pyrethroids.

The practice of utilizing a single insecticide until the appearance of resistance has become a standard practice that quickly reduces the number of insecticides available for vector control. Rotations, mosaics, and mixtures have instead been proposed as strategies for insecticide resistance management [52–54]. Mathematical models have been applied for estimating how these tools could be used in an optimal manner [55]. However, these models have been rarely tested under field conditions, especially for insect vectors, due to the difficulties in determining changes in frequencies of resistance genes in large samples of insects from resistant populations [56].

In Mexico, there was a large-scale field trial with Anopheles albimanus that used rotations or mosaics of insecticides substituting the simple use of DDT or of specific pyrethroids [56,57]. Changes in the frequency of resistance genes were monitored for 4 years [57]. The results were promising and predicted that rotations or mosaics of insecticides are viable long-term strategies for the sustainable use of insecticides in disease control programs.

With that goal in mind [58], the resistance to eight pyrethroids in collections of Ae. aegypti from the state of Veracruz located on the east coast of Mexico was examined, considering that this knowledge would facilitate the selection of viable alternative pyrethroids besides permethrin for use in a rotation program for sustained control of Ae. aegypti at the local, regional, and possibly statewide levels. The results obtained showed that the strains analyzed were resistant to δ-phenothrin, deltamethrin, cypermethrin, α-cypermethrin, z-cypermethrin, λ-cyhalothrin, bifenthrin, as well as permethrin and suggested that populations in the state of Veracruz have been exposed to strong selection pressure, resulting from the continuous application of permethrin for more than a decade. They also evaluated resistance to chlorpyrifos [59] in the same strains, and overall, the populations in this study were less resistant to chlorpyrifos than to pyrethroids, so the rotation of insecticides in the control activities is suggested to delay or minimize the occurrence of high levels of resistance to chlorpyrifos among local populations of Ae. aegypti.

Saavedra-Rodriguez et al. [60] examined changes in gene expression before, during and after five generations of permethrin laboratory selection in five strains of Ae. aegypti collections from the Yucatan Peninsula of Mexico. Changes in expression of 290 metabolic detoxification genes were measured using the Aedes Detox microarray. Selection simultaneously increased the LC50, KC50, and Ile1,016 frequency. Ten to eight genes were differentially transcribed after selection, and it was an inverse relationship between the Ile1,016 frequency and the numbers of differentially transcribed genes. Some genes were differential transcribed among field strains, but interestingly a few cytochrome P450 genes complex were overexpressed. The authors established that adaptation to permethrin in Ae. aegypti laboratory strain is conditioned presumably by geographic origin and extant target site insensitivity in the para gene. The lack of uniformity in the genes that responded to artificial selection as well as differences in the direction of their responses challenges the assumption that one or a few genes control permethrin metabolic resistance.

The selection pressure by the prolonged use of pyrethroids in Mexico had resulted in resistance to all of this kind of chemicals recommended for vector control in Mexico. All studies have shown the prevalence of cross-resistance caused by metabolic mechanisms and/or point mutations. Saavedra et al. [51] demonstrated that even in the absence of barriers to gene flow, local insecticide pressure, rather than the migration of mosquitoes with kdr-conferring mutations, is the primary determinant of the local kdr profile for Ae. aegypti. Thus, the early detection of insecticide resistance is highly relevant to establish a rotation program for insecticide resistance management in Ae. aegypti in Mexico. In an attempt to establish the importance of evaluating the strength of available techniques to assess the insecticide susceptibility in Ae. aegypti, Lopez et al. (in press) conducted a study establishing the intensity of insecticide resistance through the Resistance Intensity Rapid Diagnostic Test (I-RDT) [61]. The RDT-I consists of exposing vector populations 1, 2, 5 and ten times the diagnostic dose previously established at a diagnosis time. For this study, they used four populations of Ae. aegypti from the state of Yucatan, south of Mexico, and three population from the state of Nuevo Leon, northeastern Mexico. They were exposed to the diagnostic dose (DD) of permethrin, bifenthrin, and d-(cis-trans)-phenothrin and enhanced DD at 2, 5, and 10 times. All populations resulted resistant to the pyrethroids evaluated according to WHO recommendations for assessing the significance of detected resistance

(<90%) even when the DD was enhanced 5 times. To correlate these results with pyrethroid molecular resistance mechanisms, DNA from mosquitoes of each population were used to detect V1016I and F1534C mutations. The allelic frequency of Ile1,016 varied from 0.43 to 0.90 in the populations studied. For the 1534 locus, there was a predominance of homozygous mutant genotype in all populations with high frequencies of the mutant allele (0.75–1), showing that the F1534C mutation was more common than V1016I mutation. They also analyzed the co-occurrence of both V1016I and F1534C mutations, and results showed that more than 50% of mosquitoes genotyped expressed both mutations (double homozygous mutants).

6. CONCISE SUMMARY

Over six decades, the selection pressure from insecticides on Ae. aegypti populations in Mexico has resulted in extensive resistance to several insecticides, particularly pyrethroids in the past 15 years. It is imperative that we contemplate measures to prevent significant resistance between pyrethroids and alternative adulticides. In the future, methods must incorporate resistance monitoring, the creation of sophisticated technologies for identifying multiple pesticide resistance, and effective instruments for efficient vector control.

REFERENCES

1. Gubler DJ. 2004. The changing epidemiology of yellow fever and dengue, 1900 to 2003: full circle? Comparative Immunology, Microbiology and Infectious Diseases. 27: p. 319–330.
2. Pialoux G, Gaüzère BA, Jauréguiberry S, Strobel M. 2007. Chikungunya, an epidemic arbovirosis. Lancet Infectious Diseases 7: p. 319–327.
3. Barrett AD, Higgs S. 2007. Yellow fever: a disease that has yet to be conquered. Annual Review of Entomology 52: p. 209–229.
4. Kyle JL, Harris E. 2008. Global spread and persistence of dengue. Annual Review of Microbiology 62: p. 71–92.
5. Halstead SB. 2008. Dengue virus–mosquito interactions. Annual Review of Entomology 53: p. 273–291.
6. Garcia-Rejon J, Loroño-Pino MA, Farfan-Ale JA, Flores-Flores L, Rosado-Paredes EdP, Rivero-Cardenas N, Najera-Vazquez R, Gomez-Carro S, Lira-Zumbardo V, Gonzalez-Martinez P, Lozano-Fuentes S, Elizondo-Quiroga D, Beaty BJ, Eisen L.. 2008. Dengue virus-infected Aedes aegypti in the home environment. American Journal of Tropical Medicine and Hygiene 79: p. 940–950.
7. Scott TW, Chow E, Strickman D, Kittayapong P, Wirtz RA, Lorenz LH, Edman JD. 1993. Blood-feeding patterns of Aedes aegypti (Diptera: Culicidae) collected in a rural Thai village. Journal of Medical Entomology 30: p. 922–927.

8. Scott TW, Amerasinghe PH, Morrison AC, Lorenz LH, Clark GG, Strickman D, Kittayapong P, Edman JD. 2000. Longitudinal studies of Aedes aegypti (Diptera: Culicidae) in Thailand and Puerto Rico: blood feeding frequency. Journal of Medical Entomology 37: p. 89–101.

9. Benedictis JD, Chow-Shaffer E, Costero A, Clark GG, Edman JD, Scott TW. 2003. Identification of the people from whom engorged Aedes aegypti took blood meals in Florida, Puerto Rico, using polymerase chain reaction-based DNA profiling. American Journal of Tropical Medicine and Hygiene 68: p. 437–446.

10. Tun-Lin W, Lenhart A, Rebollar-Tellez E, Morrison AC, Barbazan P, Cote M, Midega J, Sanchez F, Manrique-Saide P, Kroeger A, Nathan MB, Meheus F, Petsold M. 2009. Reducing costs and operational constraints of dengue vector control by targeting productive breeding places: a multi-country non-inferiority cluster randomized trial. Tropical Medicine and International Health 14: p. 1143–1153.

11. Focks DA, Alexander N. 2006. A multi-country study on the methodology for surveys of Aedes aegypti pupal productivity: findings and recommendations. World Health Organization, Geneva, Switzerland.

12. Lozano-Fuentes S, Hayden MH, Welsh-Rodriguez C, Ochoa-Martinez C, Tapia-Santos B, Kobylinski KC, Uejio ChK, Zielinski-Gutierrez E, Monache LD, Monaghan AJ, Steinhoff DF, Eisen L. 2012. The dengue virus mosquito vector Aedes aegypti at high elevation in Mexico. American Journal of Tropical Medicine and Hygiene 87: p. 902– 909.

13. Gratz NG. 2004. Critical review of the vector status of Aedes albopictus. Medical and Veterinary Entomology 18: p. 215–227.

14. Rodhain F, Rosen L. 1997. Mosquito vectors and dengue virus-vector relationships. In: Dengue and Dengue Hemorrhagic Fever, D.J. Gubler and G. Kuno, editors. CABI Publishing: Cambridge, MA. p. 45–60.

15. PAHO/CDC. 2011. Preparedness and Response for Chikungunya Virus Introduction in the Americas. Pan American Health Organization/United States Centers for Disease Control and Prevention: Washington, DC.

16. Salomón-Grajales J, Lugo-Moguel GV, Tinal-Gordillo VR, Cruz-Velázquez J de la, Beaty BJ, Eisen L, Lozano-Fuentes S, Moore ChG, García-Rejón JE. 2012. Aedes albopictus mosquitoes, Yucatan Peninsula, Mexico. Emerging Infectious Diseases 18: p. 525– 527. (PMC3309596).

17. Casas-Martínez M. Torres-Estrada J. 2003. First evidence of Aedes albopictus (Skuse) in Southern Chiapas, Mexico. Emerging Infectious Diseases 9: p. 606–607.

18. Ibáñez-Bernal S, Martínez Campos C. 1994. Aedes albopictus in Mexico. Journal of the American Mosquito Control Association 10: p. 231–232.

19. Cuddehe M. 2009. Mexico fights rise in dengue fever. Lancet 374: 602–602. PMID: 19708101. doi: http://dx.doi.org/10.1016/S0140-6736(09)61509-9.

20. Secretaría de Salud México. 2012. Perfil epidemiológico del dengue en México. Available at: http://www.epidemiologia.salud.gob.mx/doctos/infoepid/publicaciones/ 2012/Monografia_Dengue_2012. pdf. Accessed October 23, 2013.

21. Hernández-Ávila JE, Rodríguez M-H, Santos-Luna R, Sánchez-Castañeda V, Román- Pérez S, Ríos-Salgado VH, Salas-Sarmiento JA. 2013. Nation-wide, web-based, geographic information system for the integrated surveillance and control of dengue fever in Mexico. PLoS One 8: e70231. doi: 10.1371/journal.pone.0070231 PMID: 23936394.

22. PAHO. Pan American Health Organization, Dengue regional information, number of cases by year. http://www.paho.org/hq/index.php?option=com_content&view=article&id=264&Itemid=363. Accessed December 11, 2014.

23. FAO. 2015. http://www.fao.org/countryprofiles/index/es/?iso3=MEX

24. Bhatt S, Gething PW, Brady OJ, Messina JP, Farlow AW, Moyes CL, Drake JM, Brownstein JS, Hoen AG, Sankoh O, Myers MF, George DB, Jaenisch T, Wint GRW, Simmons CP, Scott TW, Farrar JJ, Hay SI. 2013. The global distribution and burden of dengue. Nature 496: p. 504–507.

25. Brathwaite Dick, O. San Martín JL, Montoya RH, del Diego J, Zambrano B, Dayan GH. 2012. The history of dengue outbreaks in the Americas. American Journal of Tropical Medicine and Hygiene 87: p. 584–593. doi: 10.4269/ajtmh.2012.11-0770.

26. Kyle JL, Harris E. 2008. Global spread and persistence of dengue. Annual Review of Microbiology 62: p. 71–92.

27. Wilder-Smith A, Gubler DJ. 2008. Geographic expansion of dengue: the impact of international travel. Medical Clinics Of North America 92(6): p. 1377–1390.

28. Briseno-Garcia B, Gómez-Dantés H, Argott-Ramírez E, Montesano R, Vázquez-Martínez AL, Ibáñez-Bernal S, Madrigal-Ayala G, Ruíz-Matus C, Flisser A, Tapia-Conyer R. 1996. Potential risk for dengue hemorrhagic fever: the isolation of serotype dengue-3 in Mexico. Emerging Infectious Diseases 2(2): p. 133–135.

29. PAHO. Pan American Health Organization, Dengue regional information, number of cases by year. http://www.paho.org/hq/index.php?option=com_topics&view=rdmore&cid=6290&Itemid=40734&lang=en. Accessed October 5, 2015

30. PAHO. Pan American Health Organization, Chikungunya regional information, number of cases by year. http://www.paho.org/hq/index.php?option=com_topics&view=rdmore&cid=7928&Itemid=40931&lang=en. Accessed October 5, 2015.

31. Flores AE, Grajales JS, Fernandez IS, Ponce GG, Loaiza H, Badii MH, Lozano SF, Brogdon WG, Black WC, Beaty BJ. 2006. Mechanisms of insecticide resistance in field populations of Aedes aegypti (L.) from Quintana Roo, Southern Mexico. Journal American Mosquito Control Association 22: 672–677.

32. DOF (Diario Oficial de la Federación). 2003. NOM-032-SSA-2-2002 para la vigilancia epidemiológica, prevención y control de enfermedades transmitidas por vectores. México

33. DOF (Diario Oficial de la Federación). 2011. NOM-032-SSA-2-2010 para la vigilancia epidemiológica, prevención y control de enfermedades transmitidas por vectores. México

34. SSM. Secretaría de Salud de México. Lista actualizada de insumos recomendados por el CENAPRECE para el combate de insectos vectores de enfermedades a partir de 2015. http://www.cenaprece.salud.gob.mx/programas/interior/vectores/descargas/pdf/ListaActualizadaInsumosRecomendadosCENAPRECE2015.pdf. Accessed July 2015.

35. DOF (Diario Oficial de la Federación). 2015. NOM-032-SSA-2-2014 para la vigilancia epidemiológica, promoción, prevención y control de enfermedades transmitidas por vectores. México

36. Brown AWA, Pal R. 1971. Insecticide resistance in arthropods. WHO 38.

37. Brengues C, Hawkes NJ, Chandre F, McCarroll L, Duchon S, Guillet P, Manguin S, Morgan JC, Hemingway J. 2003. Pyrethroid and DDT cross-resistance in Aedes aegypti is correlated with novel mutations in the voltage-gated sodium channel gene. Medical and Veterinary Entomology 17: 87–94.

38. Chadwick PR, Invest JR, Bowron MJ. 1977. An example of cross resistance to pyrethroids in DDT resistant Aedes aegypti. Pesticide Science 8:618–624.

39. McDonald AE, Wood RJ. 1979. Mechanism of DDT resistance in larvae of the mosquito Aedes aegypti L.: the effect of DDT selection. Pesticide Science 10: 383–388.

40. Prasittisuk C, Busvive JR. 1977. DDT-resistant mosquito strains with cross-resistance to pyerthroids. Pesticide Science 8: 527–533.

41. Rongsriyam Y, Busvine JR. 1975. Cross resistance in DDT resistant strains of various mosquitoes (Diptera: Culicidae). Bulletin of Entomological Research 65: 459–471.

42. Malcolm CA. 1988. Current status of pyrethroid resistance in anophelines. Parasitology Today 4:S13–S15. doi: 10.1016/0169-4758(88)90081-6

43. Flores AE, Albeldaño WV, Fernandez IS, Badii MH, Loaiza H, Ponce GG, Lozano SF, Brogdon WG, Black WC, Beaty BJ. 2005. Elevated α-esterase levels associated with permethrin tolerance in Aedes aegypti (L.) from Baja California, Mexico. Pesticide Biochemistry Physiology 82: 66–78.

44. Flores AE, Reyes G, Fernandez IS, Sanchez RFJ, Ponce GG. 2009. Resistance to permethrin in Aedes aegypti (L.) in northern Mexico. Southwestern Entomology 34: 167– 177.

45. Aponte HA, Penilla RP, Dzul-Manzanilla F, Che-Mendoza A, Lopez AD, Solis F, Manrique-Saide P, Ranson H, Lenhart A, McCall PJ, Rodrıguez AD. 2013. The pyrethroid resistance status and mechanisms in Aedes aegypti from the Guerrero state, Mexico. Pesticide Biochemistry Physiology 107: 226–234.

46. Saavedra K, Urdaneta L, Rajatileka S,Moulton M, Flores AE, Fernandez I, Bisset J, Rodriguez M, Mccall PJ, Donnelly MJ, Ranson H, Hemingway J, Black IV WC. 2007. Mutations in the voltage gated sodium channel gene associated with pyrethroid resistance in Latin American Aedes aegypti. Insect Molecular Biology 16: 785–798.

47. Saavedra K, Strode C, Flores A, Fernandez I, Ranson H, Hemingway J, Black IV WC. 2008. QTL mapping of genome regions controlling permethrin resistance in the mosquito Aedes aegypti. Genetics 180: 1137–1152.

48. Ponce G, Flores A, Fernandez I, Saavedra K, Reyes G, Lozano S, Bond G, Casas M, Ramsey, Garcia J, Domínguez M, Ranson H, Hemingway J, Eisen L, Black IV WC. 2009. Recent rapid rise of a permethrin knock down resistance allele in Aedes aegypti in México. PLOS Neglected Tropical Diseases 3: 531–560.

49. Siller Q, Ponce G, Lozano S, Flores AE. 2011. Update on the frequency of Ile1016 mutation in voltage-gated channel gene of Aedes aegypti in Mexico. Journal American Mosquito Control Association 27: 357–362.

50. Flores AE, Ponce GG, Loroño MA, García JE, Machain C, Reyes GC, Lozano S, Lars E, Beaty BJ, Black IV WC. 2012. Insecticide resistance in Aedes aegypti in Mexico: implications for dengue control (abstract). DMID International Research in Infectious Diseases Meeting, Bethesda, MD. p. 49.

51. Saavedra-Rodriguez K, Beaty M, Lozano-Fuentes S, Denham S, Garcia-Rejon J, Reyes-Solis G, Machain-Williams C, Loroño-Pino MA, Flores A, Ponce G, Beaty B, Eisen L, Black IV WC. 2015. Local evolution of pyrethroid resistance offsets gene flow among Aedes aegypti collections in Yucatan State, Mexico. American Journal Tropical Medicine Hygiene 7:92(1): 201–209. doi: 10.4269/ajtmh.14-0277

52. Curtis CF. 1985. Theoretical models of the use of insecticide mixtures for the management of resistance. Bulletin Entomological Research 75: 259–265.

53. Curtis CF, Hil N, Kasim SH. 1993. Are there effective resistance management strategies for vectors of human disease? Biological Journal of the Linnean Society 48: 3– 18.

54. Roush RT. 1989. Designing resistance management programmes: how can you choose? Pesticide Science 26: 423–42.

55. Tabashnik BE. 1989. Managing resistance with multiple pesticide tactics: theory, evidence, and recommendation. Journal of Economic Entomology 82: 1263–1269.

56. Hemingway J, Penilla RP. Rodríguez AD, James BM. Edge W, Rogers H, Rodríguez MH. 1997. Resistance management strategies in malaria vector mosquito control. A large-scale field trial in Southern Mexico. Pesticide Science 51: p. 375–382.

57. Penilla R.P., Rodríguez AD, Hemingway J, Torres JL, Arredondo-Jiménez JI, Rodríguez MH. 1998. Resistance management strategies in malaria vector mosquito control. Baseline data for a large-scale field trial against Anopheles albimanus in Mexico. Medical and Veterinary Entomology 12: 217–233.

58. Flores AE, Ponce G, Silva BG, Gutierrez SM, Bobadilla C, Lopez B, Mercado R, Black IV WC. 2013. Wide spread cross resistance to pyrethroids in Aedes aegypti (L.) from Veracruz State Mexico. Journal Economic Entomology 106(2): 959:969

59. Lopez B, Ponce G, Gonzalez JA, Gutierrez SM, Villanueva OK, Gonzalez G, Bobadilla C, Rodríguez IP, Black WC IV, Flores AE. 2014. Susceptibility to chlorpyrifos in pyrethroid-resistant populations of Aedes aegypti (Diptera: Culicidae) form Mexico. Journal of Medical Entomology 51(3): 644–649. doi: http://dx.doi.org/10.1603/ME13185.

60. Saavedra-Rodriguez K, Flores-Suarez A, Salas IF, Strode C, Ranson H, Hemingway J, Black IV WC. 2012. Transcription of detoxification genes after permethrin selection in the mosquito Aedes aegypti. Insect Molecular Biology 21: 61–77.

61. CDC. 2010. Guideline for evaluating insecticide resistance in arthropod vectors using the CDC bottle bioassay. Centers for Disease Control and Prevention, Atlanta, GA. http://www.cdc.gov/parasites/education_training/lab/bottlebioassay.html

7 Chapter

MOSQUITO-BORNE DISEASES, PESTICIDES USED FOR MOSQUITO CONTROL, AND DEVELOPMENT OF RESISTANCE TO INSECTICIDES

1. INTRODUCTION: AN OVERVIEW

Mosquito-borne vectors transmit numerous pathogens responsible for infectious diseases that can be fatal to humans. In recent decades, some diseases have seen a rise in prevalence and have spread to new geographical regions. Factors that can facilitate the dissemination of disease include heightened population density, increased international travel, and the escalation of global trade in goods. The recent proliferation of notifications on mosquito-borne diseases globally is alarming, and presently, there are no efficacious vaccines available for the majority of these ailments. In several regions globally, the existence of mosquitoes poses a significant issue, as each season introduces various species that serve as vectors for diseases of medical and veterinary significance, due to their feeding on humans and other organisms. Consequently, the sole method to prevent outbreaks of mosquito-borne diseases is by controlling insect vectors and understanding their biology, behavior, and environmental factors that promote transmission. [2].

The vector qualities of mosquitoes differ based on the specific conditions of their native habitat. The mosquito undergoes four developmental stages: egg, larva, pupa, and adult, with the initial three stages requiring stagnant water for maturation. Adult mosquitoes are little insects, delicate in structure, characterized by elongated bodies, a pair of narrow wings, and three pairs of elongated, slender legs. Their lengths range from 3.16 to 1.2 inches (5 to 13 mm). They possess an extended proboscis with mouthparts specialized for skin penetration, which the female utilizes for nourishment and blood consumption. In recent decades, pest control efforts have relied heavily on the extensive application of chemical pesticides and the eradication of all containers, whether manmade or natural, that create conducive environments for the proliferation of mosquito breeding sites [3]. Nonetheless, the adverse impacts of chemicals on nontarget organism populations and the emergence of resistance to these chemicals in mosquitoes, coupled with the recent resurgence of various diseases transmitted by mosquitoes, have prompted the exploration of alternative methods that are simpler and more sustainable for mosquito control.

2. MOSQUITO-BORNE DISEASES

With the recent expansion of Aedes albopictus, there are epidemiological reports for several mosquito-borne diseases occurred in various parts of Europe [4]. These evidences show that the problem does not constitute an exclusive problem of tropical, subtropical, and/or developing countries. In North America, several mosquito vectors species of the Culex, Anopheles, and Aedes genera that transmit several diseases to humans are present [5]. Meanwhile, in Central and South America, the main disease in humans transmitted by mosquito vectors of the Anopheles, Aedes, Culex, and Ocherotatus genera are malaria, dengue, yellow fever, filariasis, St. Louis encephalitis, and western equine encephalitis [6].

Mosquito-borne diseases occur when the specific biological agent that causes the disease is transmitted to human hosts through a nonhuman carrier called vector. Therefore, the chain of transmission involves three factors: one host, usually a human, an invertebrate vector responsible for spreading the disease, and the biological agent that may be a virus, bacterium, or parasite. Vectors may act biologically or mechanically, where the mechanical vectors only transport the pathogenic agent; however, in biological vectors, agent develops and multiplies before becoming infective to the vertebrate host [7]. In that context, mosquito-borne diseases of public health importance are complex, and its occurrence will depend on the interaction of various factors such as biological, ecological, social, and economic factors [8].

2.1 Malaria

Malaria or paludism is caused by parasites of genus Plasmodium, which is transmitted to humans by the bite of female Anopheles mosquitoes. Plasmodium falciporum is responsible for most severe cases, usually causing coma or anemia in patients, which flows into death. Meanwhile, Plasmodium vivax causes recurring fevers and lesions in the brain and liver, but it rarely causes death [9]. Within the measures applied to control, Anopheles vectors, for a long time, was based on the application of DDT (1,1,1-trichloro-2,2-bis [ρ -chorophenyl] ethane), but currently it is beginning to use pyrethroids in outbreaks and transmission foci [10].

While the disease appeared to be under control in the 1950s, the infection again reappeared in many countries due to the resistance generated by vectors to insecticides of plasmodia and chloroquine. This disease is responsible for the deaths of between 700,000 and 2.7 million people [11]. Moreover, malaria causes between 400 and 900 million cases of acute fever per year in children fewer than five years in these areas. Therefore, malaria is the disease with the highest prevalence in areas with limited economic resources, causing the largest number of cases in the warm and rainy seasons. The solution to eradicate this disease would be the application of vaccine [12].

2.2 Yellow fever

This disease is caused by the yellow fever virus, an arbovirus, belonging to the Flavivirus genus is present in tropical areas of Africa and South America. Aedes aegypti mosquito is the most important vector in the transmission of the yellow fever disease in America [13]. Yellow fever virus infects both humans and monkeys, being monkeys the main reservoir of infection and transmission from monkey to monkey in woodlands and jungle. Haemagogus jantinomys and Sabethes choropterus mosquitoes are the vectors responsible for the transovarially virus transmission among the primate species [14]. On the other hand, the infection is transmitted to humans through A. aegypti mosquito bites. Yellow fever virus causes 200,000 clinical cases of disease and 30,000 deaths each year, of which 90% of the cases correspond to the African continent [15]. Unfortunately, the most of the cases and deaths are not recognized because it occurs in rural areas where surveillance and reporting are inadequate. Yellow fever distribution in America ranged from Philadelphia, in the United States, until the line connecting Bahía Blanca and Mendoza in Argentina. On the other hand, in Africa, it is located in the sub-Saharan Africa [16]. This disease can be fatal and acute or mild and inapparent. Because there is no specific antiviral treatment against it, the best strategy to prevent its spread is the prevention of infection [17]. In this context, vaccination is the best preventive measure against yellow fever.

2.3 Dengue

Dengue is a viral disease caused by infection of four viruses, known as dengue 1, 2, 3, and 4, which is endemic in more than 100 countries in Africa, America, the Eastern Mediterranean, Southeast Asia, and the Western Pacific, the latter two being the most severely affected. These viruses belong to the genus Flavivirus, Flaviciridae family. The most important mosquito vector is A. aegypti and to a lesser degree A. albopictus. Once an infected mosquito bites a human, the virus goes through an incubation period of between 3 and 14 days before disease symptoms appear. Furthermore, passive dispersion through means of transport is one of the most important factors that favor the spread of these mosquitoes and dengue virus from one region to another.

This disease is of major interest to public health because of its great impact on morbidity and mortality in the world since it is the viral disease transmitted by mosquito vectors most common and important worldwide [18]. The World Health Organization estimated that there may be 50 to 100 million dengue infections, a half-million hospitalizations, and 22,000 deaths worldwide every year [19]. Moreover, because of the absence of a vaccine to protect the population at risk, vector control is the most important method for the prevention and interruption of the transmission of the disease. The use of chemical insecticides is a key component in the control of larvae and adult mosquito vector populations. However, derived from overuse for over five decades of these insecticides to interrupt the transmission of the virus, it has generated resistance to different molecules of insecticides by part of mosquito vectors [19].

2.4 Venezuelan equine encephalitis

This disease is caused by the Venezuelan equine encephalitis virus (VEE) or encephalomyelitis in horses, donkeys, zebras, and humans. VEE virus cause acute infections in vertebrates characterized by the presence of high viremia and disease development with varying degrees of severity. Once the equines get the disease, they may die suddenly or present progressive disorders of the central nervous system. Furthermore, in humans, it causes mild to severe influenza-like symptoms, with fever and headache. Around 4–14% of cases develop neurological complications and near 1% of reported cases die [20].

The VEE complex contains a number of virus, belonging to the Alphavirus genus of the Togaviridae family, which have been classified into six viral subtypes, ranging from I to VI. VEE virus are present in all continents and most often are transmitted by the bite of infected Culex (C. vomerifer, C. pedroi, C. adamesi) and Aedes (A. taeniorhynchus) mosquito vectors and can also be transmitted through aerosols [21]. Although virtually any mosquito can be found infected with VEE virus during epizootics, it is thought that A. taeniorhynchus is the main vector responsible for transmitting VEE virus during outbreaks, whereas Culex mosquitoes species transmitted enzootic virus strains of VEE [22]. Subcutaneous injection, nasal instillation, and contact with broken skin or bedding contaminated animals are other ways to spread the virus, especially in a lab environment [23].

Epizootic and enzootic strains of VEE virus spread from northern Argentina to Florida and parts of the Rocky Mountains, being more frequent in northern South America. Since 1930, there have been 21 outbreaks of VEE throughout the American continent, being considered an emerging disease naturally due to mutations of the enzootic and endemic virus strains that circulate as a vector-borne disease among mammals host populations, especially in habitats such as forests and wetlands [24]. Some of the strategies used to reduce outbreaks of VEE in horses are through the implementation of the TC-83 vaccine and by protecting against mosquitoes by wearing protective clothing and/or insecticides. Although, the TC-83 vaccine is used in laboratories, there is still no licensed vaccine available to humans for the prevention of infection by the epizootic strains of VEE virus [25].

2.5 Japanese encephalitis

Japanese encephalitis (JE) is a viral disease transmitted to humans by mosquitoes of the Aedes and Culex genera, where main vectors are C. tritaeniorhynchus, C. annulus, C. annulirostris, C. vishnui, C. fuscocephala, and C. gelidus [26]. This disease affects horses, donkeys, pigs, and humans. Due to low qualifications and shorter duration of the viremia, humans or horses transmit the virus to other biting mosquitoes and are considered as terminal hosts [27]. The reservoirs of JE virus are pigs and wild birds. The incubation period is around 4–14 days. In pigs, JE causes reproductive losses (stillbirths and abortions), in horses, it causes encephalitis, and in humans, the disease can range from very serious to no symptoms. Neurologic sequelae occur in up to 80% of human cases and about 15,000 deaths annually [28].

JE is distributed across several areas of temperate and tropical Asia, with a higher incidence in rural areas of Southeast Asia, the Indian subcontinent and parts of Northern Asia. It also occurs, but less frequently in Japan, Taiwan, Singapore, Hong Kong, eastern Russia, and Australia [29]. Although the virus affects all age-groups, regions where there have been widespread childhood vaccination campaigns against the disease, the age distribution has changed, increasing the proportion of cases in older children and adults [27]. Therefore, the best alternative for control of JE it is the application of the vaccine, which is associated with the generation of neutralizing antibodies.

2.6 Lymphatic filariasis

Lymphatic filariasis, also know elephantiasis, is a parasitic disease caused by nematodes of the Filarioidea family, which is transmitted to humans by mosquitoes of the genus Anopheles, Culex, Aedes, and Mansonia, which inject various nematode larvae while feed on blood [30]. The disease is present in sub-Saharan Africa and Southeast Asia. It is a chronic parasitic disease in which adult worms lodge in the lymphatic vessels where they release microfilariae from the bloodstream. Filariasis can produce a wide variety of clinical manifestations whose symptoms include fever, lymphadenitis, and retrograde lymphangitis and thickening of the skin and underlying areas, caused by obstruction of the lymphatic system by the parasite. Although, the mortality associated with filariasis is low, the social and health consequences generated by their chronic manifestations are important because they cause severe disability in those subjects who suffer (blindness, deformities), which gives rise to a decrease in working capacity and in the economy

[30]. Filariasis is classified as a neglected disease by the World Health Organization due to the limited investment by pharmaceutical companies for research into new treatments and the low professional interest on the part of health personnel in the countries of first world [31]. Against lymphatic filariasis, a combination of two drugs albendazole and ivermectin is used [32]

2.7 Rift Valley fever

The disease is caused by the Rift Valley fever virus, an RNA virus belonging to the Bunyaviridae genus of the Phlebovirus family. The Rift Valley fever (RVF) is a peracute or acute, febrile, mosquito-borne disease that causes fever of viral origin that affects domestic animals like cows, buffaloes, sheep, goats, camels, and humans [33]. The route of transmission of the disease commonly is associated with mosquito-borne epidemics during the rainy season. RVF has been isolated from more than 30 mosquito species, of which it is considered that the Aedes and Culex genera are the main disease vectors [34]. The presence of RVF virus has been detected in most countries of East, West, and Southern parts of the African continent, being the disease endemic in Southern and Eastern Africa, but also epidemics have been reported in Egypt, Saudi Arabia, and Yemen [35]. The high genetic diversity observed in the RVF virus in the East, West, and Southern Africa shows the ancestral origin of the disease for these regions. Further, since outbreaks found outside these African regions have limited genetic diversity, these outbreaks are the result of the introduction of a single strain of the virus from the center of Africa, followed by its spread by mosquitoes between the vulnerable animal and human population [36]. Most cases of this disease occur in veterinarians, slaughterhouse workers, laboratory personnel, and other employees who work directly with blood and tissues of animals. Therefore, an inactivated vaccine is available against this disease; nevertheless, it is only available to laboratory personnel and veterinarians not being yet available to general public.

2.8 West Nile fever

West Nile fever (WNF) is transmitted to humans through the bite of Culex and Aedes mosquitoes vectors of West Nile virus (WNV). This disease can affect birds, horses, and humans causing an unapparent infection, mild febrile illness, meningitis, encephalitis, an acute poliomyelitis-like syndrome, or death [37]. WNV is a member of the genus Flavivirus in the Flaviviridae family. This arbovirus is maintained in nature by cycling between several bird species and mosquitoes that support the WNV replication. The resulting symptoms of WNV is variable, and the disease does not cause obvious signs in some species but often kills them after suffering a fatal systemic disease in other species [38]. Among mammals, the cyclical disease occurs mainly in horses and humans. In horses, this disease is characterized by the presence of ataxia ranging from mild to severe, besides presenting weakness, muscle twitching, and cranial nerve deficits [39]. The mortality rate is about one in three clinically affected horses unvaccinated [40]. In humans, most infection cases are asymptomatic and are usually mild and flulike if symptoms occur. The development of severe disease characterized by neurological signs and appears to be rare in most outbreaks. However, since the 1990s, this situation changed since the WNV became an important human and veterinary pathogen. WNV has a broad geographic range that includes portions of Europe, Asia, Africa, Australia (Kunjin virus), and North, Central, and South America. The incubation period of this disease ranges from 3 to 14 days with severe symptoms, in one of every 150 people infected. Symptoms may last from a few days to several weeks and have reported that approximately 80% of infected people show no symptoms at all. Although all persons are able to acquire the WNV, people older than 50 years are at higher risk of developing more severe forms of the disease. Over the past decade, the understanding gained about the clinical spectrum of the disease as well as the short- or long-term associated with human WNV infection has been substantially increased. Presently, a vaccine is available only for horses, but not for humans.[41].

2.9 Chikungunya fever

Chikungunya fever is a viral disease, manifested by fever and severe arthralgia, prevalent in Africa, Asia, and Europe, and now emerging and little studied in the Americas and the Caribbean Islands since 2013 [42]. Chikungunya virus is transmitted to humans by Aedes vectors. There are two main vectors A. aegypti and A. albopictus, which are present in the tropics and temperate zones. Although A. aegypti has always been the main vector transmitter of the disease, in most recent outbreaks, A. albopictus has become the main vector. Chikungunya word comes from the makonde dialect spoken by an ethnic group of southeast Tanzania and northern Mozambique, which means the man who walks hunched over, due to the appearance shown by patients due to the severity of joint pain that they suffer [43].

The Chikungunya virus is an enveloped positive-strand RNA virus, belonging to the Alpha-virus genus, group arbovirus A, of the Togaviridae family. The virus affects all humans without distinction, being susceptible to contracting the disease individuals not previously infected with the virus, after individuals are infected, immunity is prolonged. Chikungunya fever epidemics spread rapidly within infected community until the development of immunity in the population affected stops transmission. In general, about 3–28% of infected people are asymptomatic, although they contribute to the spread of the disease. On the other hand, in the symptomatic forms, the clinical manifestations may be acute, subacute, or chronic. After the bite, the incubation period of the disease is 1–12 days characterized by fever, severe arthralgias, back pain, incapacitating myalgia, and conjunctivitis.

Subsequently, after 2 or 3 days, maculopapular exanthem is described (sometimes only macular) in half of the cases, distributed on the trunk and extremities. Fever usually takes between 2 and 3 days, and leukopenia is the rule. Since there is no specific antiviral treatment or vaccine, treatment against the disease should be symptomatic against the symptoms presented by the patient and can be applied antipyretics, analgesics, anti-inflammatory, and ribavirin [44].

2.10 St. Louis encephalitis

St. Louis encephalitis (SLE) is a viral disease transmitted by mosquitoes of the Culex genus, where birds are the normal hosts of this disease that affects humans. Other potential vectors, which can become infected and transmit the SLE to vertebrate hosts, include the mosquito species of the Aedes, Mansonia, and Sabethes genera [45]. The SLE is a disease of greatest medical importance in North America prior the introduction of WNV in 1999. According to studies conducted in the United States, the sparrow has a key role in the chain of infection, and doves, blue jay, and robin are important too, being the young of these species most important that adults [6]. The SLEV may be enzootic in some places but eventually can occur epizootic in which a large number of birds become infected, also being able to cause infections in humans as isolated cases or epidemics, by the bite of mosquitoes infected with the virus. Under optimal conditions, the extrinsic incubation period of the virus is about 10 days, staying active throughout the life of the mosquito. Human infections with SLEV not always result in clinical manifestations of encephalitis, which is why it often goes unnoticed, except during epidemics [46]. The most susceptible population is of children and the elderly, and the mortality rate of SLE is between 5% and 20%. Currently, SLE is widely distributed throughout American continent, from Canada, Mexico, Central and South America, and the Caribbean [6]. To date, a vaccine against SLE has not been developed [41] because antibiotics are not effective against this virus. Therefore, the best form of protection against the transmission of SLEV is to avoid mosquito bites as much as possible.

2.11 Eastern equine encephalitis

Eastern equine encephalitis (EEE) is caused by the eastern equine encephalitis virus (EEEV), an RNA virus classified in the family Togaviridae. EEE infections are characterized by symptoms such as fever, headache, nausea and vomiting, malaise and weakness, confusion, myalgia, arthralgia, and neck stiffness. The main vector species involved in outbreaks of disease are Coquillettidia perturbans, Culiseta melanura, Ochlerotatus canadensis, Aedes vexans, and Culex mosquito's species [47]. The cycle of transmission of EEE occurs between wild birds and mosquito vectors, and when mosquitoes bit humans they transmit the virus. Disease incubation period varies from 4 to 10 days, and the infection can result in one of two types of disease, systemic or encephalitis, depending on the age of the person and other host factors. The illness lasts one to two weeks, and recovery is complete when the central nervous system is not affected. In humans, the rate of mortality is around 3%, and many of those recovering suffer some form of mental disability for life, and death usually occurs from 2 to 10 days after onset of symptoms. EEEV is found throughout the Western Hemisphere and North America in areas of the east of the Mississippi River in the US, southeastern Canada to Argentina, and Peru. The EEE was first recognized in humans in 1938 [48]. Since then, cases have been produced sporadically as small epidemics with a total of 223 reports between 1955 and 1993 [49], including a peak of 36 cases in 1959 [50]. Of this case, the third part of the patients affected died because of the disease. The diagnosis of the symptoms of EEE is difficult because the symptoms that occur in patients are nonspecific. Currently, the only available vaccine is for horses, and there is still no human vaccine for the general public.

2.12 Western equine encephalitis

The causal agent of Western equine encephalitis (WEE) is the WEE virus (WEEV), which is an arbovirus of the Togaviridae family, transmitted by mosquitoes of the Culex and Culiseta genres [51, 52], of which Culex tarsalis is the main vector of WEEV in western North America [53]. The natural cycle of WEEV is maintained through alternating infection between birds and ornithophilic mosquitoes. WEE primarily affects birds, mosquitoes, humans, and horses. The clinical form of WEE occurs in humans and horses, which are dead-end hosts [54]. Clinical disease can result in febrile disease of variable severity associated with neurological symptoms that range from headache to aseptic meningitis or encephalitis. The WEE has been reported in the Western United States, Canada, Mexico, and Central and South America [54]. WEEV in humans results in mild disease in adults but can become serious encephalitis in children and elderly people. The mortality rate is between 5% and 15%, of which about 50% of infants who survive have permanent brain damage. People can prevent WEE infection by avoiding outdoor activities at the primary feeding period of Culex and Culiseta mosquitoes or using mosquito repellents. For horses, a vaccine for WEE exists in the market, but there is no vaccine for humans [55].

2.13 La Crosse encephalitis

La Crosse encephalitis (LAC) is a disease caused by the La Crosse encephalitis virus (LACV). LACV is spread through the bite of infected mosquitoes of the Aedes genus, where A. triseriatus is the principal vector and reservoir of LACV in nature [56]. The vertebrate of the Sciuridae family are its normal hosts in forest habitats throughout the range of the disease [57]. Subsequently, the virus can be passed to humans by feeding on infected rodents with the virus, or transovarially from an infected female to offspring. LACV is distributed in the United States from the upper Midwestern states, primarily Illinois, Iowa, Indiana, Minnesota, Ohio, and Wisconsin to New York, and South to Texas, Alabama, and Georgia [58], with confirmed human cases of LAC in 29 of the lower 48 states [59]. The disease affects the central nervous system being able to become severe and fatal in rare cases [57]. After an incubation period of 5–15 days, LAC symptoms include fever, headache, nausea, vomiting, and fatigue and initially presents as a nonspecific summertime illness. Severe cases involving encephalitis occur more often in children under 16 years accompanied commonly of seizures and in some cases coma, paralysis, and long-term disability or death. There is no specific treatment against LACV infection, so care is based on symptoms. The best way to reduce the risk of becoming infected with LACV is to avoid mosquito bites, use insect repellents, wear clothes to avoid bites, or even stay indoors during peak hours of mosquito activity. Another preventive measure that provides good results is to eliminate mosquito breeding sites. There is no antiviral drugs or vaccine available against the LAC because antibiotics are not effective against viruses [57].

2.14 Zika fever

Zika fever (ZIKF) is caused by Zika virus (ZIKV), a flavivirus belonging to the family Flaviviridae. It is a disease of monkeys and humans spread through the bite of infected Aedes mosquito [60]. Where A. africanus and A. aegypti are the principal probable vectors [61]. Clinical symptoms are mild headache, maculopapular eruption in different parts of the body, transient fever, malaise, and joint pain. ZIKV is common in West and Central Africa but also occurs in Pakistan, India, Vietnam, Thailand, the Philippines, Malaysia, Indonesia and Micronesia. In 2014, the presence of ZIKV in some American regions was reported. There is no vaccine, drugs or specific treatment for ZIKF, so the strategies for prevention and control the disease include insect repellents and mosquito elimination [62].

3. MOSQUITO CONTROL

The prevention and control of mosquito-borne diseases globally is conducted through a comprehensive and thorough method of pest management. Where programs are not intended to completely eliminate mosquito populations but rather are aimed to reduce their number and therefore minimize the risk of disease transmission. Methods used to mosquito control include the elimination of breeding sites and the control of mosquito larvae and adults. Larvicides, by applying chemical insecticides in the breeding sites, are the best strategy to kill larvae and pupae of mosquitoes in the water. Larvicides are present in several forms ranging from powder, tablets, or liquids and include methoprene, monomolecular surface films, larvicidal oils, chemical insecticides, neurotoxic insecticides, plant-derived products, and larvicidal bacteria [19]. Adulticides technique is usually less efficient for mosquito control. However, it is the only way to kill adult mosquitoes and is the last line of defense in reducing mosquito populations. Some of the adulticide used for mosquito control include products derived from microorganisms, plants or minerals, synthetic molecules, organophosphates, some natural pyrethrins, or synthetic pyrethroids [63].

3.1 Chemical insecticides

Since its discovery, chemical insecticides have represented the most widely method used to control mosquito-borne vectors. However, the effects of chemical insecticides on mosquito vector populations are usually transitory because vectors can rapidly develop resistance against them. On the other hand, the environmental problems caused by the excessive use of chemical insecticides are a matter of current concern because it is estimated that about 2.5 million tons of pesticides are used annually, generating worldwide damage amounting to $100 billion annually [64]. Some of the disadvantages that generates when using only chemical products are (a) the selection of new insecticide resistance in pest populations; (b) the resurgence of already treated populations; (c) the generation of waste, risks, and legal complications; (d) the destruction of beneficial species; and (e) the high costs in equipment, labor, and material. In addition, the highly toxic and nonbiodegradable properties of insecticides and waste generated in soil, water, food, and crops that affect public health are additional reasons to search new methods to help solve the problems caused by chemical insecticides [64]. Consequently, the concept of integrated control arises, a method in which pest and diseases control is performed using chemicals, useful organisms, and cultural practices.

The progress of science and the chemical industry in the nineteenth century, with the discovery of DDT, made possible the development and emergence of new conventional insecticides or so-called of synthesis [65]. The most used of these insecticides of synthesis are modulators of sodium channels (organochlorines, pyrethroids, and pyrethrins), acetylcholinesterase inhibitors (carbamates and organophosphates), and the chloride channel antagonists regulated by the gamma-aminobutyric acid or also known as GABA (organochlorine cyclodiene and phenylpyrazoles).

Using these conventional insecticides gave positive results against insects disease vectors at first. However, due to its massive use, insects soon began to develop resistance to them. Thus, an insecticide that initially was effective, just being useless in the long term. In response to this problem, new-generation insecticides also called biorational insecticides have been developed, whose research strategy is based on a good understanding of the physiological processes or mechanisms specific communication of insects, and in obtaining agents that are able to affect them. These products are divided into the following: those who are analogs of juvenile hormone and molting, inhibitors tissue formation, pheromones, insecticides that prevent hatching, and biological insecticides [66].

3.1.1 Organophosphates and carbamates

Organophosphate insecticides are phosphoric acid derivatives, having activity against a wide spectrum of invertebrate. It interferes with the action of enzymes called cholinesterases that regulate the neurotransmitter acetylcholine, resulting in first instance to muscle cramps, paralysis, and eventually death [67]. Therefore, these insecticides have a toxic action that blocks an enzyme acetylcholinesterase of central and peripheral nervous system of insects, in synaptic junctions. The enzyme rapidly hydrolyzed acetylcholine, resulting in the repolarization of the membrane or the basal plate in neuromuscular connections, preparing for the arrival of a new impulse. By forming strong covalent bonds between insecticide and acetylcholinesterase, the enzyme is inhibited, causing the accumulation of acetylcholine in the synaptic junction and the interruption of normal transmission of nerve impulses [68].

However, due to the generation of resistance in vector insects to these chemical products, the use of many of these organophosphate and carbamates insecticides is no longer effective. Furthermore, because cholinesterases and neurotransmitters acetylcholine also form part of vertebrate nervous system, organophosphate pesticides are highly or moderately toxic to vertebrates [69]. In this regard, temephos are the only organophosphate pesticide that is still used to control mosquito larvae. Although temephos are not persistent in the environment being that last 7–10 days [70], it has been shown in many studies the adverse effects of temephos on a wide range of no target aquatic taxa [71]. Furthermore, carbamate pesticides, just like organophosphates, act by inhibiting the cholinesterase enzyme. Therefore, the symptoms experienced by insects per carbamate poisoning are similar to those experienced with organophosphates. However, carbamate pesticides block acetylcholinesterase enzyme hydrolyzing acetylcholine in muscle by carbamylation, which is a reversible reaction [72]. Therefore, the recovery of carbamate poisoning in humans is faster than with organophosphate intoxication since the acetylcholinesterase enzyme is able to break apart of the carbamate [73].

3.1.2 Organochlorines, pyrethroids, and pyrethrins

Organochlorine insecticides are chlorinated hydrocarbons, which are known to be effective to control mosquito populations. Its mode of action is by inhibiting GABA receptor in the nervous system through the interruption of nerve impulses due to the closure of chloride channels [72]. Therefore, when an organochlorine binds to a GABA receptor, the receptor is unable to close GABA chloride channel, which results in stimulation of the nervous system and similar symptoms to poisoning with carbamates or organophosphates [74]. However, with the Stockholm Convention on Persistent Organic Pollutants, which entered into force on May 17, 2004, the use of 12 chemicals including DDT, aldrin, dieldrin, heptachlor, mirex, chlordecone, and chlordane was prohibited because of its long average life and toxicity [75]. However, an extension clause allows countries where malaria is endemic to use DDT to control vectors that transmit the disease. Taking into account the negative effects that DDT has for the environment, malaria programs without the use of insecticides have been developed with the assistance of the Pan American Health Organization [76].

On the other hand, pyrethroids and pyrethrins used to control mosquitoes break down faster in the sunlight as opposed to chemical or microbial breakdown. However, pyrethroids are considered axonic poisons, composed of more stable substances, or degrade slower in the presence of sunlight than pyrethrins and are generally effective against most of the insect pests of agriculture. Furthermore, pyrethroids can be combined with other active ingredients, such as piperonyl butoxide, to retard its degradation and prevents the insect's system from detoxifying the pyrethroid, making it more effective [72]. Delay that allows the chemical product persists longer in the environment, requiring smaller and less frequent doses to kill pests [77]. This type of insecticidal affects the central and peripheral nervous system of insects and have a rapid knock-down effect, by interfering with the sodium channels of nerve membrane causing the interruption of the transfer of ions and transmission of impulses between nerve cells [78]. Moreover, it stimulates nerve cells to produce repetitive discharges and eventually cause paralysis and death [79]. Furthermore, because pyrethroids act on the nervous system of insects through a different pathway from the organophosphate pesticides, they generally have low toxicity in mammals and birds; however, they are toxic to fish and tadpoles [80].

3.1.3 Biorational insecticides

Biorational insecticides are those that have relatively low toxicity to humans and have few environmental effects. Among which, methoprene is an insect growth regulator insecticide with a broad spectrum of action that interferes with the insect life cycle preventing maturity or reproductive stage [81]. Meanwhile, the juvenile hormone analogue is a biorational insecticide that causes deformations in larval stage, death in the pupal stage, and sterility effect in adults [82]. Spinosad is another biorational insecticide that comes from a Saccharopolyspora spinosa neurotoxin, made by a mixture of spinosyns A and D. Spinosad act on the postsynaptic nicotinic acetylcholine receptors and GABA receptors and has proven its usefulness in the dipterans control [83]. Pyriproxyfene is another new-generation insecticide that has been tested in adult and larval mosquitoes causing a reduction in the number of sperm, egg production, blood feeding, and mating activity [84].

3.2 Plants and their derivates

For centuries, nature has created several active substances that, when applied correctly, can control insect pests such as mosquito in an efficient manner. The use of plants by man with insecticide purposes dates back to early human history. Due to their environmental advantages, the use of insecticides of vegetable origin in pest management has been increasing [85]. Among plants with potential activity against mosquitoes, Nim or Neem (Azadirachta indica) causes stunted growth, loss of appetite, reduction of fertility, molting disorders, morphological defects, and behavioral changes [86]. Moreover, it has been demonstrated that raw or partially purified plant extracts are most effective for mosquito control in place of the purified compounds or extracts [87]. The snuff (Nicotiana tabacum) is used, thanks to its insecticide and insect repellent action, where nicotine acts on the nervous system of insects through breathing, ingestion, and contact [88]. Other plants from which oils are extracted are garlic (Ocimum basilicum) and cinnamon (Cinnamomum osmophloeum), which have been shown to have insecticidal properties against larvae and adults of A. albopictus, Culex quinquefasciatus, and Armigeres subalbatus.

3.3 Biological agents

Among biological agents used for mosquito control can be mentioned derivatives of viruses, bacteria, and fungi. Entomopathogenic virus spreads from one insect generation to the next causing paralysis and eventually death on mosquito larvae being more effective in the first stage of development [89]. Within bacteria, only reports of Bacillus thuringiensis, B. sphericus, and B. popilliae with possibilities to exercise control over dipterans insects currently exist. These bacteria, during the sporulation process, produce protein crystals with insecticidal effect and/ or some toxins with the same effect [90]. Bacillus initially causes diarrhea and intestinal paralysis in mosquito, giving rise to a decrease of body movements, convulsions, and general paralysis. Internally, within the mosquito stomach, B. thuringiensis releases toxic crystals that paralyze the insect gut stopping peristalsis, causing that the insect stop feeding and die by starvation. Within the gut, bacteria multiply until they break the epithelium and invade the rest of the insect body. However, its use for mosquito control is scarce and presents some drawbacks as its duration in the environment is limited, its dispersion is rather inefficient, and the susceptibility to bacterial infection in the pest population is very heterogeneous. There are very sensitive individuals and other highly resistant. Fungi are other microorganisms that may be used to control mosquito vectors, of which 400 species are known with insecticide potential. About 20 of them have been given more attention, including those in the Lagenidium, Entomophaga, Neozygites, Entomophtora, Erynia, Aschersonia, Verticillium, Nomuraea, Hirsutella, Metarhizium, Beauveria, and

Paecilomyces genera [91]. Although, entomopathogenic fungi are not as specific as bacteria or viruses, spores persist and infect insect successive generations, so that when the infection is established, its effects can last several years. Infection occurs by adhesion of the spores on the insect cuticle, where these germinate and penetrate the cuticle leading to insect colonization by mycelium. Cuticle penetration occurs through the use of an enzyme complex that the fungi use to feed. The entomopathogenic fungus most used in controlling mosquito infestations is Beauveria bassiana, which produces various active ingredients such as beauvericin [92].

The biological control of mosquito larvae with predators and other biological control agents could be a more effective and environmentally friendly strategy, thus avoiding the use of synthetic chemicals and the consequent environmental damage [93]. Among them, some insects and vertebrates such as fish, amphibians, and some mammals have the potential to control mosquito disease-vector populations. Within vertebrates, amphibians, bats, and fish have been used to control populations of mosquito. For example, using larvivorous fish species, control of mosquito larvae in deposits used to store water has been achieved [41]. Moreover, bats are responsible for capturing flying insects such as mosquitoes at night; similarly, toads and frogs consume large numbers of insects, slugs, worms, and other invertebrates [94]. However, the use of frogs and tadpoles for disease vector control is still largely unexplored.

4. DEVELOPMENT OF RESISTANCE TO CHEMICAL INSECTICIDES

Insecticide resistance is defined as the development of the ability of a insect population to tolerate doses of an insecticide, which would be lethal to the majority of individuals in a normal population of the same species and is also the result of pressure of positive selection exerted by the insecticide on the low frequency genes initially present in the vector insect [95]. Therefore, the development of resistance by mosquito disease vectors is of international concern due to the increase worldwide exchange of plant matter that mosquitoes can spread to other parts of the world, spreading resistance genes of the plagues that they have.

Most mosquito vector control programs of diseases in humans are mainly based on the use of chemical insecticides by outdoor spraying, impregnated nets, or indoor residual spraying [96]. Thereon, the use of insecticides has helped to eradicate insect-borne diseases. In this regard, since 1950, different classes of insecticides have been successively used. Organophosphates and pyrethroid insecticides have been used to control mosquito populations in their larval and adult stages. However, more recently, the disease vector control programs are based largely on the use of synthetic pyrethroid insecticides, which are recommended by the WHO only for impregnated nets [97]. However, the massive use of pesticides has caused detrimental effects on the agroecosystem, such as the acquisition of resistance, pest resurgence, and environmental pollution. Resistance has developed in more than 84 species of mosquitoes for each of the groups of toxicological insecticides [98]. Furthermore, it was found that insecticide residues accumulated in plants often end up in water bodies where mosquito larvae feeding on such plant debris or grow in water bodies enriched with plant compounds and interactions between these xenobiotics generate tolerance to insecticides or promote detoxification pathways of these insecticides against mosquitoes [99]. In addition to abiotic factors, biotic interactions that occur among mosquitoes, the pathogens that they transmit and their microbiome (microbes living in the mosquito) may also occur [96]. These vary from symbionts to entomopathogen opportunistic organisms that are able to affect various physiological host processes, such as detoxification systems [100] or the opposite effect leading to the appearance of insecticide resistance [101]. Furthermore, allelochemicals inducing enzyme production in insects can increase their tolerance to pesticides [102]. On the other hand, other studies have shown that the degree of development of a plant can affect insecticide resistance in insects [103].

There are two main mechanisms by which mosquito vectors can develop resistance to insecticides: alterations in the target site of action and metabolic resistance, also called increased rate of detoxification of insecticides [19]. Other less common mechanisms that develop resistance in insects are the resistance per behavior and the resistance per decreased penetration through the cuticle or cross resistance [104].

4.1 Resistance mechanisms

Each insecticide triggers the selection of one or more mechanisms of resistance; in addition, an unknown number of behavioral changes in adults. For instance, changes in the target site of action are produced when no silent mutations occur in structural genes that produce an alteration of amino acids responsible for anchoring the insecticide at a specific site. For example, resistance have been reported by altering the voltage-dependent sodium channel that is the target site of action for pyrethroids and organochlorines, such as DDT, and in the insensitive acetylcholinesterase, which is the target site of action for organophosphate and carbamate [19]. Furthermore, the metabolic detoxification is an acquired resistance mechanism, which is regulated by the activity of certain oxidized enzymes such as mixed function oxidase, esterases, glutathione S-transferases, and in specific cases DDT- dehidroclorinase. Mixed function oxidase represents an important detoxification mechanism in the degradation of carbamates; moreover, esterases have an important role in the degradation of phosphorus insecticides [105]. Meanwhile, the metabolic resistance occurs through the increase in the detoxification of the insecticide. The most important form of metabolic resistance is given by detoxifying enzymes type glutathione S-transferase, mixed function oxidases, and esterases [78].

On the other hand, cross resistance can occur in two ways, positive and negative. The positive cross resistance refers to resistance to several insecticides due to expression in a single resistance mechanism [106]. Therefore, cross resistance occurs when a single gene confers resistance to a number of chemicals in the same group, such as kdr gene conferring resistance to DDT and pyrethroids [95]. Meanwhile, negative cross resistance occurs due to an increase in susceptibility to the insecticide "A," caused by the development of resistance to insecticidal "B" and vice versa. For example, in Culex pipiens quinquefasciatus larvae, it was found that resistance to organophosphorus insecticides increases susceptibility to pyrethroid insecticides [107].

Furthermore, multiple resistances occur when two or more resistance mechanisms independently selected are operating in the same insect [19]. However, the term multiple resistances not necessarily involve the cross-resistance term because an insect may be resistant to two or more insecticides, and each resistance can be attributed to different mechanisms [78]. Consequently, each additional mechanism of resistance leads to a wide cross resistance, which restricts the number of possible alternatives for the control and in extreme conditions, leading to highly resistant populations to virtually all available insecticides [108].

5. CONCISE SUMMARY

Mosquito-borne diseases are affected by biological, ecological, social, and economic variables. Regrettably, fatalities predominantly transpire in rural regions where medical care is insufficient due to resource constraints. Certain symptoms of mosquito-borne diseases are mild and easily treatable; however, for other diseases, antiviral medications and antibiotics are ineffective in treating the virus, and no vaccine is currently available for prevention. A preventive strategy to control disease transmission involves the eradication of mosquitoes and their nesting habitats. The primary technique for mosquito eradication is the application of chemical pesticides. Nonetheless, their management is challenging due to the frequent application of chemical insecticides, which induces resistance and diminishes their efficacy. The application of plants, fungi, and bacteria with potential efficacy offers certain environmental benefits; however, their effectiveness is transient, and some mosquito populations exhibit significant resistance.

A promising option involves employing chemicals and natural pesticides designed to change the normal functioning of disease-transmitting mosquitoes without harming the environment.

REFERENCES

1. Knowtlon K, Solomon G, Rotkin-Ellman M. Fever Pitch, Mosquito-Borne Dengue Fever Threat Spreading in the Americas. NRDC Issue paper. 2009; 4–22.
2. García-Gutiérrez C, Gómez-Peraza RL, López ACE, León-Váldez A. Insecticidas biorracionales para el control de mosquitos y moscas negras en Sinaloa. Ra Ximhai. 2012; 8: 47–55.
3. Henao S, Pérez FJ. Manejo de plagas sin químicos: manual para docentes. 2nd ed. San José, Costa Rica. Radio Nederland Training Center, División Internacional. 2003; 3–80.
4. Grandadam M, Caro V, Plumet S, Thiberge JM, Souarès Y, Failloux AB, Tolou HJ, Budelot M, Cosserat D, Leparc-Goffart I, Desprès P. Chikungunya virus, southeastern France. Emerging Infection Diseases. 2001; 17: 910–913.
5. DCHP. Surveillance and Control of Selected Mosquito-Borne Diseases in Florida. Guidebook. Florida Health. 2014; 8–29.
6. Almirón W, Crocco L. Mosquitos Urbanos Transmisores de Dengue y Encefalitis de San Luis, Manual de capacitación Docente. Editorial Científica Universitaria. Córdoba, Argentina. 2007; 14–51.
7. Vargas HJ. Prevención y control de la Malaria y otras enfermedades transmitidas por vectores en el Perú. Revista Peruana de Epidemiología. 2003; 11: 1–18.
8. Boratne AV, Jayanthi V, Datta SS, Singh Z, Senthilvel V, Joice YS. Predictors of knowledge of selected mosquito-borne diseases among adults of selected peri-urban areas of Puducherry. Journal of Vector Borne. Diseases. 2010; 47: 249–256.
9. Torrades S. La malaria y la controversia sobre su vacuna. OFFARM. 2001; p. 140–143.
10. Secretaría de Salud de México (SSM). Programa de Acción: Enfermedades transmitidas por Vector. México, DF.
11. Guerra CA, Gikandi PW, Tatem AJ, Noor AM, Smith DL, Hay SI, Snow RW. The limits and intensity of Plasmodium falciparum transmission: implications for malaria control and elimination worldwide. PLoS Medicine. 2008; 5: 300–311.
12. Targett ATG, Moorthy SV, Brown GV. Malaria vaccine research and development: the role of the WHO MALVAC committee. Malaria Journal. 2013; 12: 362.
13. Bisset LJA, Rodríguez MM, San Martín JL, Romero JE, Montoya R. Evaluación de la resistencia a insecticidas de una cepa de Aedes aegypti de El Salvador. Revista Panamericana de Salud Publica. 2009; 26: 229–234.

14. GVCET (Grupo de Vigilancia y Control de Enfermedades Transmisibles). Protocolo de Vigilancia y Control de Fiebre Amarilla. Subdirección de Vigilancia y Control en Salud Pública. PRO-RO22.003.0000.009. 2010; 1–18.

15. Barnett DE. Yellow fever: epidemiology and prevention. Clinical Infectious Diseases. 2007; 44: 850–856.

16. AMSE (Asociación de Médicos de Sanidad Exterior). Fiebre Amarilla. Epidemiología y Situación Mundial. [Internet]. 2012. Available: http://www.amse.es/index.php?option=com_content&view= article&id=84:fiebre-amarilla-epidemiologia-y-situacion- mundial&catid=42:inf-epidemiologica&Itemid=50 [Accessed: 2015/06/10].

17. Staples JE, Gershman M, Fischer M. Yellow fever vaccine recommendations of the Advisory Committee on Immunization Practices (ACIP). Morbidity and Mortality Weekly Report. 2010; 59: 1–27.

18. Gibbson RV, Vaughn DW. Dengue: an escalating problem. British Medical Journal. 2002; 324: 33–42.

19. Maestre SR, Gómez CD. Dengue: Epidemiología, políticas públicas y resistencia de vectores a insecticidas. Revista Ciencias Biomedicas. 2013; 4: 302–317.

20. de la Monte S, Castro F, Bonilla NJ, Gaskin de UA, Hutchins GM. The systemic pathology of Venezuelan equine encephalitis virus infection in humans. American Journal of Tropical Medicine and Hygiene. 1985; 34: 194–202.

21. Brown DT, Condreay LD. Replication of alphaviruses in mosquito cells. In: Schlesinger S, Schlesinger MJ, editors. The Togaviridae and Flaviviridae. Plenum Press, New York; 2003. p. 171–207.

22. Zacks MA, Paessler S. Encephalitic alphaviruses. Veterinary Microbiology. 2010; 140: 281–286.

23. Sidwell RW, Gebhardt LP, Thorpe BD. Epidemiological aspects of venezuelan equine encephalitis virus infections. Bacteriological Reviews. 1967; 31: 65–81.

24. Estrada-Franco JG, Navarro-Lopez R, Freier JE, Cordova D, Clements T, Moncayo A, Kang W, Gomez-Hernandez C, Rodriguez-Dominguez G, Ludwig GV, Weaver SC. Venezuelan equine encephalitis virus, Southern Mexico. Emerging Infectious Diseases. 2004; 10: 2113–2121.

25. Knipe DM, Howley PM. Fields Virology. 6th ed. Lippincot Williams & Wilkins, Philadelphia; 2003. p. 2456.

26. Hemmerter S, Slapeta J, van den Hurk AF, Cooper DR, Whelan IP, Russell CR, Johansen AC, Beebe WN. A curious coincidence: mosquito biodiversity and the limits of the Japanese encephalitis virus in Australasia. BMC Evolutionary Biology. 2007; 7: 1–11.

27. OIE—Organización Mundial de Sanidad Animal. Encefalitis Japonesa. In: Manual de la OIE sobre animales terrestres. Organización Mundial de Sanidad Animal. 2010; p. 1–9.

28. Tsai TF. Factors in the changing epidemiology of Japanese encephalitis and West Nile fever. In: Saluzzo JF, editor. Factors in the Emergence of Arboviral Diseases. Elsevier Amsterdam; 1997. p. 179–189.

29. Malaria. Mundi. Otras Enfermedades Infecciosas, Encefalitis Japonesa. [Internet]. 2014. Available: http://malariamundi.com/otras-enfermedades-infecciosas/encefalitis- japonesa [Accessed: 2015/05/25].

30. Díaz-Menéndez M, Norman F, Monge-Maillo B, Pérez-Molina JA, López-Vélez R. Enfermedades Infecciosas y Microbiología Clínica. Enfermedades Infecciosas y Microbiologia Clinica. 2011; 29: 27–37.

31. Hotez PJ, Molyneux DH, Fenwick A, Kumaresan J, Sachs SE, Sachs JD, Sachs JD, Savioli MDL. Control of neglected tropical diseases. New England Journal of Medicine. 2007; 357: 1018–1027.

32. Global Programme to Eliminate Lymphatic Filariasis (GPELF). Progress Report 2000– 2009 and Strategic Plan 2010–2020. Geneva, World Health Organization. 2010; p. 78.

33. OIE—World Organisation for Animal Health. Rift Valley Fever. OIE Terrestrial Manual 2014. OIE Reference Laboratories. 2014; p. 1–20.

34. European Food Safety Authority (EFSA). EFSA Panel on Animal Health and Welfare (AHAW); Scientific Opinion on Rift Valley fever. EFSA Journal. 2013; 11: 1–48.

35. Center for Food Security and Public Health (CFSPH). Rift Valley Fever. Institute for International Cooperation in Animal Biologics. Iowa State University. 2006; p. 1–5.

36. Ikegami T. Molecular biology and genetic diversity of Rift Valley fever virus. Antiviral Research. 2012; 95: 293–310.

37. Granwehr BP, Lillibridge KM, Higgs S, Mason PW, Aronson JF, Campbell GA, Barrett AD. West Nile virus: where are we now? Lancet Infectious Diseases. 2004; 4: 547– 556.

38. Weingartl HM, Neufeld JL, Copps J, Marszal P. Experimental West Nile virus infection in blue jays (Cyanocitta cristata) and crows (Corvus brachyrhynchos). Veterinary Pathology. 2004; 41: 362–370.

39. Snook CS, Hymann SS, Del Piero F, Palmer JE, Ostlund EN, Barr BS, Deroschers AM, Reilly LK. West Nile virus encephalomyelitis in eight horses. Journal of the American Veterinary Medical Association. 2001; 218: 1576–1579.

40. OIE—World Organisation for Animal Health. West Nile Fever. OIE Terrestrial Manual 2013. OIE Reference Laboratories. 2013; p. 1–12.

41. Connelly CR, Bolles E, Culber D, De Valerio J, Donahoe M, Gabel K, Jordi R, McLaughlin J, Neal A, Scalera S, Toro E, Walter J. Florida Resident's Guide to Mosquito Control, Integrated Pest management for Mosquito Reduction around Homes and Neighborhoods. University of Florida, USDA-NIFA. 2014; p. 22–37.

42. Frías-Salcedo JA. Actualización de aspectos clínicos epidemiológicos de la fiebre Chikungunya 2014. Revista de Sanidad Militar. 2014; 68: 313–317.

43. Porta L. Fiebre Chikungunya Amenaza para la Región de las Américas. Salud Militar. 2012; 31: 25–33.

44. Puente S, Rivas GP, Ramírez G, de Ory F, Lago M, Herrero MD, González LJ, Sánchez-Seco MP. Infección por virus de Chikungunya: 11 casos importados. In: VI Congreso de la Sociedad Española de Medicina Tropical y Salud Internacional. 5–7 Marzo 2008. Segovia, España.

45. Rodrigues SG, Nunes MRT, Casseb SMM, Prazeres ASC, Rodrigues DSG, Silva MO, Cruz ACR, Tavares-Neto JC, Vasconcelos PFC. Molecular epidemiology of Saint Louis encephalitis virus in the Brazilian Amazon: genetic divergence and dispersal. Journal of General Virology. 2010; 91: 2420–2427.

46. Reisen WK. Epidemiology of St. Louis encephalitis virus. In: Maramorosch K, Murphy FA, Shatkin AJ, editors. Advances in Virus Research. Elsevier. San Diego; 2003. p. 139–183.

47. New York State Department of Health (NYSDOH). Mosquito Borne Illness Surveillance & Response Plan 2012. 2012; p. 4–5.

48. Farber S, Hill A, Connerly ML, Dingle JH. Encephalitis in infants and children caused by the virus of the eastern variety of equine encephalitis. JAMA. 1940; 114: 1725- 1731.

49. Piliero PJ, Brody J, Zamani A, Deresiewicz RL. Eastern equine encephalitis presenting as focal neuroradiographic abnormalities: case report and review. Clinical Infectious Diseases. 1994; 18: 985–988.

50. Monath TP. Arthropod-borne encephalitides in the Americas. Bulletin World Health Organization. 1979. 57: 513–533.

51. Ray CG. Arthropod-borne and other zoonotic viruses. In: Ryan KJ, Ray CG, editors. Sherris Medical Microbiology. 4th ed. McGraw-Hill. New York. 2004; p. 585–596.

52. OIE-World Organisation for Animal Health. Equine Encephalomyelitis (Eastern and Western). OIE Terrestrial Manual 2013. OIE Reference Laboratories. 2013; p. 1–9.

53. Mahmood F, Reisen WK, Chiles RE, Fang Y. Western equine encephalomyelitis virus infection affects the life table characteristics of Culex tarsalis (Diptera: Culicidae). Journal of Medical Entomology. 2004; 41: 982–986.

54. Walton TE. Venezuelan, eastern, and western encephalomyelitis. In: Gibbs EPJ, editor. Virus Diseases of Food Animals. A World Geography of Epidemiology and Control. Disease Monographs. Vol. 2. Academic Press, New York. 1981. p. 587–625.

55. Minnesota Department of Health (MDH). Western Equine Encephalitis Fact Sheet. Minnesota Department of Health. 2006; p. 1.

56. Ohio Department of Health (ODH). La Crosse Virus Disease. Ohio Disease Reporting System. 2015; p. 1–9.

57. Rey JR. La Crosse Encephalitis. University of Florida. IFAS Extension. 2002; p. 1–2.

58. Calisher CH. Medically important arboviruses of the United States and Canada. Clinical Microbiology Reviews. 1994; 7: 89–116.

59. Moore CG, McLean RG, Mitchell CJ, Nasci RS, Tsai TF, Calisher CH, Marfin AA, Moore PS, Gubler DJ. Guidelines for Arbovirus Surveillance in the United States. Center of Disease Control, U.S. Department of Health and Human Services, Fort Collins, Colorado.1993; p. 14–16.

60. Akoua-Koffi C, Diarrassouba S, Bénié VB, Ngbichi JM, Bozoua T, Bosson A, Akran V, Carnevale P, Ehouman A. Investigation surrounding a fatal case of yellow fever in Côte d'Ivoire in 1999. Bulletin de la Societe Pathologie Exotique. 2001; 94: 227–230.

61. Centre for Disease Control (CDC). Zika Virus Infection. Northern Territory Government. Australia. 2014; p. 1–2.

62. Hayes EB. Zika Virus Outside Africa. Emerging Infectious Diseases. 2009; 15: 1347– 1350.

63. Fathy KH. Ecosmart biorational insecticides: alternative insect control strategies. In: Perveen F, editor. Insecticides—Advances in Integrated Pest Management. InTech. Rijeka; 2012. p. 17–60.

64. Koul O, Walia S, Dhaliwal GS. Essential oils as green pesticides: potential and constraints. Biopesticide International. 2008; 4: 63–84.

65. Díez DM. Efecto del Metopreno sobre la metamosfosis de la cucaracha, Blattella germanica [thesis]. Vizcaya: Universidad del País Vasco. 2013.

66. Mamatha DM, Kanji VK, Cohly H, Rajeswara RM. Juvenile hormone analogues, methoprene and fenoxycarb dose dependently enhance certain enzyme

activities in the silkworm Bombyx mori (L). International Journal of Environmental Research Public Health. 2008; 5: 120–124.

67. Ware GW, Whitacre DM. Introducción a los Insecticidas. The Pesticide Book, 6th ed.. Willoughby; 2004.

68. Rey VG. Determinación de los grados de resistencia al insecticida temefos en poblaciones de Aedes aegypti Linnaeus 1762, (Diptera: Culicidae) y su implicación en la eficacia del insecticida en los departamentos de Cauca, la Guajira, Cundinamarca y Atlántico.] Díez DM. Efecto del Metopreno sobre la metamosfosis de la cucaracha, Blattella germanica. [thesis]. Bogotá: Universidad Nacional de Colombia. 2011.

69. Mazzacano C, Hoffman BS. Ecologically Sound Mosquito Management in Wetlands. The Xerces Society for Invertebrate Conservation. 2013; p. 2–10.

70. Fortin CA, Maire A, Leclair R. The residual effect of temephos (Abate 4-E) on nontarget communities. Journal of the American Mosquito Control Association. 1987; 3: 282–288.

71. USFWS. Appendix K4. Environmental Effects of Mosquito Control. 2004; p. 1–20.

72. Gile S, Johns A, Thai V. Integrated Pest Management for Mosquito Control in Massachusetts [thesis]. Massachusetts: Worcester Polytechnic Institute. 2013.

73. Reigart RJ, Roberts JR. Recognition and Management of Pesticide Poisonings. 5th ed. U.S. Environmental Protection Agency: Washington, DC. 1999; p. 34–93.

74. Brown AE. Mode of action of insecticides and related pest control chemicals for production agriculture, ornamentals, and turf. Pesticide Information Leaflet. 2006; 43: 1- 13.

75. Reigart RJ, Roberts JR. Recognition and Management of Pesticide Poisonings. 5th ed. U.S. Environmental Protection Agency: Washington, DC. 1999; p. 34–93.

76. Pan American Health Organization (PAHO). Integrated Vector Management (IVM). [Internet]. 2011. Available: http://www.paho.org/English/AD/DPC/CD/ivm.htm [Accessed 2015/05/18].

77. Woodard V, Meléndez JL. Preliminary environmental fate and effects assessment science chapter for the registration eligibility decision of D-Phenothrin (Sumithrin)®. U.S. Environmental Protection Agency. ID: EPA-HQ-OPP-2008-0140-0005. 2008; p. 7- 16.

78. Bisset J. Uso Correcto de Insecticidas: Control de la resistencia. Revista Cubana de Medicina Tropical. 2002; 54: 202–219.

79. Guglielmone A, Castelli M, Volpogni M, Medus P, Anziani O, Mangold A. Comparación de la concentración letal 50 de diazinón y cipermetrina para

Haematobia irritans (Diptera: Muscidae) entre áreas de producción de leche o carne de Santa Fe y Entre Ríos, Argentina. Revista de Medicina Veterinaria. 2001; 82: 209–211.

80. Environmental Protection Agency (EPA). Permethrin. United States Environmental Protection Agency, Prevention, pesticides and Toxic Substances, EPA 738-F-09-110. 2009; p. 11.

81. Klowden MJ, Chambers GM. Ovarian development and adult mortality in Aedes aegypti treated with sucrose, juvenile hormone and methoprene. Journal of Insect Physiology. 1989; 35: 513–517.

82. Kawada H. New mosquito control techniques as countermeasures against insecticide resistance. In: Perveen F, editor. Insecticides. Advances in Integrated Pest Management. InTech; 2012. p. 657–682.

83. Williams CR, Leach KJ, Wilson NJ, Swart VR. The allee effect in site choice behavior of egg-laying dengue vector mosquitoes. Tropical Biomedicine. 2008; 25: 140–144.

84. Iwanaga K, Kanda T. The effects of a juvenile hormone active oxime ether compound on the metamorphosis and reproduction of an Anopheline vector, Anopheles balabacensis. Applied Entomology and Zoology. 1988; 23: 186–193

85. Leyva M, Marquetti MdelC, French L, Montada D, Tiomno O, Tacoronte JE. Efecto de un aceite de trementina obtenido de Pinus tropicalis Morelet sobre la biología de una cepa de Aedes (Stegomyia) aegypti Linnaeus, 1762 resistente a insecticidas. Anales de Biología. 2013; 35: 75–84.

86. Wandscheer CB, Duque JE, da Silva MAN, Fukuyama Y, Wohlke JL, Adelmann J, Fontana JD. Larvicidal action of ethanolic extracts from fruit endocarps of Melia azedarach and Azadirachta indica against the dengue mosquito Aedes aegypti. Toxicon. 2004; 44: 829–835.

87. Abdelouaheb A, Nassima R, Noureddine S. Larvicidal activity of a neem tree extract (Azadirachtin) against mosquito larvae in the Republic of Algeria. Biological Sciences. 2009; 2: 15–22.

88. Brechelt A. Manejo Ecológico de Plagas y Enfermedades. Red de Acción en Plagicidas y sus Alternativas para América Latina (RAP-AL). Santiago de Chile, Chile. 2004; p. 4–35.

89. Primo YE. Mundi prensa ed. Ecología química. Nuevos métodos de lucha contra insectos. Madrid; 1991. p. 191.

90. Badii MH, Abreu JL. Control biológico una forma sustentable de control de plagas. Daena: International Journal of Good Conscience. 2006; 1: 82–89.

91. Van Driesche RG, Bollows TS. Biological Control. Chapman & Hall. New York; 1996. p. 539.

92. García GC, Tamez GP. Mercado de bioinsecticidas en México. Curso de agricultura orgánica y sustentable. Fundación Produce Sinaloa. 2012; p. 99–114.

93. Kumar R, Hwang JS. Larvicidal Efficiency of Aquatic Predators: A Perspective for Mosquito Biocontrol. Zoological Studies. 2006; 45: 447–466.

94. Mahr LD, Whitaker P, Ridgway N. Biological control of insects and mites: an introduction to beneficial natural enemies and their use in pest management, UW Extension, Wisconsin Madison. 2008; p. 58.

95. World Health Organization (WHO). Seventh report Expert Committee on insecticides WHO Tech Report Ser. 1957; 125: 37.

96. Nkya TE, Akhouayri I, Kisinza W, David JP. Impact of environment on mosquito response to pyrethroid insecticides: facts, evidences and prospects. Insect Biochemistry and Molecular Biology. 2013; 43: 407–416.

97. World Health Organization (WHO). Malaria Vector Control and Personal Protection. Report of the World Health Organization. 2006; p. 1–61.

98. Falero GC. Pruebas de susceptibilidad de mosquito adulto y larvas a los insecticidas y bioensayos de las aplicaciones residuales usadas en el control de mosquitos vectores de malaria y dengue. 2001; p. 2–47.

99. David JP, Boyer S, Mesneau A, Ball A, Ranson H, Dauphin-Villemant C. Involvement of cytochrome P450 monooxygenases in the response of mosquito larvae to dietary plant xenobiotics. Insect Biochemistry and Molecular Biology. 2006; 36: 410–420.

100. Behura SK, Gomez-Machorro C, Harker BW, de Bruyn B, Lovin DD, Hemme RR, Mori A, Romero-Severson J, Severson DW. Global cross-talk of genes of the mosquito Aedes aegypti in response to dengue virus infection. PLoS Neglected Tropical Diseases. 2011; 5: 1–13.

101. Kikuchi Y, Hayatsu M, Hosokawa T, Nagayama A, Tago K, Fukatsu T. Symbiont-mediated insecticide resistance. Proceedings of National Academy Sciences. 2012; 109: 8618–8622.

102. Brewer MJ, Meade T, Trumble JT. Development of insecticide-resistant and susceptible Spodoptera exigua (Lepidoptera: Noctuidae) exposed to furanocoumarins found in celery. Environmental Entomology. 1995; 24: 392–401.

103. Attah PK, van Emden HF. The susceptibility to malathion of Metopolophium dirhodum on 2 wheat species at 2 growth stages, and the effect of plant growth regulators on this susceptibility. Insect Science and its Application. 1993; 14: 101–106.

104. Fonseca I, Quiñones M. Resistencia a insecticidas en mosquitos (Diptera: Culicidae): mecanismos, detección y vigilancia en salud pública. Revista Colombiana de Entomologia. 2005; 31: 107–115.

105. Sosa MA. Consideraciones generales sobre resistencia de los insectos a los plaguicidas con especial referencia a los "piretroides." Publicación Miscelánea. N°5. INTA Reconquista. 1992; p. 5–10.

106. Mani GS. Evolution of resistance in the presence of two insecticides. Genetics. 1985; 109: 761–783.

107. Lagunes TA. Impact of the use of mixtures and sequences of insecticides in the evolution of resistance in Culex quinquefasciatus Say (Diptera: Culicidae) [thesis]. River side California: University of California. 1980.

108. Keiding J. Resistance in the housefly in Denmark and elsewhere. In: DL Watson, editor. Pesticide management and insecticide resistance. New York: Academic Press; 1977. p. 261–302.

8
Chapter

RESISTANCE AND ITS MANAGEMENT TO MICROBIAL AND INSECT GROWTH REGULATOR LARVICIDES IN MOSQUITOES

1. INTRODUCTION: AN OVERVIEW

Mosquitoes and mosquito-borne diseases continue to be significant public health issues and socioeconomic burdens worldwide, especially in tropical and subtropical areas. Currently, the migration of humans and animals, the trade of goods, rapid demographic expansion, economic progress, and the resulting environmental consequences exacerbate the scale and severity of the issue. Mosquito control is frequently the sole or most efficacious method of integrated management to address mosquito-borne diseases. Given the stringent governmental regulations, significant environmental susceptibility, and rising demand for mosquito control due to the emergence and proliferation of mosquito-borne diseases, environmentally sustainable management strategies utilizing microbial and insect growth regulator larvicides present a promising solution due to their high efficacy, target specificity, and safety for both the environment and non-target organisms. Nonetheless, the emergence of resistance in mosquito populations to these biorational larvicides has been documented throughout the previous decades. To ensure the sustainability of mosquito control, it is essential to develop and apply susceptibility monitoring and resistance management strategies for the existing control methods.

2. BACILLUS THURINGIENSIS SUBSP ISRAELENSIS (B.T.I.)

The entomopathogenic Bacillus was first discovered in 1901 by Japanese biologist Ishiwata Shigetane, who was investigating the cause of the sotto disease (sudden-collapse disease) that was killing large populations of silkworms. Shigetane named the bacterium Bacillus sotto, but the name was later ruled invalid. In 1911, this pathogen was rediscovered in Germany by Ernst Berliner, who isolated it from Mediterranean flour moth Anagasta kuehniella (Zeller, 1879) caterpillars that suffered Schlaffsucht. He named it Bacillus thuringiensis, after the German town

Thuringia where the moth was found [1]. Up to date, at least 70 serotypes, more than 80 subspecies have been identified, among which 14 serotypes and 16 subspecies show lethal activities against mosquito larvae. Bacilllus thuringiensis israelensis (B.t.i.), a subspecies belonging to serotype H-14, was discovered from a natural mosquito habitat located in Israel desert in 1976 [2, 3]. Four synergistic endotoxins including Cry4A, Cry4B, Cry 11A, and Cyt1A are produced during sporulation of B.t.i. [4–7]. These protoxins are activated by enzymatic proteolysis activities under high pH environment in mosquito midgut. The activated toxins combine with the receptors located at the epithelium cells in midgut brush border and cause subsequent pathological consequences and death of the target species. Toxins of B.t.i. are toxic to species in nematoceran group including culicidae, simuliidae [8, 9], chironomidae [10–12], and some fungus gnats. B.t.i. is categorized as Group 11 pesticide, i.e., microbial disruptor of insect midgut membranes by the Insect Resistance Action Committee (IRAC). During the past 35 years or so, B.t.i. products have been extensively used to control mosquitoes and blackflies and occasionally chironomid midges worldwide.

2.1 Field occurrence

Generally, the risk of resistance development to wild-type B.t.i., i.e., the natural toxin complex, is very low. For example, B.t.i. products were widely used to control floodwater mosquitoes Aedes vexans (Meigen) over an area of approximately 500 km2 for more than 10 years in Rhine River area in Germany; no reduction in susceptibility was noticed [13]. One report, however, from New York, USA, showed low-level resistance in wild population of Culex pipiens L. Briefly, collections from Syracuse and Albany showed 33- to 41-fold and 6- to 14-fold resistance to B.t.i. Based on the considerable difference in resistance levels between the populations from Syracuse and Albany, it seems that there was lack of gene flow between these populations of Cx. pipiens, resulting in aggregation of resistant individuals [14]. Culex pipiens populations from Cyprus (2002–2008) showed dose–response values ranging approximately 8-fold to B.t.i., but no resistance was detected after years of application [15]. Between 1990 and 1993, the susceptibility of Cx. pipiens complex to B.t.i. was determined in 31 collections from California, USA. The samples were collected before the widespread use in California. Seven collections from the Mediterranean island of Cyprus, where no microbial insecticides have ever been used, also were tested. The collections from California during 1990–1991 exhibited 3- to 4-fold variations in susceptibility at LC50 and LC95. The collections from Cyprus in 1993 exhibited both higher mean LC values, and greater variability in those values, than the California collections. No significant geographic variation in susceptibility was observed among regions within California [16].

2.2 Laboratory studies

Multiple attempts to select resistance in laboratory colonies of Cx. pipiens complex or Aedes aegypti L. for various generations only resulted in low-level and unstable resistance. Four populations of Cx. quinquefasciatus Say collected from Southern California were subjected to different levels of selection pressure for 11–60 generations; 4.1- to 16.5-fold resistance was achieved. The resistance tended to decline in absence of selection pressure [17]. One laboratory and two wild populations of Ae. aegypti were used in selection for resistance to B.t.i. After 14 generations of selection at LC50, a statistically significant decline (2-fold) in susceptibility was observed in the F15 of one wild strain only. Regression lines, LC50 values, and slopes of parental populations between strains did not differ significantly [18]. When larvae of Cx. pipiens were subjected repeatedly to selection pressure with B.t.i. in the laboratory, only 2.78-fold tolerance was induced as a result of 20 generations of selection, which decreased by about 58% after the selection was withdrawn for 3 generations. Larval selection with B.t.i. caused a reduction in the reproductive potential in resulted survivors [19]. Another study revealed 2- to 3-fold increase in the LC50 or LC90 values to B.t.i. preparation in the larvae of Cx. quinquefasciatus after 20 generations of selection, and these values fluctuated in different generations during the selection [20]. The laboratory selection with field-persistent B.t.i. toxins led to a 3.5-fold resistance to VectoBac WG in Ae. aegypti after 30 generations, but to relatively high levels of resistance to individual Cry toxins. Bioassay procedure was developed using each Cry toxin to detect cryptic B.t.i. resistance that evolved in field mosquito populations. Although no resistance to B.t.i. was detected in three Aedes mosquito species tested, an increased tolerance to Cry4Aa (3.5-fold) and Cry11Aa toxins (8-fold) was found in one Ae. sticticus population. This study facilitated information of susceptibility status to individual Cry toxins of B.t.i. Bioassays with individual Cry toxins allow a more sensitive monitoring of B.t.i. resistance in the field [21]. When individual Cry toxins were tested against the LiTOX strain that carried low-level resistance to B.t.i., increased resistance of 68-, 9-, and 9-fold to Cry4Aa, Cry4Ba, and Cry11Aa protoxins, respectively was revealed [22].

Exposures to individual toxins of B.t.i. are conducive to resistance development, where Cx. quinquefasciatus developed high resistance to individual toxins in absence of Cyt1A toxin. Resistance became evident in the mosquitoes that were selected with a single toxin (CryIVD), reaching the highest level of 913-fold. Resistance developed at a slower pace and reached a lower level when being selected with CryIVA + CryIVB. Resistance levels were further suppressed when being selected with CryIVA + CryIVB + CryIVD or full combination of all four toxins. These results highlighted the importance of the full combination of toxins found in wild B.t.i. in resistance management [23]. This fact, i.e., target species rapidly develops resistance to individual toxins but not to the toxin complex from the wild strain, also exists in other B.t. subspecies studied, for instance, B. thuringiensis subsp jegathesan [24]. Cross- resistance occurs among the Cry toxin of B.t.i. Resistance was generally highest toward the toxin(s) that were used in the selections. The strain that was selected with CryIVD demonstrated significant cross-resistance to CryIVA + CryIVB, and vice versa. Strain that was selected with naturally occurring insecticidal toxin mixture in B.t.i. only showed a low level of resistance to this mixture, but much higher levels of resistance occurred to individual CryIV toxins and also to combinations of 2 or 3 CryIV toxins. The fact that all of the selected populations stayed susceptible to the natural toxin mixture in B.t.i. suggested that the CytA toxin with different sequence and mode of action from the CryIV toxins plays an important role in suppressing resistance to CryIV toxins [25]. Cross-resistance among Cry toxins can be extended to other B.t. subspecies. Cx. quinquefasciatus that are resistant to B.t.i. Cry toxins also show cross- resistance to Cry11B from B.t. jegathesan [26], but not to Cry19A from the same species [27].

Cyt1A, a cytolytic endotoxin of B. thuringiensis, does not possess significant larvicidal activity alone [28]. However, it plays a critical role in overcoming, preventing, and delaying resistance development to Cry toxins.

The high levels of resistance to CryIV in Cx. quinquefasciatus was reduced remarkably by combining CryIV with sublethal quantities of CytA. A 127-fold resistance to a combination of CryIVA, B, and D was completely suppressed by combining CytA in a 1:3 ratio with CryIVA, CryIVB, and CryIVD. Combining the CytA with CryIVA and CryIVB also completely suppressed 217-fold resistance to the latter toxins, whereas the combination of CytA with CryIVD reduced resistance in a CryIVD-selected mosquito strain from >1000-fold to < 8-fold [29]. The same species of mosquitoes were subjected to selection for 20 generations using the recombinant strains of B.t. that produced Cyt1Aa, Cry11Aa, or a 1:3 mixture. The resistance in the Cry11Aa strain- and Cyt1Aa strain-selected populations reached 1,237-fold and 242-fold, respectively. On the other hand, the resistance only reached 8-fold in the population that was selected with the Cyt1Aa and Cry11Aa (1:3) strain. Mosquitoes that were selected by Cyt1Aa- Cry11Aa for 48 generations only developed 9.3-fold resistance to Cry11Aa. Obviously, resistance to Cry11Aa developed at a much slower pace in the presence than in the absence of Cyt1Aa [30]. Studies on the mechanism of activity enhancement and resistance prevention of Cry toxins by Cyt1A indicated that Cyt1A functions as a receptor of Cry11A. Cyt1Aa binding to Ae. aegypti brush border membrane vesicles enhanced the binding of Cry11Aa. Two exposed regions in Cyt1Aa, namely, loop beta6-alphaE and part of beta7, bind Cry11Aa. On the other side, Cry11Aa binds Cyt1Aa proteins by means of domain II-loop alpha8 and beta-4, which are also involved in midgut receptor interaction. The key residues involved in the interaction and synergism between

Cry11Aa and Cyt1Aa were S259 and E266 in Cry11Aa and K198, E204 and K225 in Cyt1Aa [31]. Further studies revealed that binding of Cry11A to Cyt1A facilitates the formation of a Cry11A prepore oligomeric structure that is capable of forming pores in membrane vesicles [32].

It was discovered recently that the mosquitocidal toxins (Mtx) from some B. sphaericus strains not only enhance the larvicidal activity of B.t.i. Cry toxins but also mitigate resistance development to Cry toxins. Mtx-1 and Mtx-2 were active against mosquitoes that were susceptible or resistant to Cry toxins. A mixture of Mtx-1 or Mtx-2 with different Cry toxins in the ratio of 1:1 showed moderate synergism. Some combinations of Mtx and Cry toxins were highly active to kill resistant larvae and also suppressed resistance to Cry toxins [33]. There is a lack of cross- resistance to the wild type of other B.t. subspecies such as B.t. jegathesan, B.t. kyushuensis, and fukuokuensis in Cx. quinquefasciatus that are resistant to individual toxins from B.t.i. [34]. It is advised not just to express Cry toxins of B.t.i. in transgenic microbial organisms or algae from the perspectives of resistance prevention in target species.

3. BACILLUS SPHAERICUS

The mosquitocidal activity of some strains of B. sphaericus has long been recognized. Up to date, 49 serotypes over 300 strains of B. sphaericus have been identified, among which 9 serotypes 16 strains showed activity against mosquito larvae at different extent. Strains that possess high mosquitocidal activity are 2362, 1597, 2297, C3-41, and IAB-59. The mostly studied and developed strain 2362 was isolated in 1984 from adult blackfly Simulium damnosum (Diptera: Simuliidae) in Nigeria. Recently, B. sphaericus was renamed as Lysinibacillus sphaericus Meyer and Neide [35]. Active strains produce parasporal inclusions during sporulation, which contains crystal binary toxins. Some strains also synthesize noncrystal Mtx. The mode of action of the binary toxins is somewhat similar in general to B.t.i. toxins with detail differences at molecular levels. The receptor of the binary toxins is a 60-kDa alpha-glucosidase, which is anchored in the mosquito midgut membrane via a glycosyl-phosphatidylinositol (GPI) anchor. As B.t.i., B. sphaericus also belongs to IRAC Group 11. Compared with B.t.i., the target species spectrum is narrower, some Aedes spp., for example, Ae. aegypti is much less susceptible than Culex spp. to B. sphaericus. During the past decades, numerous products have been developed using various strains and applied to control Culex spp. worldwide.

3.1 Field occurrence

The earliest resistance in field populations was reported in Cx. pipiens in southern France where the resistance ratio at LC50 was 70-fold as a result of extended field applications [36]. A field- collected population of Cx. quinquefasciatus larvae from an urban area of Recife in Brazil, which has been treated for 2 years with B. sphaericus, was found to be about 10-fold less susceptible than the untreated control field populations [37]. The field resistance to strain 1593 was reported in Kochi, India, in the same year. The larvae of Cx. quinquefasciatus from the sprayed area showed LC50 and LC90 values that were 146 and 180 times higher than corresponding values for a susceptible strain from an unsprayed area after 35 rounds of application over 2 years. The subsequent laboratory selection of the collection from the treated area resulted in much higher levels of resistance, 6,223- and 31,325-fold at LC50 and LC90, respectively [38]. The similar situation also happened to strain B101 in the same mosquito species where low levels of resistance occurred in response to field applications; the population reached 52,000-fold resistance after selection for 6 generations in the laboratory [39]. Field Cx. pipiens mosquitoes that were collected after a control failure with Spherimos in southern France developed high resistance (>10,000-fold) after <8 generations of laboratory selection [40]. In southern China, a flowable formulation of strain C3-41 was continuously used for 8 years to control Cx. quinquefasciatus larvae. The resistance of field-collected larvae at LC50 was 22,672-fold [41]. More occurrences on resistance to strain 2362 were reported later in France (5,958-fold) [42] and Tunisia (750-fold) [43]. Declined efficacy and control failure of B. sphaericus was noticed within 4 months after 5 treatments using VectoLex WDG at the dosages of 50–200 mg/m2 to control Cx. quinquefasciatus in Thailand [44]. A high level of B. sphaericus resistance was documented in this population. The resistance levels at LC50, depending on reference colonies, were 21,100- to 28,100-fold against VectoLex WDG (650 ITU/mg) or >125,000- to 200,000-fold against B. sphaericus technical-grade material (2000 ITU/mg) [45]. Between 1990 and 1993, the susceptibility of Cx. pipiens complex to B. sphaericus was determined in 31 collections from California, before the registration of this agent. Variation was about 5-fold at the LC50 and LC95 [16]. Similar results were obtained for Culex spp. breeding in dairy lagoons in southern California soon after

B. sphaericus was registered and applied in California [46]. No case on resistance to B. sphaericus in the USA has ever been reported in wild mosquito populations thus far regardless of the substantial amount of B. sphaericus products that has been applied, particularly since the invasion of the West Nile virus. The resistance development in response to the field application of B. sphaericus products varies greatly, depending on prior exposure to naturally existing strains, population genetic background, and gene exchange with untreated populations, as well as product application strategies.

3.2 Laboratory studies

Resistance to B. sphaericus in laboratory colonies of Cx. pipiens complex has been reported in different countries since 1994. Larvae of laboratory colony and field-collected (southern California) Cx. quinquefasciatus developed moderate level of resistance to strain 2362 (27- to 37- fold) in response to selection at LC80 for 80 generations [47]. This moderate level of resistance to strain 2362 in laboratory colony of the same species was reconfirmed by later studies on resistance management tactics [48, 49]. A previously untreated field population of Cx. quinquefasciatus, collected near Bakersfield, California, survived the LC50 of B. sphaericus that was 7,000 times higher than in the susceptible reference colony after 12 generations of selection at LC95. Late and early instar larvae in this study showed the similar levels of resistance [50].

After 13 and 18 generations of exposure to high concentrations of C3-41 and IAB59, a field- collected low-level-resistant colony of Cx. quinquefasciatus developed >144,000-fold and 46.3- fold resistance, respectively. A field-collected susceptible colony was selected with strain 2362 and IAB59 for 46 and 12 generations and reached >162,000-fold and 5.7-fold resistance to the two agents, respectively. The slower evolution of resistance against strain IAB59 may be attributable to the presence of another larvicidal factor [51]. However, in another study, selection by treating about 15,000 of 3rd and 4th instar larvae at each generation at LC70 of IAB59 resulted in 40,000-fold resistance after a much longer selection for 72 generations [52]. The resistance development to selection depends on genetic background, size of population used, selection pressure, length of selection, etc. Resistance ratio is also dependent on the susceptibility levels of the reference population. The resistance to B. sphaericus is fairly stable in absence of selection pressure. For instance, bioassays showed that the frequency of resistant larvae did not decrease throughout 11 generations after interruption of selection, and it was associated with a similar frequency of larvae lacking the Cqm1alpha-glucosidase receptor. The frequency of the cqm1 (REC) allele remained stable throughout 11 generations without further selection [53].

Furthermore, once mosquitoes develop resistance to a given strain of B. sphaericus, they are also often resistant to other strains because of the similarity of the binary toxins in most strains. Fortunately, mosquitoes that have developed resistance to various strains of B. sphaericus remain susceptible to B.t.i. [38, 39, 44, 45, 48–52, 55–59]. The cross-resistance among different strains is mild between the strains that also produce Mtx toxins. For example, cross-resistance in strain 2362-resistant Cx. quinquefasciatus was detected toward strains 1593 and 2297, but little or no cross-resistance was observed toward strains IAB59 or ISPC5 [50]. The resistant colonies resulted from the selection with C3-41 or 2362 showed very high levels of cross-resistance to strains 2362, C3-41, 1593, and 2297 but only low-level cross-resistance to IAB59, LP1-G, and 47-6B, which all produce a major 49-kDa protein, another mosquitocidal factor. On the other hand, the IAB59- selected colonies showed high cross-resistance to both strains C3-41 and 2362 [57].

3.3 Resistance mechanism

It is mostly believed that recessive genetic mechanism is involved in resistance to B. sphaericus [20, 42, 43, 50, 52, 59]. Among the multiple theories, the predominant one is the lack of specific binding of binary toxins to alpha-glucosidase, which act as midgut receptors [37, 53, 59–63]. The main reason leading the lack of specific binding is related to the deletions of gene encoding the receptor alpha-glucosidase [64–66], where the integrity of the receptor is compromised. However, the resistance in field Cx. pipiens mosquitoes after a control failure of Spherimos in southern France is not associated with any loss of binding affinity between brush border membrane fractions and toxins [40]. The similar results were also seen in Brazil, with additional findings of slight declined receptor density [37]. Behavioral modifications such as reduced ingestion on toxins [67] and other unknown mechanisms [40, 42] are also involved.

3.4 Resistance management

can be used as a powerful tool to mitigate resistance to B. sphaericus in mosquitoes. The susceptibility to B. sphaericus was partially restored by the selection of previously resistant colony with B.t.i. alone for 10 generations. After this colony was reexposed to B. sphaericus for 20 generations, resistance to B. sphaericus surged back to a stable level. Selections of B. sphaericus-resistant colonies with B.t.i. and B. sphaericus in rotation or mixture lead to steady decline of resistance over 30 generations [48]. Resistance to B. sphaericus can be delayed or prevented by the mixture of B.t.i. and B. sphaericus because of the synergistic action among 4 toxins, particularly the presence of Cyt1A [49, 68–71]. While B. sphaericus resistance increased after F15 in response to the selection using B. sphaericus alone, the rotation of B. sphaericus and B.t.i. surprisingly resulted in much higher level and faster emergence of resistance to B. sphaericus. However, selection with mixtures of B.t.i. and B. sphaericus for 36 generations showed no emergence of resistance to B. sphaericus [49]. Recently, the recombinant that produces toxins from both B.t.i. and B. sphaericus, even at greater amount than the wild type of bacteria [72– 74], provides another path for not only mitigation of resistance but also enhancement of larvicidal activity and efficacy. The combination of B. sphaericus with botanical pesticides such as azadirachtin from the neem oil is also considered as an alternative to mitigate resistance development to B. sphaericus in mosquitoes [75].

Efforts were made to find practical strategies for controlling resistant mosquitoes and to prevent or delay the development of resistance in wild mosquito populations. In Nonthaburi Province, Thailand, the larvae of Cx. quinquefasciatus that were highly resistant (>125,000-fold) to B. sphaericus strain 2362 were successfully controlled with applications of B.t.i. alone or in combination with B. sphaericus. In order to elucidate resistance management strategy in the field, one selected site was treated with B. sphaericus 2362 alone and the other treated with a mixture of B. sphaericus 2362 and B.t.i. Moderate resistance was detected after the 9th treatment and almost complete control failure occurred by the 17th treatment in the site that was treated with B. sphaericus 2362 alone. However, no noticeable change in susceptibility to B. sphaericus was detected after 9 treatments with the mixture at another site. During this period, the site treated with B. sphaericus alone required 19 treatments, whereas the site treated with mixtures only took 9 treatments because of comparatively slower resurgence of larval populations [44]. In this resistance population, the resistance levels to the mixtures of B. sphaericus + B.t.i. increased steadily upon the increase of B. sphaericus ratios in the mixtures from 50%, 75%, 90%, 95%, to 99%. The resistance levels to the mixtures with various ratios of B. sphaericus and B.t.i., however, were substantially lower than that in B. sphaericus alone, suggesting that the addition of B.t.i. to B. sphaericus enhanced the mosquitocidal activity (synergism) against these highly B. sphaericus-resistant Cx. quinquefasciatus. Moderate tolerance and low levels of resistance to B. sphaericus/ B.t.i. recombinant (RR 7.29–12.75 at LC50 and 5.15–13.40 at LC90) were also noted in this B. sphaericus-resistant population [45]. The similar success was achieved in southern China. After 6 months of treatment with B.t.i. in the B. sphaericus-resistant populations, their susceptibility to B. sphaericus C3-41 recovered, with the resistance ratio of field- collected larvae declining from 22,672-fold to 5.67-fold [41]; the gene exchange with populations in surrounding untreated areas may also have contributed to the rapid decline of resistance levels. There is a lack of cross-resistance between binary toxins and Mtx toxins [76], indicating that Mtx could be a potential tool to manage resistance to binary toxins in the future. It was suggested that once resistance to B. sphaericus is detected in the field, its use should be discontinued until the mosquito population becomes susceptible again because of the decline in number of resistant individuals [77].

3.5 Fitness cost of resistance

In a laboratory studies [77], the resistant strains showed some disadvantages such as lower fecundity and fertility, but higher survival rates were observed at the same time. The immature stages of the females from the resistant population developed slightly faster as compared with those of the susceptible strains, which could result in a shorter generation time. The similar findings are that the resistant colony showed lower fecundity and fertility and slower development than the susceptible colony [78]. However, the opposite results were achieved in another study where the resistant colony did not display biological costs regarding fecundity, fertility, and pupal weight [53].

4. SPINOSYNS

Spinosad, a biopesticide consisting of spinosyn A ($C_{41}H_{65}NO_{10}$) and D ($C_{42}H_{67}NO_{10}$), is produced by a naturally occurring, soil-dwelling bacterium, Saccharopolyspora spinosa Mertz and Yao. As a new class of polyketide-macrolide insecticide that acts as nicotinic acetylcholine receptor (nAChR) allosteric modulator, spinosad is categorized as Group 5 insecticide by IRAC. Spinosad exerts pesticidal activity after ingestion and cuticle absorption against a broad spectrum of susceptible insect species, by stimulating nACh and γ-aminobutyric acid (GABA) receptors and causing rapid excitation of the insect nervous system. The application of spinosad products for mosquito control is relatively new; studies to evaluate resistance development risk and resistance management strategy are rather rare. The first attempt was made in Cx. quinquefasciatus. Surviving late instars and pupae were collected from a semifield evaluation on Natular® XRG (2.5% spinosad), and a laboratory colony was established. Selection pressure was applied at LC70–90 levels to 10,000–15,000 of the late 3rd and the early 4th instar larvae of each generation after initial lethal levels of Natular XRG against this colony were determined. Susceptibility changes upon selection were determined every other generation. The susceptibility to spinosad in this selected population gradually and steadily declined from generation F1 to F35. From generations F37 to F45, the susceptibility decreased at a much faster pace. For reference purposes, the susceptibility of freshly collected wild populations as well as a laboratory reference colony of the same species was also determined concurrently. By comparing with the wild populations and laboratory reference colony for resistance ratio calculation, spinosad tolerance was observed from the first 9 generations. Resistance increased gradually from generations F11 to F35 and elevated significantly from generations F37 to F45, when resistance ratios reached 1415.3- to 2229.9-fold at LC50 and 9,613.1- to 17,062.6-fold at LC90. The exponential elevation of resistance levels throughout selection indicated that a recessive mechanism might have been involved during resistance development to spinosad [79, 80]. The spinosad-resistant Cx. quinquefasciatus with various levels of resistance was found not to be cross-resistant to B.t.i., a combination of B.t.i. and B. sphaericus, methoprene, pyriproxyfen, diflubenzuron, novaluron, temephos, or imidacloprid. However, it showed various levels of cross-resistance to B. sphaericus, spinetoram, abamectin, and fipronil. On the other hand, a long-term laboratory colony of Cx. quinquefasciatus that is highly resistant to B. sphaericus [50] was as susceptible as laboratory reference colony to spinosad and spinetoram.

Field-collected and laboratory-selected Cx. quinquefasciatus that were resistant to methoprene did not show cross-resistance to spinosad and spinetoram. The presence and absence of cross- resistance to other pesticides in spinosad-resistant mosquitoes seemed to be related to their modes of actions [81].

5. INSECT GROWTH REGULATORS

5.1 Juvenile hormone analogs (methoprene and pyriproxyfen)

Methoprene, hydroprene, kinoprene, and triprene were synthesized in 1960s. These insect growth regulators interrupt juvenile hormone balance during the transition from the late 4th instar larvae to pupae and adults. Most mortality occurs at pupal stage or incompletely emerged adults. Another juvenile hormone analog pyriproxyfen was synthesized in 1970s, the IRG activity of which is much higher than methoprene [80]. The earliest experimental studies on potential of resistance development in mosquitoes to juvenile hormone analogs were in 1973 [82]. The collective results indicated low risk of resistance development [82–86]. For example, the selection of Cx. quinquefasciatus by methoprene for 10 generations only lead 3.9- to 21.3-fold of resistance [86], while the selection of Cx. pipiens for 8 generations only resulted in 8- to 13-fold resistance to methoprene and cross-resistance to triprene [83]. Higher levels of resistance to methoprene did not necessarily occur in response to longer period of selection. Culex tarsalis Coquillett developed 86-fold resistance after 62 generations of selection [84], while 218-fold of resistance was achieved in Cx. pipiens after 40 generations of selection. In the latter case, selected mosquitoes were also cross-resistant to hydroprene and triprene, but not to diflubenzuron [85]. Rapid discharge and reduced detention of methoprene in mosquito tissue played an important role during entire process of resistance development, while metabolic detoxification seemed related to development and maintenance of high level resistance [87, 88].

Data are meager with regard to resistance development in wild populations of mosquitoes. Aedes taeniorhynchus (Wiedemann) in Florida, USA, showed 15-fold resistance after applications of methoprene product during 1989 to 1994 [89]. Methoprene tolerance in Aedes nigromaculis (Ludlow) was discovered in central California, USA, after 20 years of treatment. Control failure was encountered during 1998–1999 [90], where resistance levels reached thousands of fold [91]. The documented resistance seemed not related to the metabolic detoxification by P450 monooxygenase and carboxylesterase, and treatments using B.t.i. partially restored the susceptibility to methoprene [91]. Another reports in wild populations showed that 4.7- to 16-fold in Cx. pipiens in Cypress [15] and 9- to 54-fold in Cx. quinquefasciatus in southern California [81]. Limited data showed very low risk of resistance to pyriproxyfen in mosquitoes [92].

5.2 Chitin synthesis inhibitor (diflubenzuron)

Diflubenzuron was synthesized in mid 1970s by Philips-Duphar B.V. This compound is a nonselective chitin synthesis inhibitor that interrupts formation of exoskeleton, interferes with integrity of cuticle, and causes leakage of body fluid and ultimately mortality of target organisms. Diflubenzuron acts on all stages of the mosquito life cycle, larval stages in particular, younger larvae show higher susceptibility. Up to date, studies on resistance management are limited to laboratory populations. For instance, Cx. pipiens developed 7-fold resistance to diflubenzuron in response to selection for 5 generations [85]. Culex quinquefasciatus collected from the east coastal area in Dar es Salaam, Tanzania, developed 2.4- to 6.6-fold resistance after 10 generations of selection [86]. Aedes aegypti developed 3.3-fold resistance after 10 generations of selection, of which the resistance level increased to 8- to 20-fold after this population was hybridized with a mixing collection from 35 locations and then selected for 5 generations [93]. In general, the risk of resistance development to diflubenzuron in mosquitoes is relatively low.

6. CONCISE SUMMARY

This chapter examined and evaluated historical evidence regarding resistance and resistance management in mosquitoes to biorational larvicides derived from microbial and insect growth regulator origins. Bacterial larvicide exhibits the minimal risk of resistance development, contingent upon the integrity of the endotoxin complex and the synergistic interaction among individual toxins, notably the presence of Cyt1A. B.t.i. is essential for mitigating resistance development and restoring and maintaining vulnerability in other biorational larvicides. The binary toxins from B. sphaericus offer several benefits in managing mosquito larvae; the danger of resistance development is challenging to ascertain, since numerous factors contribute to the final extent and severity of resistance. According to data from global laboratory and field studies, the integration of B.t.i. with B. sphaericus, utilizing biofuse technology or genetic engineering, is the optimal strategy to augment larvicidal activity and efficacy, mitigate resistance development, and reinstate susceptibility to B. sphaericus. Larval mosquitoes appear to rapidly develop resistance to spinosad if resistance management strategies are not strategically employed, a phenomenon linked to its method of action, specifically the activation of nACh receptors in competition with acetylcholine, together with the likelihood of sublethal exposures. It is imperative to immediately develop and implement strategies to avoid the emergence of resistance and to restore susceptibility to spinosad in mosquitoes following resistance development. The risk of resistance development to IGRs is minimal. It is important to note that juvenile hormone analogs, such as methoprene and pyriproxyfen, influence the transition from late 4th instar larvae to pupae and adults, with their efficacy mostly reliant on the internal levels of juvenile hormone. Individuals exhibiting reduced internal juvenile hormone levels, such as late 4th instar larvae and pupae, demonstrate increased susceptibility to the analogs. In natural, undeveloped mosquito populations, various instars coexist in aquatic environments, exhibiting significant variation in internal juvenile hormone levels. This mechanism would result in sublethal doses and subsequently the development of tolerance or even resistance.

There is no doubt about the consequence resulted from occurrence and spread of resistance, such as cost increase of control operations, outbreak of vector populations, and vector-borne diseases. On the other hand, there are some negative impacts of resistance development on mosquito biological fitness, such as shortened longevity and reduced fecundity [77, 78, 94], which may lower the vectorial capacity [95–97]. Therefore, evaluation on the exact impact of vector resistance to pesticides on the epidemiology of vector-borne diseases can be complicated. During the past decades, pesticide resistance development and spread promoted banning or limited applications of nonselective, long-lasting synthetic pesticides. At the same time, this situation also advanced toxicological studies and detection technology of resistance, as well as the research, development, and application of biorational pesticides, and other mosquito control techniques.

The incidence of pesticide resistance in mosquitoes, including to the biorational insecticides addressed in this chapter, has been increasing. To provide long-term advantages, susceptibility monitoring using standardized techniques must be instituted concurrently with the introduction of a pesticide into control activities. Collaboration among academic research, industry development, and field application and evaluation is essential for extending the lifespan and improving the efficacy of pesticides, while also safeguarding the environment and non-target creatures.

REFERENCES

1. Roh JY, Choi JY, Li MS, Jin BR, Je YH. Bacillus thuringiensis as a specific, safe, and effective tool for insect pest control. J. Microbiol. Biotechnol. 2007; 17:547–559.
2. Goldberg LJ, Margalit JA. Bacterial spore demonstrating rapid larvicidal activity against Anopheles sergentii, Uranotaenia unguiculata, Culex univitattus, Aedes aegypti and Culex pipiens. Mosq. News 1977; 37:355–358.
3. Margalit J, Dean D. The story of Bacillus thuringiensis var. israelensis (B.t.i.). J. Am. Mosq. Control Assoc. 1985; 1:1–7.
4. Tabashnik BE. Evaluation of synergism among Bacillus thuringiensis toxins. Appl. Environ. Microbiol. 1992; 58:3343–3346.
5. Poncet S, Délécluse A, Klier A, Rapoport G. Evaluation of synergistic interactions among the CryIVA, CryIVB, and CryIVD toxic components of B. thuringiensis subsp. israelensis crystals. J. Invertebr. Pathol. 1995; 66:131–135.

6. Crickmore N, Bone EJ, Williams JA, Ellar DJ. Contribution of the individual components of the δ-endotoxin crystal to the mosquitocidal activity of Bacillus thuringiensis subsp. israelensis. FEMS Microbiol. Lett. 1995; 131:249–254.

7. Wirth MC, Jiannino JJ, Federici BA, Walton WE. Synergy between toxins of Bacillus thuringiensis subsp. israelensis and Bacillus sphaericus. J. Med. Entomol. 2004; 41:935– 41.

8. Gibbs KE, Brautigam FC, Stubbs CS, Zibilske LM. Experimental applications of B.t.i. for larval black fly control: persistence and downstream carry, efficacy, impact on non-target invertebrates and fish feeding. Maine Agricultural Experiments Station University of Maine. Tech. Bull. 123. October 1986. 25 pp.

9. Merritt RW, Walker ED, Wilzbach MA, Cummins KW, Morgan WT. A broad evaluation of B.t.i. for black fly (Diptera: Simuliidae) control in a Michigan river: efficacy, carry and nontarget effects on invertebrates and fish. J. Am. Mosq. Control Assoc. 1989; 5:397–415.

10. Ali A. Bacillus thuringiensis serovar. israelensis (ABG-6108) against chironomids and some non-target aquatic invertebrates. J. Invertebr. Pathol. 1980; 38:264–272.

11. Mulla MS, Chaney JD, Rodcharoen J. Control of nuisance midges (Diptera: Chironomidae) with the microbial larvicide Bacillus thuringiensis var. israelensis in a man- made lake in Southern California. Bull. Soc. Vector Ecol. 1990; 15:176–184.

12. Rodcharoen J, Mulla MS, Chaney JD. Microbial larvicides for the control of nuisance aquatic midges (Diptera: Chironomidae) inhabiting mesocosms and man-made lakes in California. J. Am. Mosq. Control Assoc. 1991; 7:56–62.

13. Becker N, Ludwig M. Investigations on possible resistance in Aedes vexans field populations after a 10-year application of Bacillus thuringiensis israelensis. J. Am. Mosq. Control Assoc 1993; 9:221–224.

14. Paul A, Harrington LC, Zhang L, Scott JG. Insecticide resistance in Culex pipiens in New York. J. Am. Mosq. Control Assoc. 2005; 21:305–309.

15. Vasquez MI, Violaris M, Hadjivassilis A, Wirth MC. Susceptibility of Culex pipiens (Diptera: Culicidae) field populations in Cyprus to conventional organic insecticides, Bacillus thuringiensis subsp. israelensis, and methoprene. J. Med. Entomol. 2009; 46:881–887.

16. Wirth MC, Ferrari JA, Georghiou GP. Baseline susceptibility to bacterial insecticides in populations of Culex pipiens complex (Diptera: Culicidae) from California and from the Mediterranean Island of Cyprus. J. Med. Entomol. 2001; 94:920–928.

17. Vasquez-Gomez M. Investigations of the possibility of resistance to Bacillus thuringiensis ser. H-14 in Culex quinquefasciatus through accelerated selection pressure in the laboratory. Riverside: University of California 1983.

18. Goldman I, Arnold J, Carlton BC. Selection for resistance to Bacillus thuringiensis subspecies israelensis in field and laboratory populations of the mosquito Aedes aegypti. J. Invertebr. Pathol. 1986; 47:317–24.

19. Saleh MS, El-Meniawi FA, Kelada NL, Zahran HM. Resistance development in mosquito larvae Culex pipiens to the bacterial agent Bacillus thuringiensis var. israelensis. J. Appl. Entomol. 2003; 127:29–32.

20. Mittal P. Laboratory selection to investigate the development of resistance to Bacillus thuringiensis var. israelensis H-14 in Culex quinquefasciatus Say (Diptera: Culicidae). Natl. Acad. Lett. India 2005; 28:281–283.

21. Tetreau G, Bayyareddy K, Jones CM, Stalinski R, Riaz, MA, Paris, M David, JP, Adang MJ, Després L. Larval midgut modifications associated with B.t.i resistance in the yellow fever mosquito using proteomic and transcriptomic approaches. BMC Genomics 2012; 13:248. doi:10.1186/1471-2164-13-248.

22. Tetreau G, Stalinski R, David JP, Després L. Monitoring resistance to Bacillus thuringiensis subsp. israelensis in the field by performing bioassays with each Cry toxin separately. Mem. Inst. Oswaldo Cruz. 2013; 108:894–900.

23. Georghiou GP, Wirth MC. Influence of exposure to single versus multiple toxins of Bacillus thuringiensis var. israelensis on the development of resistance in the mosquito Culex quinquefasciatus (Diptera: Culicidae). Appl. Environ. Microbiol. 1997; 63:1095– 1101.

24. Wirth MC, Délécluse A, Walton WE. Laboratory selection for resistance to Bacillus thuringiensis subsp. jegathesan or a component toxin, Cry11B, in Culex quinquefasciatus (Diptera: Culicidae). J. Med. Entomol. 2004; 41:435–441.

25. Wirth MC, Georghiou GP. Cross-resistance among CryIV toxins of Bacillus thurin giensis subsp. israelensis in Culex quinquefasciatus (Diptera: Culicidae). J. Econ. Entomol. 1997; 90:1471–1477.

26. Wirth MC, Delécluse A, Federici BA, Walton WE. Variable cross resistance to Cry11B from Bacillus thuringiensis subsp. jegathesan in Culex quinquefasciatus (Diptera: Culicidae) resistant to single or multiple toxins from Bacillus thuringiensis subsp. israelensis. Appl. Environ. Microbiol. 1998; 64:4174–4179.

27. Wirth MC, Delécluse A, Walton WE. Lack of cross resistance to Cry19A from Bacillus thuringiensis subsp. jegathesan in Culex quinquefasciatus (Diptera: Culicidae) resistant to Cry toxins from Bacillus thuringiensis subsp. israelensis. Appl. Environ. Microbiol. 2001; 67:1956–1958.

28. Chueng, PYK, Buster D, Hammock BD. Lack of mosquitocidal activity by the cytolytic protein of the Bacillus thuringiensis subsp. israelensis parasporal crystal. Current Microbiol. 1987; 15:21–23.

29. Wirth MC, Georghiou GP, Federici BA. CytA enables CryIV endotoxins of Bacillus thuringiensis to overcome high levels of CryIV resistance in the mosquito, Culex quinquefasciatus. Proc. Natl. Acad. Sci. U.S.A. 1997; 94:10536–10540.

30. Wirth MC, Park HW, Walton WE, Federici BA. Cyt1A of Bacillus thuringiensis delays resistance to Cry11A in the mosquito Culex quinquefasciatus. Appl. Environ. Microbiol. 2005; 71:185–189.

31. Pérez C, Fernandez LL, Sun J, Folch JL, Gill SS, Soberón M. Bacillus thuringiensis subsp. israelensis Cyt1Aa synergizes Cry11Aa toxin by functioning as a membrane- bound receptor. Proc. Natl. Acad. Sci. 2005; 102:18303–18308.

32. Pérez C, Munoz-Garcia C, Portugal LC, Sanchez J, Gill SS, Soberón M, Bravo A. Bacillus thuringiensis subsp. israelensis Cyt1Aa enhances activity of Cry11Aa toxin by facilitating the formation of a pre-pore oligomeric structure. Cell Microbiol. 2007; 9:2931- 2937.

33. Wirth MC, Berry C, Walton WE, Federici BA. Mtx toxins from Lysinibacillus sphaericus enhance mosquitocidal cry-toxin activity and suppress cry-resistance in Culex quinquefasciatus. J. Invertebr. Pathol. 2014; 115:62–67.

34. Wirth MC. Mosquito resistance to bacterial larvicidal toxins. Open Toxinol. J. 2010; 3:126–140. doi: 10.2174/1875414701003010126].

35. Ahmed I, Yokota, A, Yamazoe, A, Fujiwara T. Proposal of Lysinibacillus boronitolerans gen. nov. sp. nov., and transfer of Bacillus fusiformis to Lysinibacillus fusiformis comb. nov. and Bacillus sphaericus to Lysinibacillus sphaericus comb. nov. Int. J. Syst. Evol. Microbiol. 2007; 57 (Pt 5):1117–1125.

36. Sinègre G, Babinot M, Quermel JM, Gavon B. First field occurrence of Culex pipiens resistance to Bacillus sphaericus in southern France: 1994. P17. Proceedings of the 8th European Meeting of Society for Vector Ecology, September 5–8, 1994; Barcelona, Spain. Society for Vector Ecology, Santa Ana, California, 1997.

37. Silva-Filha MH, Regis L, Nielsen-LeRoux C, Charles JF. Low level resistance to Bacillus sphaericus in a field-treated population of Culex quinquefasciatus (Diptera: Culicidae). J. Econ. Entomol. 1995; 88:525–530.

38. Rao DR, Mani TR, Rajendran R, Joseph AS, Gajanana A, Reuben R. Development of a high-level of resistance to Bacillus sphaericus in a field population of Culex quinquefasciatus from Kochi, India. J. Am. Mosq. Control Assoc. 1995; 11:1–5.

39. Adak T, Mittal PK, Raghavendra K, Subbaro SK, Ansari MA, Sharma VP. Resistance to Bacillus sphaericus in Culex quinquefasciatus Say 1823. Curr. Sci. 1995; 69:695–698.

40. Nielsen-LeRoux C, Pasquier F, Charles JF, Sinègre G, Gaven B, Pasteur N. Resistance to Bacillus sphaericus involves different mechanisms in Culex pipiens (Diptera: Culicidae) larvae. J. Med. Entomol. 1997; 34:321–327.

41. Yuan Z, Zhang Y, Cai Q, Liu EY. High-level field resistance to Bacillus sphaericus C3-41 in Culex quinquefasciatus from southern China. Biocon. Sci. Technol. 2000; 10:41–49.

42. Chevillon C, Bernard C, Marquine M, Pasteur N. Resistance to Bacillus sphaericus in Culex pipiens (Diptera: Culicidae): interaction between recessive mutants and evolution in southern France. J. Med. Entomol. 2001; 38:657–664.

43. Nielsen-LeRoux C, Pasteur N, Prètre J, Charles JF, Ben Sheik H, Chevillon C. High resistance to Bacillus sphaericus binary toxin in Culex pipiens (Diptera: Culicidae): the complex situation of west Mediterranean countries. J. Med. Entomol. 2002; 39:729- 735.

44. Mulla MS, Thavara U, Tawatsin A, Chomposri J, Su T. Emergence of resistance and resistance management in field populations of tropical Culex quinquefasciatus to the microbial control agent Bacillus sphaericus. J. Am. Mosq. Control Assoc. 2003; 19:39- 46.

45. Su T, Mulla M S. Documentation of high level Bacillus sphaericus-resistance in tropical Culex quinquefasciatus populations from Thailand. J. Am. Mosq. Control Assoc. 2004; 20:405–411.

46. Su T, Soliman BA, Chaney JD, Mulla MS, Beehler JW. Susceptibility of Culex mosquitoes breeding in dairy ponds before and after treatment with Bacillus sphaericus formulation. Proc. Pap. Mosq. Vector Control Assoc. Calif 2001; 69:110–116.

47. Rodcharoen J, Mulla MS. Resistance development in Culex quinquefasciatus (Diptera: Culicidae) to Bacillus sphaericus. J. Econ. Entomol. 1994; 87:1133–1140.

48. Zahiri NS, Su T, Mulla MS. Strategies for the management of resistance in mosquito to the microbial control agent Bacillussphaericus. J. Med. Entomol. 2002; 39:513–520.

49. Zahiri NS, Mulla MS. Susceptibility profile of Culex quinquefasciatus (Diptera: Culicidae) to Bacillus sphaericus on selection with rotation and mixture of B. sphaericus and B. thuringiensis israelensis. J. Med. Entomol. 2003; 40:672–677.

50. Wirth MC, Georghiou GP, Malik JI, Hussain G. Laboratory selection for resistance to Bacillus sphaericus in Culex quinquefasciatus (Diptera: Culicidae) from California, USA. J. Med. Entomol. 2000; 37:534–540.

51. Pei G, Oliveira CMF, Yuan Z, Nielsen-LeRoux C, Silva-Filha MH, Yan J, Regis L. A strain of Bacillus sphaericus causes slower development of resistance in Culex quinquefasciatus. Appl. Environ. Microbiol. 2002; 88:3003–3009.

52. Amorim LB, Oliveira CMF, Rios EM, Regis L, Silva-Filha MHNL. Development of Culex quinquefasciatus resistance to Bacillus sphaericus strain IAB59 needs long term selection pressure. Biol. Control 2007; 42:155–160.

53. Amorim LB, de Barros RA, Chalegre KD, de Oliveira CM, Regis LN, Silva-Filha MH. Stability of Culex quinquefasciatus resistance to Bacillus sphaericus evaluated by molecular tools. Insect Biochem. Mol. Biol. 2010; 40:311–316. doi: 10.1016/j.ibmb. 2010.02.002.

54. Poopathi S, Mani TR, Raghunatha D, Baskram G, Kabilan L. Cross-resistance to Bacillus sphaericus strains in Culex quinquefasciatus resistant to B. sphaericus 1593M. South east Asian J. Trop. Med. Public Health 1999; 30:478–481.

55. Rodcharoen J, Mulla MS. Cross-resistance to Bacillus sphaericus strains in Culex quin quefasciatus. J. Am. Mosq. Control Assoc. 1996; 12:247–250.

56. Poopathi S, Kabilan L, Mani TR, Raghunatha RD, Baskaran G. Observation of low tolerance to Bacillus thuringiensis var israelensis in Culex quinquefasciatus resistant to Bacillus sphaericus. Entomon. Ember. 2000; 25:201–208.

57. Yuan ZM, Pei GF, Regis L, Nielsen-LeRoux C, Cai QX. Cross resistance between strains of Bacillus sphaericus but not B. thuringiensis israelensis in colonies of the mosquito Culex quinquefasciatus. Med. Vet. Entomol. 2003; 17:251–256.

58. Nielsen-LeRoux C, Rao D, Rodcharoen J, Carron A, Mani TR, Hamon S, Mulla MS. Various levels of cross-resistance to Bacillus sphaericus strains in Culex pipiens (Diptera: Culicidae) colonies resistant to B. sphaericus strain 2362. Appl. Environ. Microbiol. 2001; 67:5049–5054.

59. Oliveira CMF, Silva-Filha MH, Nielsen-LeRoux C, Pei G, Yuan Z, Regis L. Inheritance and mechanism of resistance to Bacillus sphaericus in Culex quinquefasciatus (Diptera: Culicidae) from China and Brazil. J. Med. Entomol. 2004; 41:58–64.

60. Darboux I, Charles JF, Pauchet Y, Warat S, Pauron D. Transposon mediated resistance to Bacillus sphaericus in a field-evolved population of Culex pipiens (Diptera: Culicidae). Cell Microbiol. 2007; 9: 2022–2029.

61. Darboux I, Pauchet Y, Castella C, Silva-Filha MH, Nielsen-LeRoux C, Charles JF, Pauron, D. Loss of the membrane anchor of the target receptor is a mechanism of bioinsecticide resistance. Proc. Natl. Acad. Sci. U.S.A. 2002; 99:5830–5835.

62. Nielsen-LeRoux C, Charles JF, Thièry I, Georghiou GP. Resistance in a laboratory population of Culex quinquefasciatus (Diptera: Culicidae) to Bacillus sphaericus binary toxin is due to a change in the receptor on midgut brush-border membrane. Eur. J. Biochem. 1995; 228:206–210.

63. Romão TP, de Melo Chalegre KD, Key S, Ayres CF, Fontes de Oliveira CM, de-Melo- Neto OP, Silva-Filha MH. A second independent resistance mechanism to Bacillus sphaericus binary toxin targets its α-glucosidase receptor in Culex quinquefasciatus. FEBS J. 2006; 273:1556–1568.

64. Chalegre KD, Romão TP, Amorim LB, Anastacio DB, de Barros RA, de Oliveira CM, Regis L, de-Melo-Neto OP, Silva-Filha MH. Detection of an allele conferring resistance to Bacillus sphaericus binary toxin in Culex quinquefasciatus populations by molecular screening. Appl. Environ. Microbiol. 2009; 75:1044–1049.

65. Chalegre KD, Romão TP, Tavares DA, Santos EM, Ferreira LM, Oliveira CMF, de-Melo-Neto OP, Silva-Filha MHNL. Novel mutations associated with resistance to Bacillus sphaericus in a polymorphic region of the Culex quinquefasciatus cqm1 gene. Appl. Environ. Microbiol. 2012; 78:6321–6326. doi: 10.1128/AEM.01199-12.

66. Guo QY, Cai QX, Yan JP, Hu XM, Zheng DS, Yuan ZM. Single nucleotide deletion of cqm1 gene results in the development of resistance to Bacillus sphaericus in Culex quinquefasciatus. J. Insect Physiol. 2013; 59:967–973.

67. Rodcharoen J, Mulla MS. Comparative ingestion rates of Culex quinquefasciatus (Diptera: Culicidae) susceptible and resistant to Bacillus sphaericus J. Invertebr. Pathol. 1995; 66:242–248.

68. Sun F, Yuan Z, Li T, Zhang Y, Yu J, Pang Y. Reduction of resistance of Culex pipiens larvae to the binary toxin from Bacillus sphaericus by coexpression of cry4Ba from Bacillus thuringiensis subsp. israelensis with the binary toxin. World J. Microbiol. Biotechnol. 2001; 17:385–389.

69. Wirth MC, Walton WE, Federici BA. Cyt1A from Bacillus thuringiensis restores toxicity of Bacillus sphaericus against resistant Culex quinquefasciatus (Diptera: Culicidae). J. Med. Entomol. 2000; 37:401–407.

70. Wirth MC, Jiannino JA, Federici BA Walton WE. Evolution of resistance toward Bacillus sphaericus or a mixture of B. sphaericus+Cyt1A from Bacillus thuringiensis, in the mosquito, Culex quinquefasciatus (Diptera: Culicidae). J. Invertebr. Pathol. 2005; 88:154–162.

71. Chenniappan K, Ayyadurai N. Synergistic activity of Cyt1A from Bacillus thuringiensis subsp. israelensis with Bacillus sphaericus B101 H5a5b against Bacillus sphaericus B101 H5a5b-resistant strains of Anopheles stephensi Liston (Diptera: Culicidae). Parasitol. Res. 2011; 110:381–388.

72. Park HW, Bideshi DK, Federici BA. Recombinant strain of Bacillus thuringiensis producing Cyt1A, Cry11B, and the Bacillus sphaericus binary toxin. Appl. Environ. Microbiol. 2003; 69:1331–1334.

73. Park HW, Bideshi DK, Wirth MC, Johnson JJ, Walton WE, Federici BA. Recombinant larvicidal bacteria with markedly improved efficacy against Culex vectors of West Nile virus. Am. J. Trop. Med. Hyg. 2005; 72:732–738.

74. Federici BA, Park HW, Bideshi DK, Wirth MC, Johnson JJ, Sakano Y, Tang M. Developing recombinant bacteria for control of mosquito larvae. J. Am. Mosq. Control Assoc. 2007; 23:S164–175.

75. Poopathi S, Mani TR, Raghunatha RD, Baskaran G, Kabilan L. Evaluation of synergistic interaction between Bacillus sphaericus and a neem based biopesticide against Culex quinquefasciatus larvae susceptible to Bacillus sphaericus. 1593M. Insect Sci. Appl. 2002; 22:303–306.

76. Wei S, Cai Q, Cai Y, Yuan Z. Lack of cross-resistance to Mtx1 from Bacillus sphaericus in B. sphaericus-resistant Culex quinquefasciatus (Diptera: Culicidae). Pest. Manag. Sci. 2007; 63:190–193.

77. Rodcharoen J, Mulla MS. Biological fitness of Culex quinquefasciatus (Diptera; Culicidae) susceptible and resistant to Bacillus sphaericus. J. Med. Entomol. 1997; 34:5–10.

78. Oliveira CMD, Costa Filho F, Beltran JFN, Silva-Filha MH, Regis L. Biological fitness of a Culex quinquefasciatus population and its resistance to Bacillus sphaericus. J. Am. Mosq. Control Assoc. 2003; 19:125–129.

79. Su T, Cheng ML. Resistance development in Culex quinquefasciatus to spinosad: a preliminary report. J. Am. Mosq. Control Assoc. 2012; 28:263–267.

80. Su T, Cheng ML. Laboratory selection of resistance to spinosad in Culex quinquefasciatus (Diptera: Culicidae). J. Med. Entomol. 2014; 51:421–427.

81. Su T, Cheng ML. Cross resistances in spinosad – resistant Culex quinquefasciatus (Diptera: Culicidae). J. Med. Entomol. 2014; 51:428–435.

82. Schaefer CH, Wilder WH. Insect development inhibitors. 2. Effects on target mosquito species. J. Econ. Entomol. 1973; 66:913–916.

83. Brown TM, Brown AW. Experimental induction of resistance to a juvenile hormone mimic. J. Econ. Entomol. 1974; 67:799–801.

84. Georghiou GP, Lin CS, Pasternak ME. Assessment of potentiality of Culex tarsalis for development of resistance to carbamate insecticides and insect growth regulator. Proc. Papers 42nd Ann. Conf. Calif. Mosq. Control Assoc. 1974; 42:117.

85. Brown TM, Devries DH, Brown AWA. 1978. Induction of resistance to insect growth regulators. J. Econ. Entomol. 1978; 71:223–229.

86. Amin AM, White GB. Resistance potential of Culex quinquefasciatus against the insect growth regulators methoprene and diflubenzuron. Ent. Exp. Appl. 1984; 36:69–76.

87. Brown TM, Hooper GHS. Metabolic detoxication as a mechanism of methoprene resistance in Culex pipiens pipiens. Pestic. Biochem. Physiol. 1979; 12:79–86.

88. Brown TM, Brown AWA. Accumulation and distribution of methoprene in resistance Culex pipiens pipiens larvae. Ent. Exp. Appl. 1980; 27:11–22.

89. Dame DA, Wichterman GJ, Hornby JA. Mosquito (Aedes taeniorhynchus) resistance to methoprene in an isolated habitat. J. Am. Mosq. Control Assoc. 1998; 14:200–203.

90. Cornel AJ, Stanich MA, Farley D, Mulligan FS III, Byde G. Methoprene tolerance in Aedes nigromaculis in Fresno County, California. J. Am. Mosq. Control Assoc. 2000; 16:223–238.

91. Cornel AJ, M. Stanich A, McAbee RD, Mulligan FS III. High level methoprene resistnce in the mosquito Ochlerotatus nigromaculis (Ludlow) in central California. Pest Manag. Sci. 2002; 58:791–798.

92. Schaefer CH, Mulligan FS III. 1991. Potential for resistance to pyriproxyfen: a promising new mosquito larvicide. J. Am. Mosq. Control Assoc. 1991; 7:409–411.

93. Walker AL, Wood RJ. Laboratory selected resistance to diflubenzuron in larvae of Aedes aegypti. Pesti. Sci, 1986; 17:495–502.

94. Ferrari JA, Georghiou GP. Effects of insecticidal selection and treatment on reproductive potential of resistant, susceptible, and heterozygous strains of the southern house mosquito. J. Econ. Entomol. 1981; 74: 323–327.

95. Rivero A, Vézilier J, Weill M, Read AF, Gandon S. Insecticide control of vector-borne diseases: when is insecticide resistance a problem? PLoS Pathog. 2010; 6(8): e1001000. doi:10.1371/journal.ppat.1001000.

96. Alout H, Ndam NT, Sandeu MM, Djégbe I, Chandre F, Dabiré RK, Djogbénou LS, Corbel V, Cohuet A. Insecticide resistance alleles affect vector competence of Anopheles gambiae s.s. for Plasmodium falciparum field isolates. PLoS One. 2013; 8(5): e63849.

97. Liu R, Gourley SA. Resistance to larvicides in mosquito populations and how it could benefit malaria control. European J. Appl. Math. 2013; 24:415–436.

9 Chapter

ENHANCING STRATEGIC INSECTICIDE RESISTANCE MANAGEMENT FOR MALARIA VECTORS

1. INTRODUCTION: AN OVERVIEW

Malaria continues to be a vector-borne illness of significant public health importance worldwide. Approximately 198 million cases of malaria and an associated 584,000 fatalities are predicted to occur globally each year. Insecticide-based vector control, within the framework of integrated vector management (IVM), has a well-established history of preventing, mitigating, and eradicating vector-borne diseases. The ongoing effectiveness is jeopardized by the emergence of insecticide resistance in disease vectors, alongside insufficient financial resources, a dearth of necessary skills, and inadequate collaboration between health and other pertinent sectors for effective monitoring and management. Increasing evidence of pesticide resistance is systematically compromising malaria vector control and prompting policy modifications [5]. Surveillance and regulation of resistant disease vectors are crucial to curtail the emergence and dissemination of pesticide resistance and to preserve the efficacy of vector control measures [6]. Nonetheless, significant obstacles to implementation persist at the national level. Consequently, nations are urged to adopt pre-emptive insecticide resistance management (IRM) policies targeting malaria vectors in accordance with the Global Plan for IRM (GPIRM) [7]. Although Integrated Vector Management (IVM) and Insecticide Resistance Management (IRM) are the preferred methodologies for addressing vector-borne diseases and mitigating resistance proliferation, practical implementation of both techniques is confined to a restricted number of nations. IVM offers a prospective framework that might be utilized for improved national strategic IRM planning and implementation. This chapter examines the distribution, processes, and resistance management tactics of malaria vectors, along with the obstacles encountered in operational contexts. A framework of policies and strategies to enable the implementation of the GPIRM via the IVM platform is also provided, emphasizing coordinated responses among stakeholders and political commitment for effective policy execution within national health systems.

2. LITERATURE SEARCH STRATEGY

Information sources for this review included all available data and accessible archived documentary records on malaria vectors and insecticide resistance. Structured literature searches of published, peer-reviewed sources using online scientific bibliographic databases were utilised to gather pertinent data. This was conducted via systematic literature search of Library catalogues and online electronic databases, particularly PubMed [8], the WHO Library Database [9], Google Scholar [10], the African Journals Online, the Armed Forces Pest Management Board [11], and the research for life databases (AGORA, ARDi, HINARI, and OARE) were used to search for relevant literature. All digital electronic database searches for peer- reviewed, published work used a combination of key search terms: 1) Anopheles malaria vectors complex and one of the following terms; 2) insecticide resistance; 3) resistance mechanisms; 4) resistance management; 5) impact of resistance; 6) malaria vector control; and 7) malaria epidemiology. Reference sections of all relevant articles were also reviewed to identify more literature. Additional non-peer reviewed literature were examined for information related to the subject. Articles that report biochemical and molecular tools for resistance monitoring were also retrieved. The inclusion criteria considered all manuscripts and publications in English language that report on selection of insecticide resistance in malaria vectors, causes and mechanisms of resistance, vector resistance and the epidemiology of malaria, integrated vector management (IVM), resistance patterns, and the impact in malaria vectors.

3. CLASSIFICATION AND DISTRIBUTION OF MALARIA VECTORS

Mosquitoes belong to the family Culicidae in the order Diptera, class Insecta, phylum Arthropoda [12]. Culicidae is divided into three subfamilies Anophelinae, Culicinae, and Toxorhynchitinae, and comprises approximately 3,450 recognised species of mosquitoes in 38 genera. The 34 genera are in the subfamily Culicinae, 3 in Anophelinae, and only 1 in Toxorhynchitinae [13]. Malaria vectors belong to the genera Anopheles (Cellia) Myzomyia and their global distribution has been recognised in six zoo-geographical regions: Palaearctic, Oriental, Australasian, Afro-tropical, Neoarctic, and Neotropical regions [14, 15]. Globally, about 465 species have been described in the genus Anopheles with seven subgenera that vary in species composition, i.e., Anopheles (182 species), Baimaia (one species), Cellia (220 species), Kerteszia (12 species), Lophopodomyia (six species), Nyssorhynchus (39 species), and Stethomyia (five species) [16]. However, the species able to transmit parasites that cause human malaria only belong to the subgenera, Anopheles, Cellia, Kerteszia, and Nyssorhynchus [17]. Only about 80 species are capable of transmitting malaria, 70 species are vectors of malaria under natural conditions and about 45 are of major significance [13]. The distribution of major vectors of malaria is determined mainly by temperature and the capacity of the air to desiccate the insect [18].

The global distribution of principal vectors of malaria is associated with 12 epidemiological zones of malaria: North America (An. freeborni and the An. quadrimaculatus), Central America (An. albimanus, An. Aquasalis, An. pseudopunctipennis, An. argyritarsis, and An. darlingi,), South America (An. darlingi, An. albitarsis, An. Aquasalis, An. marajoara, An. nuneztovari, and An. pseudopunctipennis), Afro-tropical (Anopheles gambiae Giles 1902 and Anopheles arabiensis Patton in the An. gambiae complex, and Anopheles funestus s.s. Giles 1900 in the An. funestus complex [19, 20] with An. merus, An. melas, An. moucheti, An. pharoensis, and An. nili implicated in transmission in localised areas [21, 22]), North Eurasian (An. atroparvus), Mediterranean (An. atroparvus, An. labranchiae, An. messeae, An. sacharovi, An. sergentii, and An. superpictus), Afro- Arabian (An. arabiensis, An. pharoensis, and An. sergenti), Indo-Iranian (An. culicifacies and An. fluviatilis), Indo-Chinese Hills (An. dirus, An. fluviatilis, and An. minimus), Malaysian (An. campestris, An. donaldi, An. letifer, An. nigerrimus, An. aconitis, An. balabacencis, An. dirus, An. flavirostris, An. leucosphyrus, An. ludlowea, An. maculates, An. minimus, An. subpictus, and An. sundaicus), Chinese (An. barbirostris, An. lesteri, An. sinensis, An. aconitus, An. annularis, An. balabacensis, An. culicifacies, An. dirus, An. farauti, An. flavirostris, An. fluviatilis, An. koliensis, An. leucosphyrus, An. maculates, An. minimus, An. punctulatus, An. stephensi, An. subpictus, and An. sundaicus), and Australasian (An. farauti, An. punctulatus s.s., and An. koliensis) [15, 23]. Notably, malaria vector bionomics and their ecological variations have implications for their control.

4. INSECTICIDES AND MALARIA VECTOR CONTROL

The classes of insecticides most commonly used for contemporary malaria vector control include organochlorines, organophosphorus, carbamates, and pyrethroids [24]. The first synthetic organochlorine insecticide to be commercialised, DDT (dichlorodiphenyltrichloroethane), was central to the World Health Organisation (WHO)-led global malaria eradication campaign (1955–1969) [25]. Except in sub-Saharan Africa, this resulted in the elimination of the disease in North America, Europe, and parts of Asia [26]. While agricultural use of DDT has now ceased due to environmental persistence and reduced efficacy against resistant insects, extensive use for malaria control continues as a cost-effective and safe insecticide for indoor residual spraying (IRS). More recently, pyrethroids have been widely used for malaria control. They are the only class of insecticides recommended by the WHO for impregnation of long-lasting insecticidal bed nets (LLINs) [24] and are also available for IRS. Pyrethroids, such as permethrin and deltamethrin, and the pseudo-pyrethroid etofenprox, including DDT and its analogues, share a similar mode of action of targeting the sodium channels of the nerve membranes. Carbamates, such as bendiocarb, share the same mode of action as organophosphates, such as pirimiphos methyl, malathion, and temephos, binding to acetylcholinesterase at the nerve junction [27]. Chlorfenapyr has a different mode of action involving disruption of oxidative phosphorylation and consequently the disruption of the conversion of ADP to ATP in mitochondria [28]. Although insecticides from different chemical classes are available as larvicides (e.g., temephos), the arsenal of insecticides recommended for IRS is limited to four classes only [29]. The selection of DDT resistance in malaria vectors resulted in the declining political and financial support for the Global Malaria Eradication Campaign launched by the WHO [30].

Vector control, personal protection, and community participation are the pillars of WHO strategies for insect-transmitted disease control. IVM has been advocated for as a recommended approach for combating vector-borne diseases in the past decade [31]. IVM is defined as "a rational decision-making process for optimal use of resources for vector control". The objective of vector control is to reduce and/or interrupt transmission of malaria by preventing human contact with malaria-bearing mosquitoes, eliminating breeding sites, killing the mosquito larvae, or reducing the longevity of adult mosquitoes [30]. The use of IRS and LLINs are the mainstream contemporary malaria vector control interventions [32, 33]. The efficacy of these two methods has been evaluated in different epidemiological settings [34] at experimental field trial [35, 36] and community-wide levels [37, 38]. In reducing abundance and infectivity of malaria vectors, these tools reduce overall transmission and protect all individuals within a community [35], albeit with variation in responsiveness amongst vector populations. Present ly, there is mounting evidence that combining IRS and insecticide treated nets (ITNs) affords enhanced protection to exposed populations compared to using one method alone [39]. As such, deployment of these interventions together in high malaria risk areas has been advocated [35, 40, 41]. Although these two interventions have been critical in providing community protection, the optimal policy for their co-implementation still remains to be determined. Moreover, the growing resistance of malaria vectors to available insecticides is a major cause for concern and an increasing threat to such essential and effective interventions [24, 42, 43].

In light of the inherent heterogeneity in the responsiveness of malaria vectors to control, the core interventions can be supplemented in specific locations by larval source management (LSM) strategies (e.g., larviciding, biological control, and environmental management) in the context of IVM [44, 45]. Larvivorous fishes or bacterial pathogens such as Bacillus thuringiensis israelensis and Bacillus sphaericus are examples of biological agents that are used to kill larvae [46]. Temephos is the commonly used chemical larvicide. The environmental methods to prevent malaria include elimination of breeding sites by drainage or by applying locally grown plants. With the selection of resistance, new insecticides, and novel approaches to vector control must be developed. Effective and sustained malaria vector control requires clear commitment from national authorities, including long-term support from funding partners [47]. Several malaria control programmes have fragmentary empirical evidence to inform policy formulation for rational vector control. For this reason, malaria control programmes are encouraged to adopt the WHO-led IVM strategy [48], which should be an evidence-based decision making process that requires coherent monitoring and evaluation component [49]. This should include routine surveillance of resistance profiles of major malaria vectors and potential resistance mechanisms to facilitate informed decisions and policy changes, such as the incorporation of insecticide resistance management operations into control programmes [29].

5. INSECTICIDE RESISTANCE IN MALARIA VECTORS

The selection of insecticide resistance in malaria vectors has the potential to compromise effective control of vector-borne diseases. Resistance is defined as "the development of an ability in a strain of some organism to tolerate doses of a toxicant that would prove lethal to a majority of individuals in a normal population of the same species" [27, 29]. Alternatively, a resistant phenotype has been defined as an insect that survives a dose of insecticide that would normally have killed it [50, 51]. This heritable change in the sensitivity of a vector population is reflected in the repeated failure of a product to achieve the expected level of control when used according to the label recommendation for that disease vector species [52]. The biological phenomenon is a genetically inherited characteristic that develops as a result of selective effects of the relevant insecticidal compound or its analogue and increases in the vector population [15]. In mosquitoes, genetic and phenotypic resistance results from a mutation or gene duplication leading to the alteration of a normal physiology, morphology, or behaviour of the individual phenotype. In this regard, the sensitivity of the nervous system to the insecticide is reduced or the process of detoxification of the insecticide is enhanced. When an insecticide is applied, susceptible individuals are less likely to survive relative to the resistant individuals. The consequence is the propagation and exponential increase in the frequency of the resistance gene within the population over time [29].

Resistance is a multidimensional biological phenomenon that depends for its development on the interaction of multiple influences [27]. The evolution of insecticide resistance is complex and depends on several genetic, biological, and operational factors [53, 54]. The genetic factors include the intrinsic characteristics of the resistant genes (e.g., mono versus polygenic resistance, dominance, fitness cost, and gene interaction), while the biological factors relate to the life cycle of the insect (e.g., rate of reproduction, number of generation/offspring, and rate of migration and isolation). Behavioural/ecological encompass, migration in and out of exposed population, avoidance of the insecticide, effects of age and natural inducers on degradative enzymes, and endophagy/exophagy. Operational factors concern the treatment itself, including the method and frequency of application, dosage, and residual activity of the insecticides as well as insecticide coverage [55]. Among known and potential factors affecting the evolution of resistance, the operational factors are the only ones open to manipulation by man. Therefore, investigation on the development of resistance should ideally take into account of all these factors.

The level of resistance in insect populations is dependent on the amount and frequency of insecticides used, and the inherent characteristics of the insect species selected. Mosquitoes, for instance, are endowed with all attributes suited for rapid resistance development including high reproductive potentials and short life cycles producing several generations per season with abundant progeny [27, 52]. Mostly resistance in a particular species is considered to occur throughout the control area, but in reality, insecticide resistance can be focal in nature and is very heterogeneous even over very small distances. It often develops within a small part of the population of one species of Anopheles and assumes different patterns depending on the type of selection pressure [27]. In Guatemala, sampling sites for Anopheles albimanus only a few kilometres apart varied not only by presence or absence of resistance, but also by level of resistance and the mix of mechanisms responsible for resistance [56]. The WHO Global Technical Strategy for Malaria 2016–2030 highlights insecticide resistance as a major obstacle to achieving malaria control targets [57]. The current major emphasis in research into vector resistance is double pronged. The first approach strives towards understanding the molecular mechanisms underlying resistance with the view of developing novel vector-control methods that avoid or minimise resistance problems. The second approach to research involves rational resistance management, which is developing and implementing control methods that minimise the likelihood that vectors will evolve strong resistance to important insecticides [58].

6. METHODS FOR DETECTING INSECTICIDE RESISTANCE

Information on insecticide resistance is important to inform effective vector control policy formulation. As such, detection and monitoring of insecticide resistance in malaria vectors is crucial and has to be conducted together with other entomological surveys [29]. Insecticide resistance can be detected and investigated at many levels; from dose-related phenotypical observations and genotypic approaches ranging from molecular characterisation of genes conferring resistance and their biochemical products, to the role these gene products play in overcoming the toxic effects of insecticides. In order to detect resistance, one needs to be continually looking for it. Measuring phenotypic resistance using bioassays is the recommended initial step in establishing resistance levels before genotyping for target-site and metabolic resistance and biochemical assays [59, 60]. Establishing an effective resistance management plan requires simple assays to monitor and evaluate insecticide resistance and its underlying mechanisms. To this end, different biochemical and molecular approaches have been devised, some of which are amenable for field operations. The contemporary applicable methods for resistance monitoring of field populations of mosquitoes are outlined below:

1. **WHO Diagnostic Assays:** A bioassay is used to determine the relationship between a physiologically active agent and the effect that it produces in a living organism [61, 62]. Bioassays with the dosage or the exposure time as the variable are carried out to test the resistance status of insect populations. The WHO diagnostic assay is a useful and handy approach to detect resistance. Insecticide impregnated filter paper is used as a contact surface for exposed mosquitoes. The assay uses insecticide discriminating dosages twice the LD99 that kills 100% of non-blood-fed, adult-susceptible Anopheline mosquitoes of known age [59, 63, 64]. The demerits associated with the assays include: the assay is only able to detect high levels of resistance, i.e., if more than 5% of insects survive the exposure, with the potential exception of dieldrin bioassays, they cannot monitor resistance gene frequencies accurately; cannot give an indication of the underlying mechanisms of resistance; and cannot be used to predict cross- resistance between insecticides [29]. Therefore, the resistance status detected using bioassays, can then be further studied by looking at the mechanisms responsible for resistance using biochemical and molecular assays.
2. **CDC Bottle Assay:** These are similar to the WHO discriminating dose assays. However, the CDC bottle assay relies on time mortality data, which are measures of the time it takes an insecticide to penetrate a vector, traverse its intervening tissues, get to the target site, and act on that site. Mosquitoes are exposed to glass bottle surfaces coated with an acetone- or alcohol-based formulation of insecticides. The advantage of this assay is that the rate of insecticide knock down can easily be scored during the course of the exposure period. It is able to give predictive kdr-type resistance mechanism with rapid acting insecticides, such as pyrethroids. However, caution should be exercised, as metabolic resistance mechanisms are capable of eliciting reduced knock-down phenotype without any concomitant change in sensitivity at the sodium channel target site [65].
3. **Synergists:** Synergists are enzyme inhibitors of insecticide detoxification enzymes. The synergists, piperonyl butoxide (PBO) and S, S, S-tributylphosphorotrithioate (DEF) are inhibitors of monooxygenases

and esterases respectively [64, 66]. Glutathione s-transferase activity is inhibited by Ethacrynic acid (EA), diethyl maleate (DM), and chlorfenethol (CF). By inhibiting specific detoxification enzymes, insecticide synergists can reduce or eliminate the selective advantage of individuals possessing over-expressed or mutated enzymes [67]. Therefore, they are used to suggest the type of metabolic resistance mechanisms present in insect populations [68]. For example, they are used in bioassays to counteract or inhibit the enzymes responsible for resistance to the insecticide. Some are used in control to reduce the dose or rate of application [64]. For example, piperonyl butoxide is commonly added to pyrethroid-based aerosol formulations to decrease the time to knock down and increase the time to recover from the insecticide.

4. **Biochemical Assays:** There are two ways that metabolic enzymes can produce resistance [69]; overproduction of the enzyme, which leads to either increased metabolism or sequestration of the insecticide and an alteration in the catalytic centre activity of the enzyme, which increases the rate of insecticide metabolism by the enzyme. Sequestration occurs when the overproduced enzyme rapidly binds and slowly metabolises the insecticide, therefore preventing it from reaching the target site within the insecticide [70]. With sequestration, the resistance level is proportional to the increase in the quantity of the enzyme produced because of the slow insecticide turn-over rate [70]. Biochemical assays are used to give a first indication of the enzyme system involved in resistance [69]. A number of simple biochemical assays [71] are available to detect increased activity of three enzyme systems, esterases [72, 73], glutathione- S-transferases (GST) [74, 75], and cytochrome P450-dependent monoxygenases (P450s) [76] involved in insecticide metabolism. Many of these assays detect increased enzymatic activity against model substrates in resistant individuals. While simple microtitre plate assays to measure AChE insensitivity using a carbamate or an oxon analogue of a phosphorothioate insecticide exist [77], the applicability of biochemical assays for the GSTs and P450s are not easily amenable in the field [29].

5. **Molecular Assays:** Molecular techniques can be used to detect some well-characterised resistance mechanisms. Most techniques employ the method of polymerase chain reaction (PCR). Mutations in the insecticides' target site lend themselves to detection through simple PCR assays, which can readily be used in many field settings. Allele-specific PCR assays have been developed for three major target sites, the GABA receptors [78, 79], the sodium channels (kdr) [80, 81], and AChE. The challenge is to adapt these assays for high-throughput field applications as they have the advantage of

detecting heterozygous-resistant individuals that may be missed by other assays [29]. Recent advances in genomics have allowed a much more rapid identification of genes that are up or down regulated in insecticide resistant insects using microarray technology [82]. The detoxification microarray chip, developed for An. gambiae, contains all potential insecticide resistance genes. The detoxification chips have been developed for An. stephensi, pyrethroid resistant An. funestus, and A. aegypti [29].

Bioassay data as measured by either the CDC or WHO technique have a common limitation of inability to accurately determine gene frequency or predict the epidemiological impact of resistance [83]. Resistance management requires that resistance is detected at very low frequencies, but the two approaches are not sensitive enough to achieve this. While contemporary techniques diverge in their sophistication and ease of use, there is still a need for approaches to enable measurement of the frequency of different mechanisms of resistance in mosquito populations [29].

7. INSECTICIDE RESISTANCE MECHANISMS

In mosquitoes, resistance is mediated through complex mechanisms, including behavioural and/or physiological changes resulting in insecticide avoidance (reduced contact with insecticide), reduced cuticular penetration (of the insecticide through the cuticle), increased sequestration (i.e., stored in the body where it is not harmful), target site insensitivity (i.e., the target site is altered and not affected by the insecticide), or increased bio-degradation (so that it is detoxified before it reaches the target site) [84], and possible increased excretion. The molecular basis of insecticide resistance has been attributed to the existence of mutations in target site genes or metabolic alterations at the level of the activity of the detoxification proteins [56, 85]. Insecticide resistance mechanisms have a biochemical basis and target-site resistance and detoxification enzyme-based resistance remain the two major forms of biochemical resistance [52, 56]. Alone or in combination, target site resistance, which results from the inability of the insecticide to bind to its target, and metabolic resistance resulting from failure of the insecticide to reach its site of action due to enhanced levels of modified activities of detoxification enzymes, confer various levels of resistance to all classes of insecticides [84]. Though long- and well-recognised, the importance of behavioural and cuticular resistance in malaria vectors has been largely overlooked. While resistance arises through Darwinian selection in a population, it is often a combination of factors that results in the overall expression of the phenomenon [15]. Thus, gene over-expression, amplification, and structural mutations have been linked to insecticide resistance mechanisms in some insects [84]. Cognizant of detailed elaborations given elsewhere [86], a brief description of the key mechanisms is outlined below:

Target site resistance (also called phenotypic resistance) is based on alterations of amino acids in the site of action where the insecticide is supposed to bind, rendering them less sensitive to the active ingredient [58]. Majority of insecticides used in vector control are nerve poisons and target: the acetylcholinesterase (AChE) that hydrolyses the neurotransmitter acetylcholine in the synapses particularly in carbamates and organophosphates; the sodium channels responsible for raising the action potential in the neurons during the nerve impulses involved in the resistance in organochlorines and pyrethroids; or the c-aminobutyric acid (GABA) receptors responsible for chloride-ion neurotransmission in the nervous system, specific for cyclodienes [87, 88]. Mutations have been observed in neuronal enzymes and receptors, leading to well- defined target site alteration and resistance to chemical insecticides [89]. Knock down resistance (Kdr) occurs due to a single or multiple substitutions/mutations in the para-gated sodium channel gene [29]. In An. gambiae, two mutations (Leu-Phe [90] and Leu-Ser [91]) have been identified at the same codon. As this is the target site of DDT and pyrethroids, this mechanism produces cross-resistance to the two insecticide classes. Organophosphate and carbamate insecticides inhibit acetylcholinesterase (AChE). Many insect vectors have developed resistance through structural alterations of this target site [58]. These point mutations may act individually or in combination.

Metabolic resistance, on the other hand, usually involves over-expression/over-production of a complex array of specific enzymes capable of detoxifying insecticides or modifications in the amino acid sequences that cause alterations in the levels and activity of detoxifying proteins [58, 90]. The mechanism of increased detoxification contributes to a decrease in the effective dose of insecticides available at the target site [67]. The overproduction of these endogenous detoxifying enzymes may be achieved via two nonexclusive mechanisms: 1) increase in the gene's copy number of available molecules (by gene amplification or expression activation); or 2) mutation in the enzyme coding portion of the gene, so that its product metabolises the insecticide more efficiently [92], preventing it from reaching its target in the nervous system. Metabolic resistance occurs through increased biodegradation of the insecticide, usually through overproduction of detoxification enzymes such as P450s, GST, and carboxylesterases (EST). The P450 cytochromes primarily metabolises pyrethoids and to a lesser extent, of carbamates and organophosphates, carboxylesterases largely detoxify organophosphate and carbamate and to a lesser extent in pyrethroid resistance [93]. GSTs are involved in the detoxification of a wide range of xenobiotics, including the organochloride insecticide DDT [94]. In A. gambiae, metabolic resistance to insecticides can be conferred by elevation in the activity of these three classes of detoxifying enzymes. In contrast, there are few examples in literature regarding insect behavioural changes and tegument alterations.

Behavioural resistance involves behaviour changes in response to prolonged exposure to an insecticide resulting in avoidance and reduced contact with lethal doses of an insecticide [95].

Behavioural resistance does not have the same "importance" as physiological resistance but may be considered to be a contributing factor [96, 97]. It remains unclear whether adaptation of malaria vectors species to insecticidal-based vector control interventions may result from a phenotypic plasticity or from selected behavioural traits [98]. Notably, behavioural resistance is characteristically difficult to quantify [99].

Penetration resistance: Reduced penetration involves changes that decrease the rate of penetration or absorption of insecticide through the insect cuticle or digestive tract linings and confers low levels of resistance [68, 100]. This resistance mechanism is not specific and can affect a broad range of insecticides. Reduced uptake of insecticide, often referred to as cuticular resistance, is frequently described as a minor resistance mechanism. More effort is required to identify the significance of cuticular resistance in phenotypic resistance [86].

The understanding of the development of resistance and the design of novel strategies to manage it and to effectively control disease vectors is greatly owed to the characterisation of genes and the molecular mechanisms involved in insecticide resistance [101]. However, the mechanisms of insecticide resistance are generally far less well-understood. Particularly, the contribution these enzymes make towards pyrethroid resistance and their biochemical relationships with P450-mediated resistance is still unclear [84].

8. DISTRIBUTION OF INSECTICIDE RESISTANCE

The emergence and spread of insecticide resistance to all four classes of insecticides useful in public health invariably threatens the effectiveness of malaria vector control as most programmes rely heavily on insecticide usage [60]. Resistance has been observed in more than 500 insect species worldwide, among which over 50 Anopheles species (Diptera: Culicidae) are responsible for the transmission of malaria parasites to humans [13, 58]. Globally, resistance to at least one insecticide has been identified in 64 countries with on-going malaria transmission [7]. Currently, 27 countries in sub-Saharan Africa have reported pyrethroid resistance in Anopheles vectors [102]. The real figure could very well be higher, as a lack of in-country resistance monitoring prevents accurate assessment [60]. Insecticide resistance is a focal phenomenon and as such is not evenly distributed among vector species and varies markedly from one place to the other. Several platforms are available online with vast information on the distribution of insecticide resistance in malaria vectors such as: Anobase (http:// anobase.vectorbase.org/

ir/), Arthropod Pesticide Resistance Database (http://www.pesticideresistance.org), MARA (http://www.mara.org.za), and IR mapper (http://www.irmapper.com). Persuasive evidence for the presence of resistance in primary vector species to all available insecticides has been presented from Africa, Southeast Asia and India, and South and Central America. Corbel and N'Guessan present a detailed description of the country by country situation analysis of resistance in these regions [86]. A summary is presented below:

In Africa, target-site and metabolic-mediated resistance has been detected in An. gambiae s.l. malaria vectors across the continent south of the Sahara. Most of the documented evidence comes from west Africa where pyrethroid resistance is predominant in An. gambiae s.s. compared to An. arabiensis [102, 103]. High levels of resistance have also been extensively reported in the two major vectors in Central, East, Austral, and South African countries [104, 105]. Two kdr alleles exist in An. gambiae s.s. and An. arabiensis, the L1014S gene originally from East Africa [90] and the 1014F gene of west African origin alleles [91]. Both mutations have also been reported to co-exist in Gabon and Cameroon [106] and in Uganda [107]. The western kdr was also detected in Kenya [108], Tanzania [109], and also in Zambia [110, 111]. In pyrethroid-resistant An. gambiae s.l. metabolic resistance involving increased levels of P450 has been reported in several countries with CYP6P3 and CYP6M2 genes over-expressed [112, 113]. Most data on resistance in An. funestus comes from South African countries where the species seems to be the predominant malaria vector [101]. In An. funestus, pyrethroid resistance involving increased activity of P450 monooxygenase and/or GST was demonstrated in Southern Africa [114–116] with over-expressed CYP6M7, CYP6P9a, and CYP6P9b genes [117]. Pyrethroid resistance in An. funestus has also been detected in East [118] and West Africa [119, 120]. Recently, the spatial scale of the problem in sub-Saharan Africa was brought to the fore through the IRMapper [105].

Southeast Asia and India insecticide resistance has been detected in the main malaria vector species [87]. In the Mekong region, Anopheles dirus s.s. [121] and Anopheles minimus s.l. [122]. No kdr mutation has been observed so far in these species [123] and pyrethroid resistance seems to result from increased detoxification by esterases and/or P450 monooxygenases [124]. Esterase-mediated pyrethroid detoxification in both An. epiroticus and An. subpictus and GST- mediated DDT resistance in An. subpictus have been reported [121]. An. vagus and An sinensis are resistant to pyrethroids with high 1014S kdr alleles [124–127]. The presence of the 1014F allele has been revealed in An. sundaicus, An. aconitus, An. subpictus, and An. vagus [128] (http:// www.itg.be/malvecasia/). In India, An. culicifacies s.l. has developed strong resistance to pyrethroids [129], DDT [130, 131], dieldrin/HCH [132], and malathion [131]. Both 1014F and 1014S kdr phenotypes have been detected in pyrethroid and DDT-resistant An. culicifacies s.l. [130] and An. stephensi [134] with elevated activities of GST in DDT resistance in this mosquito species [135]. An. annularis, An. subpictus, and An. philippinensis are resistant to pyrethroid, DDT, and/or dieldrin/HCH [135]. In Sri Lanka, metabolic resistance involving carboxylesterases (malathion) or monooxygenases and GSTs (DDT) has been detected in An. culicifacies s.l. and An. subpictus [136, 137]. In Bangladesh, An. philippinensis, An. maculatus s.l., and An. aconitus have all developed resistance to DDT [138]. An. stephensi and An. sacharovi in Iran and Turkey are resistant to DDT and dieldrin [139–141]. While An. maculatus s.l. and An. Aconitus have developed resistance to DDT in Nepal, An. stephensi is resistant to malathion in Pakistan [142].

In Central and South America, the primary malaria vectors are An. darlingi and An. albimanus. In Mexico, An. albimanus exhibits high levels of DDT and pyrethroid-resistance with elevated levels of GST, P450, and esterases, and iAChE-mediated carbamate and organophosphate resistance [143, 144]. In Peru, An. Albimanus is resistant to pyrethroids [145]. In Colombia, DDT resistance has been reported in An. darlingi [146, 147] and pyrethroid resistance in both An. darlingi and An. albimanus [148]. In An. darlingi, both multi-function oxidase (MFO)- and non-specific esterase (NSE)-based metabolic resistance were reported in a deltamethrin and DDT-resistant population [149]. An. Nuneztovari, a secondary malaria vector, is resistant to organophosphate and pyrethroids [150].

9. RESISTANCE AND VECTOR CONTROL

Insecticide resistance has been perceived to have the potential to undermine efforts to control vector-borne diseases including malaria [151, 152]. However, the impact of resistance on the ability of malaria control intervention to reduce disease transmission is poorly understood [153]. Insecticide resistance triggers a chain reaction that through deteriorated efficacy leads to vector control failure and disease control failure may be expected [153]. Evidence linking the potential of ITNs increasing phenotypic resistance and kdr frequency [154, 155] that threaten to compromise their effectiveness exists [86]. However, whether these various forms of resistance have an impact on the effectiveness of ITNs in malaria control remains a topic of debate among policy makers and researchers [60]. Conclusive evidence of insecticide resistance impacting on the efficacy of vector control interventions in decreasing disease transmission is by large still absent. However, minimal evidence of an effect of resistance on entomological indicators having an impact on disease transmission exist [60]. The number of studies aimed at evaluating the operational significance of insecticide resistance on epidemiological outcomes of malaria remains nominal. This could be ascribed to multiple confounding factors capable of complicating the interpretation of data. The most available evidence is laboratory or experimental huts-based and harnessing entomological outcomes to assess the impact resistance on mosquito biting rates, blood feeding rates, or insect mortality [153]. Conflicting findings on the impact of resistance on vectorial capacity has been reported with some results indicating an increasing effect [156–160] while others present decreasing outcomes [159, 160].

Mostly, the impact of pyrethroid resistance is not clearly observable in entomological and epidemiological terms. For instance, in areas with detected kdr resistance the distribution of LLINs has been shown to successfully reduce malaria transmission [161]. Insecticide resistance has only been directly implicated in operational control failure of pyrethroids in An. funestus in South Africa [162]. In 1996, pyrethroid resistance compromised malaria control in KwaZulu Natal following a switch of IRS insecticides from using DDT to deltamethrin [163]. The re- introduction of IRS with DDT controlled the pyrethroid resistant An. funestus population and malaria cases dropped by 91% [164]. In Bioko Island, IRS with pyrethroid had no impact on kdr-mediated resistant An. gambiae population, but had significant impact on transmission index and malaria cases [39, 42]. After switching to IRS with a carbamate, the mosquito population declined [42]. In Burundi, programmatic IRS with pyrethroids and ITNs markedly reduced Anopheles density by 82% and transmission intensity by 90% and occurrence of clinical episodes by 43% in children despite high kdr frequencies in An. gambiae s.s. [40, 165]. In Côte d'Ivoire, ITN-randomised controlled trials demonstrated a significant reduction on the entomological inoculation rate (55%) [166] and on malaria incidence in children <5 (56%) [167] despite the presence of kdr-based pyrethroid resistance.

The current information gathered across Africa indicates that there is rapid loss of efficacy of most pyrethroids against malarial vectors [109, 168]. In Malawi, pyrethroid resistance did not trigger an operationally significant epidemiological impact on malaria parasite prevalence in children [114]. To compromise insecticide vector control, the level of resistance must be high enough to adversely affect disease transmission [169]. Despite the observed decline of vector abundance after the use of the pyrethroid derivatives [170, 171], the reported loss of efficacy of these widely used insecticides should be taken as a major threat for potential resurgence of malarial transmission in areas where gains have already been achieved against malaria vectors [172]. In many cases, vector control may not be affected by the level of resistance but enhanced surveillance and monitoring would be required [56]. This has refocused attention on the production of chemicals that are efficient and cost-effective [161]. The impact of resistance on the ability of the vector to transmit malaria is underexplored due to the scanty published literature available. However, most studies use kdr alleles frequency as a proxy for resistance due to the lack of molecular markers for alternative resistance mechanisms. This can be misleading if metabolic or other resistance mechanisms are the predominant drivers of the phenomenon. There is a need for additional attention to investigate on evolution and development of resistance to insecticides by disease vectors and consequently the epidemiological impacts of malaria and other vector-borne diseases [153].

10. RESISTANCE MANAGEMENT STRATEGIES

The long-term control of vectors is threatened by insecticide resistance, which is occurring at a faster pace than new insecticides are being developed. Pre-emptive action to mitigate the development and spread of insecticide resistance is critical in preserving the limited arsenal of insecticides available for public health [143]. With only four classes of insecticides currently recommended for vector control, implementation of effective resistance management strategies remains inevitable [29]. Resistance management can be defined as "the containment of the frequency of resistance genes below an acceptable threshold by means of strategic choices of insecticide, dosage, mode of application, or frequency of use"[173]. The mutant alleles that confer insecticide resistance are generated by random events. In the absence of insecticide selection pressure, resistance management strategies take advantage of the adverse fitness costs of resistance genes, to the insects carrying them. Though generally selected against in the absence of selection pressure, alleles with strong pleiotropic effects increase in frequency when insecticide selection pressure is applied. However, the outcome of resistance management strategies can be affected by dominance status of the trait [29, 143].

Resistance management entails the development and implementation of control interventions that minimise the likelihood that vectors will evolve strong resistance to important insecticides [169]. The aim is to prevent or delay the onset of resistance in populations exposed to an insecticide, or develop management programmes that cause existing resistance in populations to decline, through rotating or alternating insecticides as a resistance management strategy before resistance reaches measurable levels [174]. The use of combined classes of insecticides, rotations of insecticides, or mosaic design has shown to overcome resistance problems effectively than using a single class of insecticide [175, 176]. Temporal rotation over time of two, or preferably more, insecticide classes with different modes of action applied in an alternating sequence is also based on the assumption that an individual mosquito does not carry two resistant alleles [177]. Rotations are particularly effective if the resistance gene has an associated fitness cost [87] and assumes that if resistance to each insecticide is rare, then multiple resistances will be extremely rare [178]. The "mosaic" approach refers to applications of different compounds against the same insect in spatially segregated locations [55] and aim to preserve susceptibility by spatial restriction of insecticides [7]. Larger scale mosaics have been shown to be effective for the management of pyrethroid resistance in An. albimanus in Mexico [143, 179]. An alternative is simultaneous utilisation of a mixture of two or more insecticides of unrelated mode of action, the aim being that resistance will evolve more slowly to both insecticides than if either had been used on its own [180]. Unlike rotations, the effectiveness of mixtures is not directly related to the degree of fitness cost. Mixtures of insecticides require the expected frequency of resistant alleles at two different genetic loci to be low and that individual mosquitoes carrying both alleles are rare [181]. The other approach is through combinations of two vector control tools, such that a mosquito that survives contact with one (e.g., LLIN) is exposed to the other one (e.g., IRS) [182]. The success of combinations in effectively managing resistance depends on the ability to kill the vector despite the existence of resistance by using another intervention or insecticide [183]. However, caution should be exercised not to increase selection pressure by combining insecticides with same mode of action (e.g., avoid pyrethroids for both IRS and LLINs) [86].

Ideally, insecticide resistance management should be undertaken using insecticide-based approaches in conjunction with other non-insecticidal vector control methods, in the context of IVM [3]. However, resistance surveillance is a fundamental step and insecticide susceptibility an indispensable resource of resistance management; it provides baseline data for program planning and insecticide selection before the commencement of control operations, facilitates detection of resistance at an early stage so that timely management can be implemented, and enables continuous monitoring of the effect of control strategies on resistance. Establishing international, multi-disciplinary technical working groups with a clear reporting system and defined responsibilities to facilitate data collation and rational policy transformation is critical for optimal IRM strategies. This would require the presence of a multiplicity of partners with vested interest in insecticide resistance, demand close collaboration and sustained coordination of local and external technical experts, and require good stewardship for them to succeed. Availability of entomological resources provides an ideal opportunity to develop a rational IRM plan underpinned by entomological and epidemiological baseline data to facilitate tracking of spatial and temporal resistance profiles of malaria vectors and evaluating its impact on the efficacy of control interventions. There remains a paucity of evidence on the utility of conventional resistance management strategies (e.g., insecticide rotations, mosaics, mixtures, and combinations) in restoring the susceptibility of malaria vectors. There is also a need for well-designed assessments of the operational impact of combinations of insecticidal and non-insecticidal interventions, including larval source management approaches [87].

11. OPERATIONAL CHALLENGES OF INSECTICIDE RESISTANCE MANAGEMENT

The WHO has developed the GPIRM to help member states mitigate the development and spread of resistance [7]. However, countries continue to experience substantial constraints for effective deployment. First, there is limited country-level technical resource capacity to support entomological intervention monitoring and evaluation, minimal essential physical infrastructure and logistical resources to support implementation of the plan, including insufficient qualified vector control workforce. Second, gaps in availability of reliable routine monitoring data on vector bionomics, spatial distribution, insecticide resistance, underlying resistance mechanisms, including operational cost of insecticide resistance from epidemiologically representative sites, makes decision-making on resistance management difficult. Third, deficiency in local financial support and sustainability that is threatened by donor dependency. Fourth, timely scale up has been constrained by paucity in coordinating in-country entomological resources, coupled with scepticism surrounding scientific findings by some key national and international implementing and funding organisations. Fifth, skilled international technical assistance is a scarce resource that is overstretched. Sixth, there is limited data on malaria transmission and its correlation to epidemiological indices to guide the targeting of tools and monitoring of their impact. Seventh, poor data quality, management, and willingness to share data by different partners is usually nominal and remain a challenge to documenting insecticide resistance. Generally, there are limited resources and both human and institutional capacity to fill these gaps. However, the potential of IVM provides a window of opportunity that could be exploited for enhanced IRM activities.

Eight countries, Equatorial Guinea, Eritrea, Mozambique, Namibia, Rwanda, South Africa, United Republic of Tanzania, and Zambia, currently have plans of implementing the GPIRM, representing less than 10% of countries that need them. These plans are mainly reactive rather than proactive. Two examples of countries with well-developed plans are Bioko, Equatorial Guinea and Zambia [184]. Despite having good plans, the operational implementation of these plans remains challenging. In Bioko, large-scale LLIN distribution and island-wide pyrethroid-based IRS were conducted before a switch to bendiocarb IRS for eight years, after the detection of kdr-based pyrethroid resistance. Despite kdr, there is evidence that pyrethroids remain operationally effective. Therefore a bendiocarb-deltamethrin annual rotation has been implemented. Pirimiphos-methyl remains a reserve option should this rotation fail, but was considered too expensive to include initially despite the greater treatment longevity. In Zambia, two major vectors (Anopheles funestus and An. gambiae sensus stricto) are resistant to carbamates and pyrethroids and pyrethroids alone, respectively. A mosaic pattern of insecticide use, driven by the prevalence of the different vectors has therefore been implemented. However, due to the increased cost, coverage has been reduced in a format that may adversely impact disease transmission. Widespread pyrethroid resistance is now a major problem. Getting new active ingredients to market quickly is imperative; large-scale randomised control trials over many years to document efficacy may be unrealistic given the urgency [184].

12. POLICY IMPLICATIONS OF INSECTICIDE RESISTANCE

To maintain the effectiveness of vector control, countries are encouraged to deploy tools within the context of IVM [3] and to pre-emptively implement suitable IRM strategies against malaria vectors [7]. To help control programmes re-orient to IVM and IRM, strategic direction and technical assistance have been provided for the two approaches. WHO guidance on IVM includes: the Global Strategic Framework for IVM (2004) [3], the Report of the WHO Consultation on Development of a Global Action Plan for IVM [4], Guidance on Policy Making for IVM [185], Core Structure for Training Curricula on IVM [186], Handbook for IVM [187], and Monitoring and Evaluation Indicators for IVM [188]. Yet, only 62% of 113 endemic countries globally and 53% of countries in Africa have national IVM policies and implemented the strategy [184]. Moreover, resistance to at least one insecticide has been identified in 64 countries with on-going malaria transmission [7]. The threat posed by insecticide resistance is highlighted in the GPIRM consisting of five key pillars including: 1) planning and implementation of IRM strategies in malaria-endemic countries; 2) ascertaining proper, timely entomological and resistance monitoring coupled with effective data management; 3) the development of new, innovative vector control tools; 4) filling of gaps in knowledge on mechanisms of insecticide resistance and the impact of current IRM approaches; and 5) making available enabling mechanisms such as advocacy and human and financial resources [7]. The current monitoring of insecticide resistance is inadequate and inconsistent in most settings in which vector control interventions are used. Often, monitoring is performed reactively or ad hoc, depending on local research projects being conducted [7].

With the view to operationalise the GPIRM and optimise resistance monitoring and management, the WHO has developed a framework document for countries to use as a template for their insecticide resistance monitoring and management plans [189]. However, very few countries have established rational IRM strategies and incorporated them into operational IVM-based vector control programmes. Notably, an emergency approach needs to be adopted for IRM with continued advocacy for the GPIRM, similar to that given to Artemisinin resistance management plans is essential. Incorporating other vector-borne disease (i.e., dengue, leishmaniasis, etc.) in the GPIRM and emphasising biological agents, housing improvement, and larviciding as IRM tools is crucial. For example, larviciding uses differ ent classes of chemical insecticides and biological agents with different modes of action to the four classes available for adult vector control and can reduce overall density [184]. The current areas of focus within IVM include: redesigning programs in the context of insecticide resistance response and climate change; reorientation of programs with capacity building and career pathways; encouraging intersectoral work; and IVM in emergency situations [184]. The WHO should address resistance and entomological capacity challenges via support to countries for developing IRM plans, the inclusion of additional mechanism data in the global database, bi-regional training, the development of a global insecticide resistance response plan, and advocacy for action and resource mobilisation. In attempting to control and contain the spread of insecticide resistance, multi-country cross-border reporting systems and proactive planning is also crucial to preserve new tools and should be considered to inform policy at this level, especially in light of the malaria elimination efforts that many countries have embarked upon [168].

13. USING IVM FOR OPTIMAL IRM IMPLEMENTATION

Given the backdrop of escalating resistance and limited vector control tools, as well as global finances that continue to fall short of estimated requirements for malaria control and elimination [2] and restricted entomological capacity [190], there has been some progress in the implementation of the GPIRM [191]. A successful IVM programme includes actions along five key strategic elements that can be harnessed for addressing the pillars of the GPIRM pertinent to country-level strategic planning and implementation. First, Advocacy, social mobilisation, and legislation: to strengthen national insecticide legislation and regulatory mechanisms for their safe and judicious use; ensure insecticide resistance advocacy and communications to effectively target policy makers, implementers, communities, and other stakeholders. Second, Collaboration within the health sector and partners: to establish technical support linkages with insecticide manufacturers and distributors for joint entomological monitoring, insecticide selection, and resistance management; establish partnerships with the ministry of agriculture and ministry of environment for supervision and pesticide management. Third, Capacity building: to identify competencies and staffing levels essential for effective IRM; strengthen human resource capacity through training for entomological resistance monitoring; establish requisite infrastructure including insectaries, entomology labs; establish vector control data management systems. Fourth, Evidence-based decision-making: clarify information needs and data collection methods; establish entomological and epidemiological monitoring plans to help target and evaluate interventions; select insecticides based on local data regarding vector susceptibility and transmission ecology, ensure insecticide selection is based on an IRM plan as outlined in the GPIRM; ensure vector control and vector data collection are completed in a timely and rigorous manner; manage and utilise evidence for decisions and strategy refinement, including annual reassessment. Fifth, Integrated approach: ensure there is adequate, evidence-based guidance on the impact of resistance on malaria vector control interventions; evaluate whether agricultural use and other vector-borne diseases have an impact on resistance; explore additional non-insecticide complementary malaria vector control measures where they may be appropriate [3, 4].

14. CONCISE SUMMARY

The formulation and execution of national Insecticide Resistance Monitoring and Management Plans for malaria control is essential for operationalizing the GPIRM. IVM can be utilized as a framework for strategic IRM planning. Consequently, sensible Integrated Resistance Management techniques should be an essential component of Integrated Vector Management-based malaria vector control programs. Nonetheless, a substantial coordinated response among stakeholders and political commitment is essential for the prompt and successful execution of policies within the framework of a national health system.

REFERENCES

1. Van den Berg H, Takken W. A framework for decision-making in integrated vector management to prevent disease. Trop Med Int Health. 2007; 12: 1230–1238.
2. WHO. World malaria report. Geneva: World Health Organization; 2014. Available from: http://www.who.int/malaria/publications/world_malaria_report_2014/ report/en/ [Accessed on 5 April 2015].
3. WHO. Global strategic framework for integrated vector anagement. WHO/CDS/CPE/PVC/2004.10. Geneva: World Health Organization; 2004. http://whqlibdoc.who.int/hq/2004/WHO_CDS_CPE_PVC_2004_10.pdf.
4. WHO. Development of a global action plan for integrated vector management (IVM). Report of a WHO Consultation. Geneva: World Health Organization; 2009. http:// whqlibdoc.who.int/hq/2009/WHO_HTM_NTD_VEM_2009.1_eng.pdf
5. Coleman M, Casimiro S, Hemingway J, Sharp B. Operational impact of DDT reintroduction for malaria control on Anopheles arabiensis in Mozambique. J. Med. Entomol. 2008; 885–890.
6. Kelly-Hope L, Ranson H, Hemingway J. Lessons from the past: Managing insecticide resistance in malaria control and eradication programmes. Lancet Infect Dis. 2008; 8: 387–389.

7. WHO. Global plan for insecticide resistance management in malaria vectors (GPIRM). WHO Global Malaria Programme. [Accessed on 5 April 2015]. http://www.who.int/malaria/vector_control/ivm/gpirm/en/index.html.
8. PubMed. www.ncbi.nlm.nih.gov/pubmed.
9. WHO. http://www.who.int/.
10. Google Scholar. http://www.google.com.
11. Armed forces pest management board (US). [http://www.afpmb.org/].
12. Darsie R, Ward R. Systematics. In Darsie R, Ward R, eds., Identification and Geographical Distribution of the Mosquitoes of North America, north of Mexico, 2nd ed. 1-6. 2005. University Press of Florida, Gainesville.
13. Foster WA, Walker ED. Mosquitoes (Culicidae). In Mullen G, Durden L, eds., Medical and Veterinary Entomology. 2002. Academic Press, San Deigo. Hackett LW. Malaria in Europe. 1937. Oxford, Oxford University Press.
14. Bruce-Chwatt. Essential Malariology. 2nd ed. John Willey and Sons. New York. 1985.166-179.
15. Harbach RE. Mosquito Taxonomic Inventory. http://mosquito-taxonomic-inventory.info/ [Accessed on 11 July 2015].
16. Harbach RE. The Phylogeny and Classification of Anopheles. In Anopheles mosquitoes - New insights into malaria vectors. Published by InTech. Edited by Sylvie Manguin. 2013. pp 3–55.
17. Lindsay SW, Parson L, Thomas CJ. Mapping the ranges and relative abundance of the two principal African malaria vectors, Anopheles gambiae sensu stricto and An. arabiensis, using climate data. Proceedings of the Royal Society of London. Series B. Biological Sciences. 1998; 265: 847–854.
18. Gillies MT, De Meillon BA. The Anophelinae of Africa south of the Sahara (Ethiopian Zoogeographical Region). 2nd ed. South African Institute for Medical Research, Johannesburg. Publication of the South African Institute for Medical Research. 1968; 54: 131– 132.
19. White GB. Anopheles gambiae complex and disease transmission in Africa. Trans. R. Soc. Trop. Med. Hyg. 1974; 68(4): 278–301.
20. Krafsur ES. Anopheles nili as a vector of malaria in a lowland region of Ethiopia. Bull. World Health Organ. 1970; 42(3): 466–471.
21. Gillies MT, Coetzee M. A supplement to the Anophelinae of Africa South of the Sahara (Afro-tropical region). Publication of the South African Institute for Medical Research, Johannesburg. 1987; 55: 78–143.

22. Macdonald G. The epidemiology and control of malaria. Oxford University Press, London. 1957.

23. N'guessan R, Corbel V, Akogbeto M, Rowland M. Reduced efficacy of insecticide-treated nets and indoor residual spraying for malaria control in pyrethroid resistance area, Benin. Emerg. Infect. Dis. 2007; 13: 199–206.

24. Najera JA. 1999. Malaria control: Achievements, problems and Startegies. WHO/MAL/99.1087. World Health Organization.

25. Gravitz L. The last bite: Preventing mosquitoes from transmitting the malaria parasite is a crucial piece of the eradication puzzle. Nature. 2012; 484: S26–S27.

26. WHO. Insecticides. 7th report of the expert committee on insecticides. WHO Technical Report Series 125. Geneva: World Health Organization; 1957.

27. Raghavendra K, Barik TK, Sharma P, Bhatt RM, Srivastava HC, Sreehari U, Dash AP. Chlorfenapyr: A new insecticide with novel mode of action can control pyrethroid resistant malaria vectors. Malar J. 2011; 10: 16.

28. Coleman M, Hemingway J. Insecticide resistance monitoring and evaluation in disease transmitting mosquitoes. J. Pestic. Sci. 2007; 32: 69–76.

29. van den Berg H. Global status of DDT and its alternatives for use in vector control to prevent disease. Environ Health Perspect. 2009; 117: 1656–1663.

30. WHO. Position statement on integrated vector management. Geneva: World Health Organization; 2008. http://whqlibdoc.who.int/hq/2008/ WHO_HTM_NTD_VEM_2008.2_eng.pdf[Accessed on 5 April 2015].

31. Pluess B, Tanser FC, Lengeler C, Sharp BL. Indoor residual spraying for preventing malaria. Cochrane. Database. Syst. Rev. 2010; 4: CD006657.

32. Lindsay SW, Snow RW, Broomfield GL, Janneh MS, Wirtz RA, Greenwood BM. Impact of permethrin-treated bednets on malaria transmission by the Anopheles gamiae complex in The Gambia. Med Vet Entomol. 1989; 3(3): 263–271.

33. Lengeler C, Sharp B. Indoor residual spraying and insecticide-treated nets, In Reducing Malaria's Burden: Evidence of Effectiveness for Decision Makers. Global Health Council Technical Report - December 2003. pp 17–24.

34. Lengeler C. Insecticide-treated bed nets and curtains for preventing malaria. Cochrane. Database. Syst. Rev. 2004; (2): CD000363.

35. Mabaso ML, Sharp B, Lengeler C. Historical review of malarial control in southern African with emphasis on the use of indoor residual house-spraying. Trop Med Int Health. 2004; 9(8): 846–856.

36. Lines J, Lengeler C, Cham K, de Savigny D, Chimumbwa J, Langi P, Carroll D, Mills A, Hanson K, Webster J, Lynch M, Addington W, Hill J, Rowland M, Worrall E, Mac Donald M, Kilian A. Scaling-up and sustaining insecticide-treated net coverage. Lancet Infect. Dis. 2003; 3(8): 465–466.

37. Curtis CF, Mnzava AE. Comparison of house spraying and insecticide-treated nets for malaria control. Bull World Health Organ. 2000; 78(12): 1389–1400.

38. Kleinschmidt I, Schwabe C, Shiva M, Segura JL, Sima V, Mabunda SJ, Coleman M. Combining indoor residual spraying and insecticide-treated net interventions. Am. J. Trop. Med. Hyg. 2009; 81(3): 519–524.

39. Kleinschmidt I, Sharp B, Benavente LE, Schwabe C, Torrez M, Kuklinski J, Morris N, Raman J, Carter J. Reduction in infection with Plasmodium falciparum one year after the introduction of malaria control interventions on Bioko Island, Equatorial Guinea. Am. J. Trop. Med. Hyg. 2006; 74(6): 972–978.

40. Protopopoff N, Van BW, Marcotty T, Van HM, Maes P, Baza D, D'Alessandro U, Coosemans M. Spatial targeted vector control is able to reduce malaria prevalence in the highlands of Burundi. Am J Trop Med Hyg. 2008; 79(1): 12–18.

41. Pages F, Orlandi-Pradines E, Corbel V. Vectors of malaria: Biology, diversity, prevention, and individual protection. Med Mal Infect. 2007; 37(3): 153–161.

42. Sharp BL, Ridl FC, Govender D, Kuklinski J, Kleinschmidt I. Malaria vector control by indoor residual insecticide spraying on the tropical island of Bioko, Equatorial Guinea. Malar. J. 2007; 6: 52.

43. Townson H, Nathan MB, Zaim M, Guillet P, Manga L, Bos R, Kindhauser M. Exploiting the potential of vector control for disease prevention. Bull. World Health Organ. 2005; 83(12): 942–947.

44. Keiser J, Singer BH, Utzinger J. Reducing the burden of malaria in different eco-epidemiological settings with environmental management: A systematic review. Lancet Infect. Dis. 2005; 5(11): 695–708.

45. Walker K, Lynch M. Contributions of Anopheles larval control to malaria suppression in tropical Africa: Review of achievements and potential. Med Vet Entomol. 2007; 21(1): 2–21.

46. Komatsu R, Korenromp EL, Low-Beer D, Watt C, Dye C, Steketee RW, Nahlen BL, Lyerla R, Garcia-Calleja JM, Cutler J, Schwartlander B. Lives saved by Global Fund- supported HIV/AIDS, tuberculosis and malaria programs: Estimation approach and results between 2003 and end-2007. BMC. Infect. Dis. 2010; 10: 109.

47. Chanda E, Masaninga F, Coleman M, Sikaala C, Katebe C, MacDonald M, Baboo KS, Govere J, Manga L. Integrated vector management: The Zambian experience. Malar J. 2008; 7: 164.

48. Beier JC, Keating J, Githure JI, Macdonald MB, Impoinvil DE, Novak RJ. Integrated vector management for malaria control. Malar J. 2008; 7 (Suppl 1): S4

49. Zlotkin E. The insect voltage-gated sodium channel as target of insecticides. Annual Review of Entomology. 1999; 44; 429–455.

50. Hemingway J, Field L, Vontas J. An overview of insecticide resistance. Science. 2002; 298(5591): 96–97.

51. Nauen R. Insecticide resistance in public health pests: A challenge for effective vector control. Public Health Journal. 2006; 18: 8–15.

52. Georghiou GP, Taylor CE. Genetic and biological influences in the evolution of insecticide resistance. Journal of Economic Entomology. 1977; 70(3): 319–323.

53. Denholm I, Rowland MW. Tactics for managing pesticide resistance in arthropods: Theory and practice. Annu Rev Entomol. 1992; 37: 91–112.

54. IRAC: Prevention and Management of Insecticide Resistance in Vectors of Public Health Importance In: Resistance Management for Sustainable Agriculture and Improved Public Health: Second Edition 2010 Insecticide Resistance Action Commitee; 2010. p 72.

55. Brogdon WG, McAllister JC. Insecticide resistance and vector control. Emerging Infectious Diseases. 1988; 4: 605–613.

56. WHO. 2015. Global technical strategy for malaria (2016–2030). http://www.who.int/ malaria/areas/global_technical_strategy/draft_strategy/en/.

57. Hemingway J, Ranson H. Insecticide resistance in insect vectors of human disease. Annual Review of Entomology. 2000; 45: 371–391.

58. WHO. Test procedures for insecticide resistance monitoring in malaria vectors, bio- efficacy and persistence of insecticide on treated surfaces. Geneva: World Health Organization. 1998.

59. Strode C, Donegan S, Garner P, Enayati AA, Hemingway J. The Impact of Pyrethroid Resistance on the Efficacy of Insecticide-Treated Bed Nets against African Anopheline Mosquitoes: Systematic Review and Meta-Analysis. PLoS Med 2014; 11(3): e100 1 619.

60. Hoskins WM, Craig R. Uses of bioassay in entomology. Annual. Rev. Entomol. 1962; 7: 437–464.

61. WHO. Criteria and meaning of tests for determining susceptibility or resistance of insects to insecticides. WHO Technical Report series. 1963; 265: 135–138.

62. WHO. Resistance of vectors of disease to pesticides. 5th report of expert committee on vector biology and control. WHO Technical Report series. 1980; 655.

63. Devine GJ, Denhom I. An unconventional use of piperonyl butoxide for managing the cotton whitefly, Bemisia tabci (Hemiptera: A byrodidae). Bulletin of Entomological Research. 1998; 88: 601–610.

64. Brogdon WG. Managing the emergence of pesticide resistance in vectors. In The resistance phenomenon in microbes and infectious disease vectors: implications for human health and strategies for containment. National Academic Press. 2003.

65. Soderlung DM, Bloomquist JR, Roush RT, Tabashnik BE. Molecular mechanisms of insecticide resistance. In: Pesticide Resistance in Arthropods. Chapman and Hall, New York and London. 1990.

66. Matsumura F. Toxicology of insecticides. Plenum Press, New York and London. 1975.

67. Scott JG, Cochran DG, Siefried BD. Insecticide toxicity, synergism, and resistance in the Germany cockroach, Blattella germanica (Dictyoptera: Blattellidae). Journal of Economic Entomology. 1990; 83; 1698–1703.

68. Hemingway J. Genetics and Biochemistry of insecticide resistance in Anophelines. Ph.D. thesis. London University. 1981.

69. Aldridge WN. The esterases: Perspectives and problems. Chemico Biological Interactions. 1993; 87: 5–13.

70. WHO. Techniques to detect insecticide resistance mechanisms (field and laboratory manual), WHO, Geneva. 2000.

71. Brogdon WG, Hobbs JH, St JY, Jacques JR, Charles LB. Microplate assay analysis of reduced fenitrothion susceptibility in Haitian Anopheles albimanus. J Am Mosq Control Assoc. 1988; 4(2): 152–158.

72. Dary O, Georghiou GP, Parsons E, Pasteur N. Microplate adaptation of Gomori's assay for quantitative determination of general esterase activity in single insects. J Econ. Entomol. 1990; 83(6): 2187–2192.

73. Brengues C, Hawkes NJ, Chandre F, McCarroll L, Duchon S, Guillet P, Manguin S, Morgan JC, Hemingway J. Pyrethroid and DDT cross-resistance in Aedes aegypti is correlated with novel mutations in the voltage-gated sodium channel gene. Med.Vet. Entomol. 2003; 17: 87–94.

74. Brogdon WG, Barber AM. Fenitrothion-deltamethrin cross-resistance confered by esterases in Guatemalan Anopheles albimanus. Pestic. Biochem. Physiol. 1990; 37: 130– 139.

75. Brogdon WG, McAllister JC, Vulule J. Heme peroxidase activity measured in single mosquitoes identifies individuals expressing the elevated oxidase mechanism for insecticide resistance. J. Am. Mosq. Control. Assoc. 1997; 13: 233–237.

76. Hemingway J, Smith C, Jayawarden KJI, Earth PRJ. Field and laboratory detection of the altered acetylcholinesterase resistance genes which confer organophosphate and carbamate resistance in mosquitoes (Diptera: Culicidae). Bulletin of Entomological Reserch. 1986; 76: 559–565.

77. French-Constant RH, Steichen JC, Shotkoski F. Polymerase chain reaction diagnostic for cyclodiene insecticide resistance in the mosquito Aedes aegypti. Med Vet Entomol. 1994; 8: 99–100.

78. Du W, Awolola TS, Howell P, Koekemoer LL, Brooke BD, Benedict MQ, Coetzee M, Zheng L. Independent mutations in the Rdl locus confer dieldrin resistance to Anopheles gambiae and An. arabiensis. Insect Mol. Biol. 2005; 14(2): 179–183.

79. Martinez-Torres D, Chandre F, Williamson MS, Darriet F, Berge JB, Devonshire AL, Guillet P, Pasteur N, Pauron D. Molecular characterization of pyrethroid knockdown resistance (kdr) in the major malaria vector Anopheles gambiae s.s. Insect Mol. Biol. 1998; 7(2): 179–184.

80. Lynd A, Ranson H, McCall PJ, Randle NP, Black WC, Walker ED, Donnelly MJ. A simplified high-throughput method for pyrethroid knock-down resistance (kdr) detection in Anopheles gambiae. Malar. J. 2005; 4(1): 16.

81. David JP, Strode C, Vontas J, Nikou D, Vaughan A, Pignatelli PM, Louis C, Hemingway J, Ranson H. The Anopheles gambiae detoxification chip: a highly specific microarray to study metabolic-based insecticide resistance in malaria vectors. Proc. Natl. Acad. Sci. U. S. A. 2005; 102(11): 4080–4084.

82. Casimiro SL, Hemingway J, Sharp BL, Coleman M. Monitoring the operational impact of insecticide usage for malaria control on Anopheles funestus from Mozambique. Malar. J. 2007; 6: 142.

83. David J-P, Ismail HM, Chandor-Proust A, Paine MJI. Role of cytochrome P450s in insecticide resistance: impact on the control of mosq uito-borne diseases and use of insecticide s on Earth. Phil Trans R Soc B. 2013; 368: 20120429.

84. Hemingway J. Taking aim at mosquitoes. Nature. 2004; 430(7002): 936.

85. Corbel V, N'Guessan R. Distribution, Mechanisms, Impact and Management of Insecticide Resistance in Malaria Vectors: A Pragmatic Review. In Anopheles mosquitoes - New insights into malaria vectors. Published by InTech. Edited by Sylvie Manguin. 2013. pp 579–633.

86. Himeidan YE, Temu EA, Kweka EJ. Insecticides for Vector-Borne Diseases: Current Use, Benefits, Hazard and Resistance, Insecticides - Advances in Integrated Pest Management, Dr. Farzana Perveen (Ed.). pp. 683–708. 2012.

87. Ranson H, Abdallah H, Badolo A, Guelbeogo WM, Kerah-Hinzoumbe C, Yangalbe- Kalnone E, Sagnon N, Simard F, Coetzee M. Insecticide resistance

in Anopheles gambiae: Data from the first year of a multi-country study highlight the extent of the problem. Malar. J. 2009; 8(1): 299.

88. Donnelly MJ, Corbel V, Weetman D, Wilding CS, Williamson MS, Black IV WC. Does kdr genotype predict insecticide-resistance phenotype in mosquitoes? Trends Parasitol. 2009; 25: 213–219.

89. Martinez-Torres D, Chandre F, Williamson MS, Darriet F, Berge JB, Devonshire AL, Guillet P, Pasteur N, Pauron D. Molecular characterization of pyrethroid knockdown resistance (kdr) in the major malaria vector Anopheles gambiae s.s. Insect Mol. Biol. 1998; 7(2): 179–184.

90. Ranson H, Vulule JM, Ang A, Hemingway J, A. C. F. H. Identification of a point mutation in the voltage-gated sodium channel gene of Kenyan Anophelse gambiae associated with resistance to DDT and pyrethroids. Insect Molecular Biology. 2000; 9: 491–497.

91. Hemingway J, Karunaratne SHPP. Mosquito carboxylesterases: A review of the molecular biology and biochemistry of a major insecticide resistance mechanism. Med. Vet. Entomol. 1998; 12: 1–12.

92. Félix R, Silveira H. The Role of Anopheles gambiae P450 Cytochrome in Insecticide Resistance and Infection, Insecticides - Pest Engineering, Dr. Farzana Perveen (Ed.), 2012. pp 503–518.

93. Enayati AA, Ranson H, Hemingway J. Mini Review: Insect glutathione transferases and Insecticide resistance. Insect Molecular Biology. 2005; 14: 3–8.

94. Sparks TC, Lockwood JA, Byford RL, Graves JB, Leonard BR. The role of behaviour in insecticide resistance. Pestic. Sci. 1989; 26: 383–399.

95. Roberts DR, Chareonviriyaphap T, Harlan HH, Hshieh P. Methods of testing and analyzing excito-repellency responses of malaria vectors to insecticides. J Am Mosq Control Assoc. 1997; 13(1): 13–17.

96. Chandre F, Darriet F, Duchon S, Finot L, Manguin S, Carnevale P, Guillet P. Modifications of pyrethroid effects associated with kdr mutation in Anopheles gambiae. Medical and Veterinary Entomology. 2000; 14(1): 81–88.

97. Durnez L, Coosemans M. Residual Transmission of Malaria: An Old Issue for New Approaches. In Anopheles mosquitoes - New insights into malaria vectors. Published by InTech. Edited by Sylvie Manguin. 2013. pp 671-704.

98. Miller TA, Salgado VL. The mode of action of pyrethroids on insects. In: The pyrethroid insecticides. Leahey JP (Ed). Taylor & Francis Publ, London & Philadelphia. 1985. pp 440.

99. Oppenoorth FJ. Biochemical and genetic in insecticide resistance. In: Comprehensive Insect Physiology Biochemistry and Pharmacology. Kerkut GA and Gilbert LI (Eds.). Pergamon Press. 1985. pp 731–773.

100. Nannan L. Pyrethroid Resistance in Insects: Genes, Mechanisms, and Regulation. Insecticides - Advances in Integrated Pest Management, 2012.

101. Ranson H, N'Guessan R, Lines J, Moiroux N, Nkuni Z, et al. Pyrethroid resistance in African anopheline mosquitoes: What are the implications for malaria control? Trends Parasitol. 2011; 27: 91–98.

102. Mitchell SN, Stevenson BJ, Muller P, Wilding CS, Egyir-Yawson A, Field SG, Hemigway J, Paine MJ, Ranson H, Donnelly MJ. Identification and validation of a gene causing cross-resistance between insecticide classes in Anopheles gambiae from Ghana. Proc Natl Acad Sci U S A. 2012; 109(16): 6147–6152.

103. Mahande AM, Dusfour I, Matias JR, Kweka EJ. Knockdown Resistance, rdl Alleles, and the Annual Entomological Inoculation Rate of Wild Mosquito Populations from Lower Moshi, Northern Tanzania. J Glob Infect Dis. 2012; 4(2): 114–119.

104. Knox TB, Juma EO, Ochomo EO, Jamet HP, Ndungo L, Chege P: An online tool for mapping insecticide resistance in major Anopheles vectors of human malaria parasites and review of resistance status for the Afrotropical region. Parasites & Vectors. 2014; 7: 76.

105. Etang J, Fondjo E, Chandre F, Morlais I, Brengues C, Nwane P, Chouaibou M, Ndjemai H, Simard F. First Report of Knockdown Mutations in the Malaria Vector Anopheles gambiae from Cameroon. American Journal of Tropical Medicine and Hygiene. 2006; 74: 795–797.

106. Verhaeghen K, Van Bortel W, Roelants P, Backeljau T, Coosemans M. Detection of the East and West African kdr Mutation in Anopheles gambiae and Anopheles arabiensis from Uganda Using a New Assay Based on FRET/ Melt Curve Analysis. Malaria Journal. 2006; 5: 16.

107. Stump AD, Atieli FK, Vulule JM, Besansky NJ. Dynamics of the Pyrethroid Knock down Resistance Allele in Western Kenyan Populations of Anopheles gambiae in Response to Insecticide-Treated Bed Net Trials. American Journal of Tropical Medicine and Hygiene. 2004; 70: 591–596.

108. Kulkarni MA, Rowland M, Alifrangis M, Mosha FW, Matowo J, et al. Occurrence of the Leucine-to-Phenylalanine Knockdown Resistance (kdr) Mutation in Anopheles arabiensis Populations in Tanzania, Detected by a Simplified High-Throughput SSOP ELISA Method. Malaria Journal. 2006; 5: 56.

109. Chanda E, Hemingway J, Kleinschmidt I, Rehman AM, Ramdeen V, Phiri FN, et al. Insecticide resistance and the future of malaria control in Zambia. PLoS One. 2011 Jan; 6(9): e24336.

110. Thomsen EK, Strode C, Hemmings K, Hughes AJ, Chanda E, Musapa M, et al. Underpinning Sustainable Vector Control through Informed Insecticide Resistance Management. PLoS One. 2014 Jan; 9(6): e99822.

111. Muller P, Donnelly MJ, Ranson H. Transcription profiling of a recently colonised pyrethroid resistant Anopheles gambiae strain from Ghana. BMC Genomics. 2007; 8: 36.

112. Muller P, Warr E, Stevenson BJ, Pignatelli PM, Morgan JC, Steven A, Yawson AE, Mitchell SN, Ranson H, Hemingway J, et al. Field-caught permethrin-resistant Anopheles gambiae overexpress CYP6P3, a P450 that metabolises pyrethroids. PLoS Genet. 2008; 4(11): e1000286.

113. Brooke BD, Kloke G, Hunt RH, Koekemoer LL, Temu EA, Taylor ME, Small G, Hemingway J, Coetzee M. Bioassay and biochemical analyses of insecticide resistance in southern African Anopheles funestus (Diptera: Culicidae). Bulletin of Entomological Research. 2001; 91(4): 265–272.

114. Amenya DA, Naguran R, Lo TC, Ranson H, Spillings BL, Wood OR, Brooke BD, Coetzee M, Koekemoer LL. Over expression of a cytochrome P450 (CYP6P9) in a major African malaria vector, Anopheles funestus, resistant to pyrethroids. Insect Mol Biol. 2008; 17(1): 19–25.

115. Wondji CS, Coleman M, Kleinschmidt I, Mzilahowa T, Irving H, Ndula M, Rehman A, Morgan J, Barnes KG, Hemingway J. Impact of pyrethroid resistance on operational malaria control in Malawi. Proc Natl Acad Sci U S A. 2012; 109(47): 19063– 19070.

116. Riveron JM, Ibrahim SS, Chanda E, Mzilahowa T, Cuamba N, Irving H, Barnes KG, Ndula M and S Wondji CS. The highly polymorphic CYP6M7 cytochrome P450 gene partners with the directionally selected CYP6P9a and CYP6P9b genes to expand the pyrethroid resistance front in the malaria vector Anopheles funestus in Africa. BMC Genomics. 2014; 15: 817.

117. Morgan JC, Irving H, O kedi LM, Steven A, Wondji CS. Pyrethroid Resistance in an Anopheles funestus Population from Uganda. PLoS ONE. 2010; 5(7): e11872.

118. Djouaka R, Irving H, Tukur Z, Wondji CS. Exploring mechanisms of multiple insecticide resistance in a population of the malaria vector Anopheles funestus in Benin. PLoS ONE. 2011; 6(11): e27760.

119. Dabire KR, Baldet T, Diabate A, Dia I, Costantini C, Cohuet A, Guiguemde TR, Fontenille D. Anopheles funestus (Diptera: Culicidae) in a humid savannah

area of western Burkina Faso: bionomics, insecticide resistance status, and role in malaria transmission. J Med Entomol. 2007; 44(6): 990–997.

120. Van Bortel W, Trung HD, Thuan le K, Sochantha T, Socheat D, Sumrandee C, Baimai V, Keokenchanh K, Samlane P, Roelants P, et al. The insecticide resistance status of malaria vectors in the Mekong region. Malar J. 2008; 7: 102.

121. Somboon P, Prapanthadara LA, Suwonkerd W. Insecticide susceptibility tests of Anopheles minimus s.l., Aedes aegypti, Aedes albopictus, and Culex quinquefasciatus in northern Thailand. Southeast Asian J Trop Med Public Health. 2003; 34(1): 87– 93.

122. Verhaeghen K, Van Bortel W, Trung HD, Sochantha T, Coosemans M. Absence of knockdown resistance suggests metabolic resistance in the main malaria vectors of the Mekong region. Malar J. 2009; 8: 84.

123. Chareonviriyaphap T, Rongnoparut P, Chantarumporn P, Bangs MJ. Biochemical detection of pyrethroid resistance mechanisms in Anopheles minimus in Thailand. J Vector Ecol. 2003; 28(1): 108–116.

124. Verhaeghen K, Van Bortel W, Trung HD, Sochantha T, Keokenchanh K, Coosemans M. Knockdown resistance in Anopheles vagus, An. sinensis, An. paraliae and An. peditaeniatus populations of the Mekong region. Parasit Vectors. 2011; 3(1): 59.

125. Kang S, Jung J, Lee S, Hwang H, Kim W. The polymorphism and the geographical distribution of the knockdown resistance (kdr) of Anopheles sinensis in the Republic of Korea. Malar J. 2012; 11:151.

126. Tan WL, Wang ZM, Li CX, Chu HL, Xu Y, Dong YD, Wang ZC, Chen DY, Liu H, Liu DP, et al. First report on co-occurrence knockdown resistance mutations and susceptibility to beta-cypermethrin in Anopheles sinensis from Jiangsu Province, China. PLoS ONE. 2012; 7(1): e29242.

127. Syafruddin D, Hidayati AP, Asih PB, Hawley WA, Sukowati S, Lobo NF. Detection of 1014F kdr mutation in four major Anopheline malaria vectors in Indonesia. Malar J. 2010; 9: 315.

128. Raghavendra K, Verma V, Srivastava HC, Gunasekaran K, Sreehari U, Dash AP. Persistence of DDT, malathion & deltamethrin resistance in Anopheles culicifacies after their sequential withdrawal from indoor residual spraying in Surat district, India. Indian J Med Res. 2010; 132: 260–264.

129. Singh OP, Dykes CL, Das MK, Pradhan S, Bhatt RM, Agrawal OP, Adak T. Presence of two alternative kdr-like mutations, L1014F and L1014S, and a novel mutation, V1010L, in the voltage gated Na+ channel of Anopheles culicifacies from Orissa, India. Malar J. 2010; 9: 146.

130. Mishra AK, Chand SK, Barik TK, Dua VK, Raghavendra K. Insecticide resistance status in Anopheles culicifacies in Madhya Pradesh, central India. J Vector Borne Dis. 2012; 49(1): 39–41.

131. Sharma SN, Shukla RP, Raghavendra K. Susceptibility status of An. fluviatilis and An. culicifacies to DDT, deltamethrin and lambdacyhalothrin in District Nainital, Uttar Pradesh. Indian J Malariol. 1999; 36(3-4): 90–93.

132. Singh OP, Dykes CL, Lather M, Agrawal OP, Adak T. Knockdown resistance (kdr)- like mutations in the voltage-gated sodium channel of a malaria vector Anopheles stephensi and PCR assays for their detection. Malar J. 2011; 10: 59.

133. Gunasekaran K, Muthukumaravel S, Sahu SS, Vijayakumar T, Jambulingam P. Glutathione S transferase activity in Indian vectors of malaria: A defense mechanism against DDT. J Med Entomol. 2011; 48(3): 561–569.

134. Tikar SN, Mendki MJ, Sharma AK, Sukumaran D, Veer V, Prakash S, Parashar BD. Resistance status of the malaria vector mosquitoes, Anopheles stephensi and Anopheles subpictus towards adulticides and larvicides in arid and semi-arid areas of India. J Insect Sci. 2011; 11: 85.

135. Perera MD, Hemingway J, Karunaratne SP. Multiple insecticide resistance mechanisms involving metabolic changes and insensitive target sites selected in anopheline vectors of malaria in Sri Lanka. Malar J. 2008; 7: 168.

136. Surendran SN, Jude PJ, Weerarathne TC, Parakrama Karunaratne SH, Ramasamy R. Variations in susceptibility to common insecticides and resistance mechanisms among morphologically identified sibling species of the malaria vector Anopheles subpictus in Sri Lanka. Parasit Vectors. 2012; 5: 34.

137. Mittal PK, Wijeyaratne P, Pandey S. Status of Insecticide Resistance of Malaria, Kala- azar and Japanese Encephalitis Vectors in Bangladesh, Bhutan, India and Nepal (BBIN). In. Edited by Project EH. Washington 2004.

138. Abai MR, Mehravaran A, Vatandoost H, Oshaghi MA, Javadian E, Mashayekhi M, Mosleminia A, Piyazak N, Edallat H, Mohtarami F, et al. Comparative performance of imagicides on Anopheles stephensi, main malaria vector in a malarious area, southern Iran. J Vector Borne Dis. 2008; 45(4): 307–312.

139. Lak SH, Vatandoost H, Entezarmahdi MR, Ashraf H, Abai MR, Nazari M. Monitoring of Insecticide Resistance in Anopheles sacharovi (Favre, 1903) in Borderline of Iran, Armenia, Naxcivan and Turkey, 2001. Iranian J Publ Health. 2002; 31(3–4): 96– 99.

140. Vatandoost H, Mashayekhi M, Abaie MR, Aflatoonian MR, Hanafi-Bojd AA, Sharifi

141. Monitoring of insecticides resistance in main malaria vectors in a malarious area of Kahnooj district, Kerman province, southeastern Iran. J Vector Borne Dis. 2005; 42(3): 100–108.

142. Rowland M. Location of the gene for malathion resistance in Anopheles stephensi (Diptera: Culicidae) from Pakistan. J Med Entomol. 1985; 22(4): 373–380.

143. Hemingway J, Penilla RP, Rodriguez AD, James BM, Edge W, Rogers H, Rodriguez MH. Resistance management strategies in malaria vector mosquito control. A large- scale field trial in Southern Mexico. Pesticide Science. 1997; 51(3): 375–382.

144. Dzul FA, Patricia Penilla R, Rodriguez AD. Susceptibility and insecticide resistance mechanisms in Anopheles albimanus from the southern Yucatan Peninsula, Mexico. Salud Publica Mex. 2007; 49(4): 302–311.

145. Zamora Perea E, Balta Leon R, Palomino Salcedo M, Brogdon WG, Devine GJ. Adaptation and evaluation of the bottle assay for monitoring insecticide resistance in disease vector mosquitoes in the Peruvian Amazon. Malar J. 2009; 8: 208.

146. Quinones ML, Suarez MF. Irritability to DDT of natural populations of the primary malaria vectors in Colombia. J Am Mosq Control Assoc. 1989; 5(1): 56–59.

147. Suarez MF, Quinones ML, Palacios JD, Carrillo A. First record of DDT resistance in Anopheles darlingi. J Am Mosq Control Assoc. 1990; 6(1): 72–74

148. Fonseca-Gonzalez I. Estatus de la resistencia a insecticidas de los vectores primarios de malaria y dengue en Antioquia, Chocó, Norte de Santander y Putumayo, Colombia. Universidad de Antioquia, Colombia; 2008.

149. Fonseca-Gonzalez I, Quinones ML, McAllister J, Brogdon WG. Mixed-function oxidases and esterases associated with cross-resistance between DDT and lambda-cyhalothrin in Anopheles darlingi Root 1926 populations from Colombia. Mem Inst Oswaldo Cruz. 2009; 104(1): 18–26.

150. Fonseca-Gonzalez I, Cardenas R, Quinones ML, McAllister J, Brogdon WG. Pyrethroid and organophosphates resistance in Anopheles (N.) nuneztovari Gabaldon populations from malaria endemic areas in Colombia. Parasitol Res. 2009; 105(5): 1399- 1409.

151. Karunamoorthi K, Sabesan S. Insecticide Resistance in Insect Vectors of Disease with Special Reference to Mosquitoes: A Potential Threat to Global Public Health. Health Scope. 2013; 2: 4–18.

152. Kelly-Hope L, Ranson H, Hemingway J. Lessons from the Past: Managing Insecticide Resistance in Ma-laria Control and Eradication Programmes. The Lancet Infectious Disease. 2008; 8: 387–389.

153. Philbert A, Lyantagaye SL, Nkwengulila G. A Review of Agricultural Pesticides Use and the Selection for Resistance to Insecticides in Malaria Vectors. Advances in Entomology. 2014; 2: 120-128.

154. Ndiath MO, Sougoufara S, Gaye A, Mazenot C, Konate L, et al. Resistance to DDT and pyrethroids and increased kdr mutation frequency in An. gambiae after the implementation of permethrin -treated nets in Senegal. PLoS ONE. 2012; 7: e31943.

155. Norris LC, Norris DE. Insecticide resistance in Culex quinquefasciatus mosquitoes after the introduction of insecticide-treated bednets in Macha, Zambia. J Vector Ecol. 2011; 36: 411–420.

156. Kumar S, Christophides GK, Cantera R, Charles B, Han YS, et al. The Role of Reactive Oxygen Spe-cies on Plasmodium Melanotic Encapsulation in Anopheles gambiae. Proceedings of the National Academy of Sciences of the United States of America. 2003; 100: 14139–14144.

157. Vontas J, Blass C, Koutsos AC, David JP, Kafatos FC, Louis C, Hemingway J, Christophides GK, Ranson H. Gene Expression in Insecticide Resistant and Susceptible Anopheles Gambiae Strains Constitutively or after Insecticide Exposure. Insect Molecular Biology. 2005; 14: 509–521.

158. Vontas J, David JP, Nikou D, Hemingway J, Christophides GK, Louis C, Ranson H. Transcrip-tional Analysis of Insecticide Resistance in Anopheles stephensi Using Cross-Species Microarray Hybridization. Insect Molecular Biology. 2007; 16: 315–324.

159. McCarroll L, Hemingway J. Can Insecticide Resistance Status Affect Parasite Transmission in Mosquitoes? Insect Biochemistry and Molecular Biology. 2002; 32: 1345- 1351.

160. McCarroll L, Paton MG, Karunaratne SHPP, Jayasuryia HTR, Kalpage KSP, Hemingway J. Insecticides and Mosquitoborne Disease. Nature. 2000; 407: 961–962.

161. Maharaj R. 2011. Global trends in insecticide resistance and impact on disease vector control measures. Open Access Insect Physiology. 2011; 3: 27–33.

162. [162[Brooke BD, Kloke G, Hunt RH, Koekemoer LL, Temu EA, Taylor ME, Small G, Hemingway J, Coetzee M. Bioassay and biochemical analyses of insecticide resistance in southern African Anopheles funestus (Diptera: Culicidae). Bulletin of Entomological Research. 2001; 91: 265–272.

163. Maharaj R, Mthembu DJ, Sharp BL. Impact of DDT re-introduction on malaria transmission in KwaZulu-Natal. S. Afr. Med. J. 2005; 95: 871–874.

164. Sharp BL, Kleinschmidt I, Streat E, Maharaj R, Barnes KI, Durrheim DN, Ridl FC, Morris N, Seocharan I, Kunene S, et al. Seven years of regional malaria control collaboration—Mozambique, South Africa, and Swaziland. Am J Trop Med Hyg. 2007; 76(1): 42–47.

165. Protopopoff N, Verhaeghen K, Van Bortel W, Roelants P, Marcotty T, Baza D, D'Alessandro U, Coosemans M. A significant increase in kdr in Anopheles gambiae is associated with an intensive vector control intervention in Burundi highlands. Trop Med Int Health. 2008; 13(12): 1479–1487.

166. Dossou Yovo J, Henry M, Chandre F, Assi SB, Guillet P, Doannio J, Diarassouba A, Koffi A, Mouchet J, White GB, et al. Anti-malaria efficacy of lambda-cyalothrin treated bednets where Anopheles gambiae is pyrethroid resistant in Ivory Coast In: XVth Internationnal Congress for Tropical Medicine and Malaria: 20-25 August 2000; Cartagena de Indias, Colombia; 2000; 58.

167. Henry MC, Assi SB, Rogier C, Dossou-Yovo J, Chandre F, Guillet P, Carnevale P. Protective efficacy of lambda-cyhalothrin treated nets in Anopheles gambiae pyrethroid resistance areas of Cote d'Ivoire. Am J Trop Med Hyg. 2005; 73(5): 859–864.

168. Trape JF, Tall A, Diagne N, et al. Malaria morbidity and pyrethroid resistance after the introduction of insecticide-treated bednets and artemisinin-based combination therapies: A longitudinal study. Lancet Infect Dis. 2011; 11: 925–932.

169. Himeidan YE, Temu EA, Kweka EJ. Insecticides for Vector-Borne Diseases: Current Use, Benefits, Hazard and Resistance, Insecticides - Advances in Integrated Pest Management, Dr. Farzana Perveen (Ed.). 2012. pp 683–708.

170. Bayoh MN, Mathias DK, Odiere MR, et al. Anopheles gambiae: Historical population decline associated with regional distribution of insecticide-treated bed nets in western Nyanza Province, Kenya. Malar J. 2010; 9: 62.

171. Meyrowitsch DW, Pedersen EM, Alifrangis M, et al. Is the current decline in malaria burden in sub-Saharan Africa due to a decrease in vector population? Malar J. 2011; 10: 188.

172. Kweka EJ, Mazigo HD, Munga S, Magesa SM, Mboera LEG. Challenges to malaria control and success stories in Africa. Global Health Perspectives. 2013; 1: 71–79.

173. Georghiou GP. Principles of insecticide resistance management. Phytoprotection. 1994; 75: 51–59.

174. Curtis CF, Hill N, Kasim SH. Are there effective resistance management strategies for vectors of human disease? Biological Journal of the Linnean Society. 1993; 48: 3–18.

175. Guillet P, N'Guessan R, Darriet F, et al. Combined pyrethroid and carbamate 'two-in- one' treated mosquito nets: field efficacy against pyrethroid-resistant Anopheles gambiae and Culex quinquefasciatus. Med Vet Entomol. 2001; 15: 105–112.

176. Mosha FW, Lyimo IN, Oxborough RM, et al. Experimental hut evaluation of the pyrrole insecticide chlorfenapyr on bed nets for the control of Anopheles arabiensis and Culex quinquefasciatus. Trop Med Int Health. 2008; 13: 644–652.

177. Georghiou GP. Insecticide resistance and prospects for its management. Residue Reviews. 1980; 76: 131–144.

178. Curtis CF. Theoretical models of the use of insecticide mixtures for management of resistance. Bull Ent Res. 1985; 75: 259–265.

179. Penilla RP, Rodriguez AD, Hemingway J, Torres JL, Arredondo Jimenez JI, Rodriguez MH. Resistance management strategies in malaria vector mosquito control. Baseline data for a large-scale field trial against Anopheles albimanus in Mexico. Medical and Veterinary Entomology. 1998; 12(3): 217–233.

180. Mani GS. Evolution of resistance in the presence of two insecticides. Genetics. 1985; 109(4): 761–783.

181. Curtis CF, Lines JD. Impregnated fabrics against malaria mosquitoes. Parasitol Today. 1985; 1(5): 147.

182. Ngufor C, N'Guessan R, Boko P, Odjo A, Vigninou E, Asidi A, Akogbeto M, Rowland M. Combining indoor residual spraying with chlorfenapyr and long-lasting insecticidal bed nets for improved control of pyrethroid-resistant Anopheles gambiae: An experimental hut trial in Benin. Malar J. 2011; 10: 343.

183. Corbel V, Akogbeto M, Damien GB, Djenontin A, Chandre F, Rogier C, Moiroux N, Chabi J, Banganna B, Padonou GG, et al. Combination of malaria vector control interventions in pyrethroid resistance area in Benin: a cluster randomised controlled trial. Lancet Infect Dis, 2012; 12(8): 617–626.

184. RBM. Roll Back Malaria Vector Control Working Group (VCWG) 10th Annual Meeting, 28th–30th January 2015. Geneva, Switzerland.

185. WHO. Guidance on policy-making for Integrated Vector Management. Geneva: World Health Organization; 2012.

186. WHO. Core Structure for training curricula on Integrated Vector Management. Geneva: World Health Organization; 2012. http://whqlibdoc.who.int/publications/ 2012/9789241502788_eng.pdf.

187. WHO. Handbook for Integrated Vector Management WHO/HTM/NTD/VEM/ 2012.3. Geneva: World Health Organization; 2012. http://whqlibdoc.who.int/ publications/ 2012/9789241502801_eng.pdf.

188. WHO. Monitoring and Evaluation indicators for Integrated Vector Management. Geneva: World Health Organization; 2012. http://www.who. int/iris/bitstream/ 10665/76504/1/9789241504027_eng.pdf.

189. WHO. Framework for the development of national insecticide resistance monitoring and management plans. Geneva. Switzerland: World Health Organization; 2014.

190. Mnzava AP, Macdonald MB, Knox TB, Temu EA, Shiff CJ. Malaria vector control at crossroads: Public health entomology and the drive to elimination. Trans R Soc Trop Med Hyg. 2014; 108: 550–4.

191. Mnzava AP, Knox TB, Temu EA, Trett A, Fornadel C, Hemingway J, Renshaw M. Implementation of the global plan for insecticide resistance management in malaria vectors: Progress, challenges and the way forward. Malaria Journal. 2015; 14: 173.

10 Chapter

INSECTICIDE RESISTANCE IN EAST AFRICA - HISTORICAL CONTEXT, GEOGRAPHIC DISTRIBUTION, AND IMPLICATIONS FOR MALARIA VECTORS AND DISEASE MANAGEMENT

1. INTRODUCTION: AN OVERVIEW

East Africa comprises six nations: Kenya, Uganda, Ethiopia, Tanzania, Rwanda, and Burundi, all of which are susceptible to malaria transmission via established effective vectors. The primary malaria vectors in the region include Anopheles gambiae s.s., Anopheles arabiensis, and Anopheles funestus. [1–4]. These vectors breed in different habitats ranging from temporary rain pools to permanent water bodies [5–8]. Vector species distribution in East Africa are governed by several factors which include anthropogenic activities [4, 5, 7], such as development projects [9–12]. Also, climate, particularly temperature and rainfall, has been regarded as the function of habitats for vector abundance and distribution between low- and high-altitude areas [13, 14]. Human migration and movement from high land to low land have facilitated the distribution of parasites [15]. Topography has influenced the abundance and distribution of vector in all areas [16–19]. Thus, the abundance and distribution of efficient vectors have led to the wide use of control tools and intensive interventions across the sub-region. The main tools used for the control of malaria vectors are long-lasting insecticidal nets (LLINs) and indoor residual spraying (IRS) [20]. The pyrethroids are the only insecticides which have been used for treating LLINs while organochlorides, organophosphates, carbamates and pyrethroids are used for IRS [1, 21]. Currently, organochlorides (especially DDT) are banned in most of the East African countries for IRS use due to resistance developed by the major malaria vectors and environmental concern [20]. The development of resistance is influenced by many factors [22]. These include genetic factors including the number and frequency of resistance alleles in the insect population, fitness cost and relative dominance of the characters; biological factors including the insect life history parameters, the fitness of the heterozygous and homozygous resistant phenotypes and initial population size; reproductive factors including the rate of increase and fluctuations in population size; and operational factors including application methods of the insecticide and properties of an insecticide in use, previous selection with other insecticides, proportion of population exposed to selective doses, dosage of insecticide taken up by exposed insects and the life stage of the mosquito selected [22, 23].

Insecticide resistance is not new in insect vectors, and it is a genetically inherited characteristic which increases in the populations of vectors as a result of increased resistance selection pressure and also a trait capable of rapid spread. Malaria vector control in East Africa relies principally on the use of insecticides that can be applied either as an indoor residual deposit or can be used to treat mosquito nets and curtains. However, the long-term vector control program based on prolonged and frequent insecticide application faced the problem of resistance. Vector control subjects mosquito populations to selection and survival of the fittest. Evolution of insecticide resistance in an insect population arises when there is an increase in the frequency of one or more resistance genes in the population following exposure to insecticides. Attempts to kill the tolerant individuals lead to ever increasing doses and eventually resistant pest populations. This is an unavoidable constraint in the application of any novel or established class of insecticides. Malaria control operations implemented DDT from 1945 to 1948 during the Second World War to eliminate malaria. Since then, DDT has shown efficient in controlling malaria vectors; nevertheless, resistance has developed in endemic regions, notably East Africa.

2. MALARIA PREVENTION AND CONTROL STRATEGIES IN EAST AFRICA

In East Africa, several intervention strategies are set to reduce morbidity and mortality from malaria. Effective measures of malaria control have been achieved mainly through the use and high coverage of IRS and scaling up of LLINs in Tanzania [24, 25], Kenya [26], Uganda [27], Ethiopia [28] and Rwanda [29]. Community involvement has been another strategy in malaria prevention in different parts of East Africa [10, 27, 28].

In the recent past, house modification through window screen and blocking of eaves has been practiced in the prevention of malaria vectors in Tanzania [30, 31], Kenya [32] Ethiopia [33] and Uganda [34]. More innovative vector control strategies including control of resistant vector populations in the sub-region are use of entomopathogenic fungi [35, 36], larval source management [7] and vector trapping [37-39]. Plant-based derivatives have also been used in vector control in Uganda [40], Tanzania [41–43], Kenya [44, 45] and Ethiopia [46–48].

3. HISTORY, DISTRIBUTION AND CURRENT STATUS OF INSECTICIDE RESISTANCE IN EAST AFRICA

Insecticides (chemicals) which have been used for the control of vector-borne diseases and crop protection is believed to enhance the evolution of resistance in insects [49]. The intensive use of DDT in agriculture and public health programs and the introduction of pyrethroids in 1970s and its increased utilization since 1990s caused resistance to have been detected in malaria vectors from different sites in different countries of East Africa. Moreover, the long- term use of a single class of insecticide or combination of different classes of insecticides have led to the emergence of single resistance mechanism or multiple resistance mechanisms in different areas of East Africa [50]. Thus, insecticide resistance intensity and its distribution are increasing in East Africa (Figure 1).

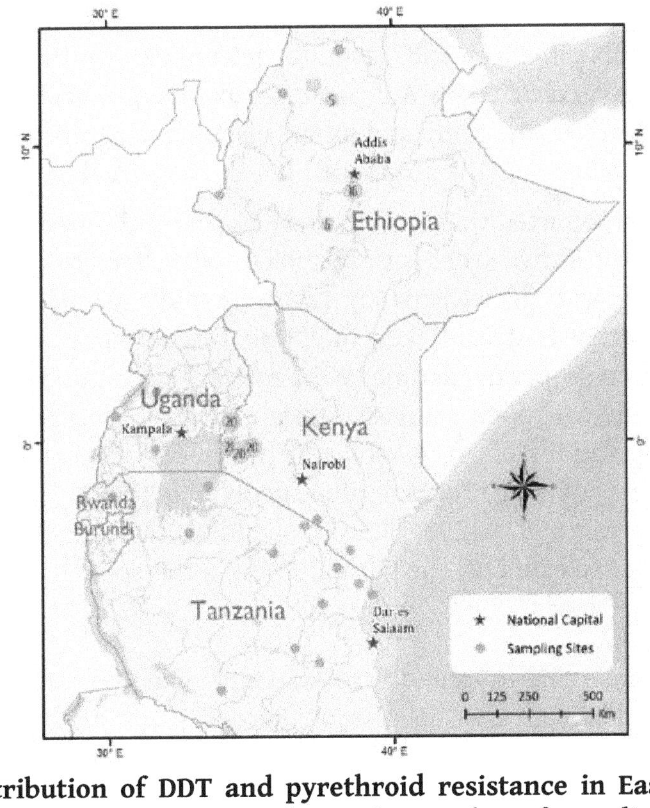

Figure 1. Distribution of DDT and pyrethroid resistance in East Africa. Note: Numbers in orange spots indicate the number of sampling sites.

3.1 Kenya

The major malaria vectors in Kenya are An. gambiae complex (An. gambiae s.s., An. arabiensis, An. merus) and An. funestus while other vector species in the country include An. melas, An. nili, An. paludis, An. pharoensis and An. coustani [20]. From peer-reviewed sources, other anopheline species documented included: An.

christyi, An. demeilloni, An. gibbinsi, An. harperi, An. implexus, An. maculipalpis, An. marshalli, An. pretoriensis, An. rufipes, An. squamosus, An. swahilicus, An. theileri, An. wilsoniand An. ziemanni, none of which are considered as important or primary vectors in Africa [51]. The malaria vector distribution in the country is not uniform due to variation in climatic factors, particularly temperature and rainfall.

In Kenya, the first reported case of resistance was in the context of insecticide-treated net use in Western Kenya where reduced knockdown rates have been observed [1]. Complete susceptibility of populations of An. arabiensis to DDT, fenitrothion, bendiocarb, lambdacyhalothrin and permethrin was documented from Mwea rice irrigation scheme, Central Kenya [52]. Widespread resistance against pyrethroids and DDT was observed across western Kenya [53]. An. gambiae s.l. showed different levels of resistance to deltamethrin, lambdacyhalothrin and bendiocarb Kilifi, Malindi and Taveta districts in coastal Kenya. Pyrethroid resistance has been reported in An. gambiae s.s and An. arabiensis from four districts of Western Kenya. Stump and others also found significant differences in kdr gene frequency between the large-scale insecticides treated nets [54].

Kamau and Vulule reported that An. gambiae s.l. and An. funestus from western, coastal and central Kenya were susceptible to DDT, fenitrothion, bendiocarb, lambdacyhalothrin and permethrin [52]. The same study also showed the presence of Leucine-Serine (East African) kdr mutation in An. gambiae s.s. of western Kenya, but the leucine–phenylalanine (West African) mutation was absent in this mosquito population. Though the East African kdr mutation was detected from west Kenyan populations of An. gambiae, it has never occurred at homozygous state. The frequency of the L1014S kdr allele doubled in the ITN test village and its nearest neighbor from 1987 to 2001, but not outside of this area. This suggests that ITN use has further selected for the kdr mutation in the population.

3.2 Uganda

The main malaria vectors in Uganda are An. gambiae and An. funestus, with An. arabiensis involved in local transmission. Recent study also showed that An. funestus and An. gambiae are the widely distributed vectors in Uganda. Other less dominant anophelines which were implicated in malaria transmission in the country include: An. coustani, An. listeri, An. marshalli and An. kingi [55, 56].

There is widespread insecticide resistance in the main malaria vectors, An. gambiae, An. funestus and An. arabiensis. In Uganda, resistance to pyrethroid insecticides has been reported in the three main malaria vectors, An. gambiae, An. arabiensis [55, 57, 58] and An. funestus [59]. A reduced susceptibility by An. gambiae s.l. to three pyrethroid insecticides, deltamethrin, cyfluthrin and cypermethrin, has been observed [60]. An.gambiae s.l. was DDT- and pyrethroid- resistant in central and eastern Uganda [58]. There are currently no reports of organophosphate resistance, but resistance to carbamates including propoxur has been documented. Mawejje and co-workers observed high pyrethroid resistance in An. gambiae and An. Arabiensis, but both species were fully susceptible to bendiocarb and fenitrothion from eastern Uganda [55]. Resistance to DDT and deltamethrin has also been reported in populations of An. funestus and An. gambiae s.l. from southwestern Uganda [56]. An. funestus in Tororo, eastern Uganda, was resistant to pyrethroids, permethrin and deltamethrin. Suspected DDT resistance was also observed in An. funestus. However, this population was completely susceptible to bendiocarb (carbamate), malathion (organophosphate) and dieldrin. Recently, widespread resistance against pyrethroids and DDT was observed across Uganda [53, 61]. Mutations which confer resistance to DDT and pyrethroids, West African (L1014F) and East African (L1014S) mutations, have been reported from the Ugandan An. gambiae. Increased esterase activity was also detected in pyrethroid- and DDT-resistant An. gambiae populations. The presence of the East African kdr mutation (L1014S) is shown for the first time in An. arabiensis from Uganda [62]. The resistance in this species was due to both target site (kdr) and metabolic mechanisms and there was also cross-resistance between DDT and pyrethroids. Resistance to pyrethroids is present, and apparently increasing, in An. arabiensis from Jinja, eastern Uganda [55], but it is not mediated by known 'knockdown resistance' target-site mechanisms (L1014F and L1014S) in the voltage-gated sodium channel, which are extremely rare in this species in this area [55]. In the absence of a known target-site mechanism, metabolic mechanisms are strongly implicated in the resistance phenotype. However, knockdown resistance mutation conferring pyrethroid/DDT resistance has also been suggested to occur in other axons of the sodium channel gene in An. gambiae. Biochemical assays suggest that resistance in this population is mediated by metabolic resistance with elevated level of GSTs, P450s and pNPA. The low frequency of L1014S and L1014F mutations and complete restoration of susceptibility to permethrin and deltamethrin by the two species after synergist assay using PBO indicate involvement of other mechanisms such as P450s in the same study. Populations of An. gambiae s.l. from eastern Uganda tested for the presence of knockdown resistance (kdr) and altered acetylcholinesterase (ace-1R) alleles showed the presence of kdr L1014S allele, while ace-1R and kdr L1014F alleles were absent [57]. All populations from the same area remain highly susceptible to carbamate, organophosphate and dieldrin insecticides. Metabolic resistance through elevated expression of cytochrome P450s has been implicated in these mosquito populations.

3.3 Ethiopia

Forty-two anopheline species have been recorded in Ethiopia [63]. There are only four anopheline mosquito species reported as malaria vectors. Anopheles arabiensis is the primary vector of malaria and it is widely distributed throughout the country [64], while An. funestus, An. pharoensis and An. nili are secondary vectors with localized distribution [65]. An. arabiensis belongs to the An. gambiae complex of sibling species. Only two member species of the An. gambiae complex, An. arabiensis and An. Amharicus (formerly known as An. quadriannulatus B), are reported to exist in Ethiopia. An. quadriannulatus species B had been described as a new species from southwestern Ethiopia [66]. This species was reported to be zoophilic and exophilic and is assumed to have no role in malaria transmission in Ethiopia [67]. Anopheles arabiensis is responsible for most of malaria infections in Ethiopia. Indoor residual spraying (IRS) and long-lasting insecticidal nets (LLINs) are pillars in malaria prevention and control strategy in Ethiopia. For over five decades, the main vector control strategy by the national malaria control program has been indoor residual spraying (IRS), using DDT with a limited application of malathion as an alternate insecticide. However, DDT use for IRS was replaced in favor of deltamethrin in 2009, and after 2 years of use deltamethrin was also replaced with bendiocarb in 2011 due to the reduced susceptibility of the principal vector to the mentioned insecticides. Insecticide susceptibility tests carried out in different parts of the country have shown different levels of resistance by the principal vector to insecticides in use for IRS and/or to treat LLINs. Insecticide resistance by An. arabiensis to DDT was reported during the early 1990s [64, 68]. Balkew and others reported resistance by An. arabiensis to permethrin and DDT [69]. Another study by Yewhalaw and his colleagues from southwestern Ethiopia indicated that An. arabiensis developed resistance to DDT, permethrin, deltamethrin and malathion. In contrast, An. arabiensis was susceptible to bendiocarb and Propoxur [70] and primiphos methyl (PMI/USID unpublished data). Abate and Haddis also reported high level of DDT and pyrethroid resistance in populations of An. gambiae s.l, presumably An. arabiensis from different parts of the country [71]. Another recent report by Massebo and others showed that populations of An. arabiensis from southwest Ethiopia developed resistance against lambda-cyhalothrin, alpha-cypermethrin, cyfluthrin, deltamethrin and DDT [33]. Moreover, high knockdown resistance mutation (West African kdr) was detected in populations of An. arabiensis from northwestern, central and southwestern Ethiopia [70, 72]. Bottle bioassay studies using synergists also revealed possible involvement of metabolic resistance in addition to kdr mutations in these populations of An. Arabiensis, which could further complicate the current malaria vector control program in the country [73]. The development of resistance by malaria vectors against insecticides used for public health could potentially jeopardize the malaria vector control strategy in Ethiopia, and hence it is imperative to monitor the level and distribution of insecticide resistance to develop new effective vector control tool and/or plan sound insecticide resistance management (IRM) strategy in the country.

3.4 Tanzania

The principal vectors of malaria in Tanzania are mosquitoes of the An. gambiae s.s, An. arabiensis and An. funestus. Other vectors which have limited role in malaria transmission include: A. merus, A. nili, A. paludis, A. pharoensis, An. coustani. An. leesoni, An. parensis, An. merus, An. marshallii and An. rivulorum [74, 75]. Recent entomological data indicate that An. funestus is prevalent on the mainland as well, particularly in the Kagera Region. Moreover, in coastal areas of north-eastern Tanzania and Zanzibar, high coverage of ITNs and IRS has resulted in a shift in the malaria vector population from An. gambiae to An. arabiensis. Resistance to pyrethroids by An. gambiae s.s. and An. arabiensis has been reported from several districts of the mainland of Tanzania [76–78]. Okumu and his colleague reported that An. arabiensis from southeastern Tanzania showed 100% susceptibility to DDT but 95.8% to deltamethrin, 90.2% to lambda cyhalothrin and 95.2% to permethrin [79]. In Zanzibar, An. arabiensis was resistance to pyrethroids (lambda-cyhalothrin, deltamethrin and permethrin), but was susceptible to carbamates (bendiocarb) and organochlorides (DDT). Moreover, in a similar study, resistance was documented in An. gambiae s.s to the same pyrethroid insecticides but was susceptible to bendiocarb, DDT and malathion [80]. In Pemba, resistance was detected in sites monitored for lambdacyhalothrin, permethrin, deltamethrin and DDT, but no resistance was detected for bendiocarb and pirimiphos-methyl CS. Similarly in Unguja, lambda-cyhalothrin resistance was detected in four of the five sites tested and permethrin resistance in one of the two sites tested. However, insecticide resistance was not detected for bendiocarb, pirimiphos-methyl CS and DDT. Anopheles gambiae s.s showed reduced susceptibility to the carbamate insecticide, bendiocarb [81]. An. arabiensis collected from Lower Moshi showed complete susceptibility to pirimiphos-methyl and malathion, but reduced susceptibility to permethrin [82, 83]. In northwestern Tanzania, there was cross-resistance between pyrethroids and DDT. In Zanzibar, resistance is not homogeneously expressed across islands, and pyrethroid resistance is stronger in Pemba than Unguja.

West African leucine phenylalanine kdr mutation was detected in two heterozygous individuals field-collected An. arabiensis from Tanzania [84]. A study also showed that a low frequency of permethrin resistance mediated by mixed function oxidases and esterases are present in An. arabiensis from Lower Moshi. The permethrin resistance is probably caused by the agricultural use of insecticides, especially in the rice fields, as permethrin-treated nets were not widely used in Lower Moshi [76]. The kdr-eastern variant was present in homozygous form in 97% of An. gambiae s.s but was absent in An. arabiensis. Synergist assays with PBO showed to restore susceptibility to pyrethroids, indicating that the resistance is in part due to an oxidase enzyme mechanism. Knockdown resistance mutation (target site insensitivity) was also detected in Pemba [84, 85].

3.5 Burundi

The primary vector of malaria in Burundi is Anopheles gambiae s.s, while secondary vectors An. funestus, An. arabiensis and An. nili. The most predominant members of vector species complex in the highlands of Burundi are An. gambiae s.s and An. funestus s.s [86, 87]. Insecticide susceptibility study in Karusi for An. gambiae s.l. showed reduced mortality to permethrin, DDT and deltamethrin. There was complete susceptibility of An. funestus to DDT and pyrethroids. A high frequency of East African kdr allele was detected in An. gambiae s.l., leading to cross resistance between DDT and permethrin in mosquito population. As there is little information on the frequency and distribution of insecticide resistance and the status of the susceptibility level of malaria vectors to insecticides used for vector control in the country, there is an urgent need for a nationwide and systematic evaluation of vector susceptibility level to current WHOPES-approved insecticides for malaria vector control, to inform ongoing interventions and control program.

3.6 Rwanda

Earlier entomological studies indicate that Anopheles gambiae s.l. and An. funestus are the main vectors responsible for malaria transmission in Rwanda. An. arabiensis is also a locally important vector of malaria. The main malaria foci are in the east and southeast areas where the altitude is generally below 1,500 m and surrounded by marshy plains.

Insecticide susceptibility studies conducted in 2012 in several sites indicated signs of resistance to DDT in some areas, possible emergence of resistance to some pyrethroid compounds, and complete susceptibility to bendiocarb and fenitrothion. A similar insecticide susceptibility study conducted by the national malaria control program in the same year showed established resistance to pyrethroids in Mimuri, a sentinel site in the country. A high frequency of the kdr gene in An. gambiae s.l. has been attributed to explain the new established resistance to pyrethroids in one district. A countrywide resistance monitoring also showed resistance to pyrethroids, DDT and bendiocarb and higher resistance was reported from eastern province, southern province and Kigali city. A continuous monitoring of resistance and resistance mechanisms is required in order to guide program for the best strategies to prevent the development and spread of resistance in the country.

4. POSSIBLE CAUSES OF INSECTICIDE RESISTANCE

Emergence of resistance in disease vectors in particular mosquitoes have been associated with different factors and sources. One of the factors is the intensive use of some classes of insecticides such as pyrethroids both in public health and in agriculture, which led to its reduced efficacy of insecticides [88–90]. Agricultural use of pesticides plays a role on the development of resistance and cross-resistance in malaria vectors has been implicated in literature. Resistance of An. arabiensis to pyrethroids in Tanzania [76] and Ethiopia [72] was attributed to use of insecticides in agriculture and livestock.

Insecticide resistance selection pressure in malaria vectors in East Africa region has also been attributed to wide coverage of LLINs and/or IRS [91] and use of agricultural pesticides [92]. As the most commonly used pesticides in agriculture and IRS are pyrethroids, organophosphate and organochlorides and for the treatments of LLINs are pyrethroids, cross-resistance is common between pyrethroids and DDT [85, 93].

5. FREQUENCY AND MECHANISMS OF RESISTANCE IN MALARIA VECTORS IN EAST AFRICA

There is variation in the frequency of resistance in malaria vectors and the mechanisms conferring resistance in different sites of different countries in East Africa (Table 1). The frequency and mechanism of resistance in insects depend on the degree of selection pressure and the mode of action of the insecticide, respectively. Insecticides target the nervous system of an insect. Organophosphate insecticides are cholinesterase inhibitors. Cyclodienes insecticides affect the chloride channel by inhibiting the gamma amino butyric acid (GABA) receptor. Pyrethroids and DDT act on the sodium channel preventing those channels from closing, resulting in continual nerve impulse transmission which eventually leads to the death of an insect [94].

Target site insensitivity is the most frequently reported mechanism conferring resistance to several insecticides used for vector control by altering the target site of the insecticides. The mode of action of each insecticide on insects is site-specific. For instance, the mode of action of organophosphate and carbamate insecticides is mainly by inhibition of the enzyme acetyl-cholinesterase (ACHE). Insects develop resistance to these insecticides through structural modification of ACHE due to large number of point mutations that occurs in gene encoding the protein for acetyl cholinesterase (ACHE), an active target site for carbamates and organophosphates which operates in the nerve cell synapses. These mutations result in altered ACHE, which reduces the sensitivity of target site to an insecticide. Another common site insensitivity mechanism is referred as knockdown resistance (kdr): insects usually get paralyzed rapidly following exposure to DDT and pyrethroids, and this is expressed as 'knockdown resistance' (kdr). However, knockdown is absent in insects exposed to DDT and pyrethroids due to mutations in the para-gated sodium channel gene, whose protein sub-units make up the voltage-sensitive sodium channels on the nerve membranes. Voltage-gated sodium channels are the target for both pyrethroid insecticides and DDT by which insecticides alter the function of the sodium channels in nerve membranes. Knockdown resistance mutation results from a single nucleotide polymorphism in the domain II, segment

6 of the sodium channel gene. Lucine (TTA) to serine (TTT) and leucine (TTT) to phenylalanine (TCA) amino acid substitutions at this position result in West and East African kdr mutations, respectively, which confer resistance to DDT and/or pyrethroids in the East African malaria vectors An.gambiae s.s and An. arabiensis [95, 96]. Another mutation of methionine to threonine, known as the super-kdr mutation, occurs between segment 4 and segment 5 of domain II of the sodium channel gene and results in a much higher resistance than kdr. The super-kdr mutation is mostly occurring together with the kdr mutation. The kdr resistance mechanism produces cross-resistance between DDT and pyrethroids and it is a genetically recessive mechanism.

Table 1. Insecticide resistance mechanisms conferring resistance to different insecticide families in the major malaria vectors and other mosquitoes in East Africa

Country	Insecticide	Mosquito species	Mechanism	Reference(s)
Kenya	DDT & pyrethroids	An. gambiae, An. arabiensis	Kdr (L1014S, L1014F)	[21, 52, 54, 104, 105, 116, 130–132]
	Pyrethroids	An. gambiae, An.arabiensis, An.funestus	Cytochrome P450s monooxygenases, esterases	[53, 61, 130, 132]
Uganda	DDT & pyrethroids	An. gambiae, An. arabiensis	Kdr (L1014S, L1014F)	[55, 57, 58, 62, 114, 133]
	DDT & pyrethroids	An. gambiae, An. arabiensis , An. funestus	cytochrome P450s, GSTs, pNPA	[53, 55, 59, 91]
Ethiopia	DDT & pyrethroids	An. arabiensis	kdr (L1014F)	[67, 70, 72, 134] [73]
Tanzania	DDT & pyrethroids	An. arabiensis	Cytochrome P450s monooxygenases	[76, 84, 92]
	DDT & pyrethroids	An. gambiae, An. arabiensis	Kdr (L1014F, L1014S), rdl	[76, 92, 135, 136]
Burundi	DDT & pyrethroids	An. arabiensis, Cx. quinquefasciatus	Mixed function oxidases, b-esterases, P450s, cuticle proteins, GABA, sulfotransferase	[133]
Rwanda	DDT & pyrethroids	An. gambiae s.l.	Kdr (L1014S)	[137]

Metabolic resistance is another important mechanism conferring resistance to insect vectors which is associated with the production of increased quantities of families of enzymes involved in insecticide metabolism. It resulted from structural change in the enzyme molecule that enhances its ability to detoxify or bind the insecticides which alter the affinity of the enzyme to insecticides. In the latter case, this mechanism enhances insecticide tolerance status in insects. Some of the common enzymes involved in detoxifying or sequestrating insecticides in insects are monooxygenases which include the cytochrome P450 enzymes. These large groups of enzymes confer resistance mainly to pyrethroids and carbamates and to a lesser extent to organochlorines and organophosphates. Another extremely important group of enzymes which confer resistance to organophosphate, carbamates and to some extent pyrethroid insecticides are esterases. Elevated level of esterases results in sequestration and metabolism of the target insecticides. Elevated glutathione S-transferases (GSTs) also play a role in the detoxification and excretion of organophosphates and DDT in insects. Cross-resistance between DDT and organophosphates is often caused by GSTs.

Behavioral resistance in mosquito vectors tends to change their behavior due to long-term exposure to insecticide-treated surfaces such as walls and LLINs. This behavior has been found to be associated with avoidance of exposure to lethal doses of insecticides due to reduced contact with the insecticide [97, 98]. The behavior is known to increase the longevity of insects in an environment where there is insecticide application through IRS, LLINs or both for vector control. Insects show limited tendency to enter sprayed houses or in houses with LLINs. For example, the evaluation of LLINs or IRS compounds in East African experimental huts have shown avoidance behavior by An. gambiae, An. arabiensis and An. funestus [98]. This also results in irritancy and excito-repellency, which keeps the mosquitoes away from different treated surfaces before contact with the host [99–102]. Shifting of vector species composition (from An. gambiae to An. arabiensis) due to implementation of LLINs or IRS has also been observed in Tanzania and Kenya [61, 103, 104]. In Africa, there is high proportion of An. gambiaes.l and An. funestus in areas with high coverage of LLINs [105]. The host-seeking behavior of vectors have changed from endophagic to exophagic due to intensive LLINs coverage [106].

Moreover, mutation in the GABA-gated chloride channel, which leads to dieldrin resistance other than DDT, has been described in different species of mosquitoes. The role of cuticular resistance mechanism is not yet known in the phenotypic resistance in the East African malaria vectors.

6. IMPACT AND IMPLICATIONS OF INSECTICIDE RESISTANCE ON THE EFFICACY OF LLINS, IRS AND MALARIA TRANSMISSION

The increased bed nets ownership and its utilization have significantly reduced malaria- related cases and mortality in Kenya [107, 108], Tanzania [109, 110], Uganda [111] and Ethiopia [112]. However, the high coverage of IRS and scaling up of LLINs is believed to induce the development of resistance in vector species to various classes of insecticides. These have been documented in Kenya [21],

Tanzania [79, 113], Uganda [114] and Ethiopia [70, 73, 115].

After the successful reduction of the malaria vector and disease transmission, the increased resistance among potential vector populations has been witnessed across East Africa [1, 21, 116]. The current status of pyrethroid resistance in malaria vectors and an increase in malaria incidence shows the compromised vector control system due to insecticide resistance, which calls the need for the development of new tools for malaria control. Insecticide resistance has shown to compromise the effectiveness of malaria control efforts in Kenya and other West African countries. The use of non-pyrethroid insecticides for IRS is a potential option as the ITN are mainly pyrethroid-based. It has been observed that pyrethroid resistance mosquitoes are entering and surviving exposure to LLINs, which may quantify the indoor transmission resurgence in areas with high level of pyrethroid resistance.

7. PROSPECTS OF PREVENTION ON DEVELOPMENT AND SPREAD OF IR IN MALARIA VECTORS IN EAST AFRICA

In vector control, constant use of the same insecticide induces resistance selection pressure in small vector population which subsequently spread to the large population. The spread of resistance depends on the frequency of the resistance genes within the vector population. In operational programs, the coverage of LLINs and IRS are most critical to be considered in the prevention of the development and spread of the insecticide resistance. Further, insecticide- resistant monitoring plan and management strategy should be developed and implemented to delay the development or spread of resistance.

8. INSECTICIDE MONITORING AND INSECTICIDE RESISTANCE MANAGEMENT OPTION

8.1 Mixture

In resistance management strategies, four tools (rotation, mosaic, mixture and combination) have been suggested [117] either to slow down resistance or reduce the rate of insecticide selection pressure. In control programs, simultaneous use of two or more insecticide compounds with different modes of action within a single product or formulation is preferred to manage resistance in insects. Two mixed insecticides with different modes of action can lead to reduced chance of double resistance by killing an insect which is resistant to one of the insecticide compounds [71, 82]. The use of mixture of insecticides relies on the assumption that the number of insects carrying a resistant allele at both loci is rare if the frequency of resistant allele at two loci is low [23, 118]. This approach may have a reduced efficacy if resistance in insects is at detected level to one of the mixed insecticide compounds. The major aim of insecticide mixture is to overcome the resistance selection pressure rather than maintaining the high susceptibility status of the insect population. The mixture should be up to the standard application concentration ratio of two insecticide compounds for effective control. Mixture of insecticides has usually high cost implication which may not be affordable in community protection against malaria. The mixture of insecticide compounds has practically shown to be effective when applied in small scale [82].

8.2 Mosaic

This approach is the use or application of two different classes of insecticides to control the same disease vector in the same area [119]. The mosaic approach is effective if application takes into consideration the spatial pattern. This technique helps in restoring the susceptibility status of the vector to an insecticide. It is a method for control of resistance secured to be working if properly done and monitored [120]. In some malaria endemic countries, large-scale mosaic application has shown to effectively control resistant populations of An. albimanus [119]. It has been observed that resistance developed fast in areas with pyrethroid alone than in areas with mosaic application along organophosphate, pyrethroids and carbamates [119]. Recently, industries have developed mosaic LLINs (PermaNet 3.0 and OlysetPlus) containing a pyrethroid insecticide and a synergist (piperonyl butoxide), an oxidase inhibitor on the fabric to increase the bio-efficacy against pyrethroid-resistant vectors [73, 114, 121]. Further research is needed in the future to use mosaic in LLINs and IRS.

8.3 Rotation

This is employing two or more insecticide compounds of different insecticide classes with different modes of action by switching the insecticide of choice each round or in alternating sequences. This approach is based on the assumption that resistance genes have a selective disadvantage in the absence of an insecticide used in operational program. If vector resistance to each insecticide is low, then the occurrence of multiple insecticide resistance is minimal or practically impossible [122]. The rotational use of insecticides plays a major role in killing resistant insects when the switch is made to a second insecticide. The defined rotation time should be as short as possible to reduce the risk of resistance development against the insecticide in use. It also slows down the evolution of the resistance [119]. For LLINs, it is difficult to implement rotation technique as only pyrethroids are used for the treatment of nets [123]. This method has higher financial implications for the implementation in vector control.

8.4 Combination of tools

In monitoring and management of resistance, the use of two or more tools or combinations of interventions simultaneously is an option in insecticide resistance management. The use of tools targeting adults such as LLINs and implementation of IRS or vise versa or combined with larviciding or larval source reduction is shown to have effects on vector control in Kenya [26, 124], Tanzania [43, 79, 113], Uganda [125] and Ethiopia [126].

Combination of tools is appreciated as it is cost-effective, prohibits mosquito feeding and causes mortality instead of reducing resistance alone. In this approach, using insecticides which share the same resistance mechanism should be avoided as resistance in malaria vectors develops faster. Combination tools have shown to increase the protection efficiency against vectors and maintain reduced susceptibility status of the vectors for longer period.

9. CHALLENGES IN INSECTICIDE RESISTANCE MANAGEMENT

The growing and widespread of insecticide resistance among vector species have been a major challenge in vector control and managing resistance. The resistant vectors have developed different mechanisms to tolerate the insecticides [53]. Each of the mechanism has its own target site for an insecticide [53]. The main insecticides used for the treatment of LLINs are pyrethroids, to which the major malaria vectors have shown tolerance [21]. The main challenge is that there is no other new class of insecticide to be used for LLINS and IRS [53, 61, 127]. Malaria control programs in East Africa and most sub-Saharan Africa rely heavily on donor-funded programs for LLINs distribution and IRS implementation. Insecticide resistance monitoring and management and operational research were not the primary agenda for the main donors. The control programs of East African countries have also not yet established a mechanism (s) for generating local funds to foster malaria control efforts [128]. This makes the whole effort of vector control program more challenging with the risk of malaria resurgence in some foci along the emergence and widespread of resistance in large areas of East Africa [59, 129, 130]. In general, insecticide resistance data in East Africa are patchy, and in some countries such as Burundi and Rwanda nearly non-existent. Therefore, countries need to create a national insecticide resistance data base for insecticide resistance monitoring data to understand the trend of insecticide resistance for timely decision-making and sharing of information..

10. CONCISE SUMMARY

Resistance to insecticides among malaria vectors has proliferated across East African nations. Countries such as Tanzania and Rwanda have implemented a national insecticide resistance monitoring and management strategy, while others are in the midst of building such plans primarily to avert the establishment of resistance or in response to identified resistance. Effective implementation of the plan necessitates national capabilities in trained personnel and infrastructure to conduct surveillance and monitoring of resistance to policy recommendations, explore alternative control options or novel vector control tools, and ensure that existing interventions remain viable choices in vector control programs. Successful execution of pesticide resistance monitoring and management requires coordination and inter-sectoral collaboration within the individual countries.

The insufficient financial resources in East African nations may postpone the execution of resistance monitoring and management measures. This may impede the initiation of resistance monitoring or response, the application of the recommended pesticide for vector control, and the alteration of control strategies once robust evidence of resistance is obtained. Internal financial sources, self-directed resource mobilization, and the allocation of sufficient resources are essential.

Crucial significance in executing national pesticide resistance monitoring and management policies within the framework of integrated vector management. The National Malaria Control Program of each country must also formulate operational guidelines with donors and other pertinent stakeholders to execute pesticide resistance monitoring and control methods.

REFERENCES

1. Ochomo E, Bayoh N, Kamau L, Atieli F, Vulule J, Ouma C, Ombok M, Njagi K, Soti D, Mathenge E, Muthami L, Kinyari T, Subramaniam K, Kleinschmidt I, Donnelly M, Mbogo C. Pyrethroid susceptibility of malaria vectors in four Districts of western Kenya. Parasite Vectors 2014;7(1):310.
2. Mnzava AE, Kilama WL, Kasigwa PF. Application of a biochemical key to study transmission of malaria and Bancroftian filariasis in sibling species of the Anopheles gambiae complex in north-eastern Tanzania. Acta Trop 1989; 46(5–6):323–33.
3. Braack L, Hunt R, Koekemoer L, Gericke A, Munhenga G, Haddow A, Becker P, Okia M, Kimera I, Coetzee M. Biting behaviour of African malaria vectors: 1. where do the main vector species bite on the human body? Parasite Vectors 2015; 8(1):76.
4. Animut A, Gebre-Michael T, Balkew M, Lindtjorn B. Abundance and dynamics of anopheline larvae in a highland malarious area of south-central Ethiopia. Parasite Vectors 2012;5(1):117.
5. Sattler M, Mtasiwa D, Kiama M, Premji Z, Tanner M, Killeen G, Lengeler C. Habitat characterization and spatial distribution of Anopheles sp. mosquito larvae in Dar es Salaam (Tanzania) during an extended dry period. Malar J 2005; 4(1):4.

6. Kweka EJ, Kamau L, Munga S, Lee M-C, Githeko AK, Yan G. A first report of Anopheles funestus sibling species in western Kenya highlands. Acta Trop 2013;128(1):158–61.

7. Kweka EJ, Zhou G, Munga S, Lee M-C, Atieli HE, Nyindo M, Githeko AK, Yan G. Anopheline Larval Habitats Seasonality and Species Distribution: A Prerequisite for Effective Targeted Larval Habitats Control Programmes. PLoS ONE 2012;7(12):e52084.

8. Fletcher M, Teklehaimanot A, Yemane G. Control of mosquito larvae in the port city of Assab by an indigenous larvivorous fish, Aphanius dispar. Acta Trop 1992;52(2–3):155– 66.

9. Yewhalaw D, Legesse W, Van Bortel W, Gebre-Selassie S, Kloos H, Duchateau L, Speybroeck N. Malaria and water resource development: the case of Gilgel-Gibe hydroelectric dam in Ethiopia. Malar J 2009;8(1):21.

10. Alemayehu T, Ye-ebiyo Y, Ghebreyesus TA, Witten KH, Bosman A, Teklehaimanot A. Malaria, schistosomiasis, and intestinal helminths in relation to microdams in Tigray, northern Ethiopia. Parassitologia 1998;40(3):259–67.

11. Afridi MK. Control of malaria in parties engaged on road and railway construction. Ind J Malariol 1947;1(2):275–7.

12. Covell G. Note on malaria problems created during road and railway construction. India J Malariol 1947;1(2):258–61.

13. Stern DI, Gething PW, Kabaria CW, Temperley WH, Noor AM, Okiro EA, Shanks GD, Snow RW, Hay SI. Temperature and Malaria Trends in Highland East Africa. PLoS ONE 2011;6(9):e24524.

14. Omumbo J, Lyon B, Waweru S, Connor S, Thomson M. Raised temperatures over the Kericho tea estates: revisiting the climate in the East African highlands malaria debate. Malar J 2011;10(1):12.

15. Lynch CA, Bruce J, Bhasin A, Roper C, Cox J, Abeku TA. Association between recent internal travel and malaria in Ugandan highland and highland fringe areas. Trop Med Int Health 2015;20(6):773–80.

16. Himeidan Y, Zhou G, Yakob L, Afrane Y, Munga S, Atieli H, El-Rayah E-A, Githeko A, Yan G. Habitat stability and occurrences of malaria vector larvae in western Kenya highlands. Malar J 2009;8(1):234.

17. Minakawa N, Munga S, Atieli F, Mushinzimana E, Zhou G, Githeko AK, GY. Spatial distribution of anopheline larval habitats in Western Kenyan highlands: effects of land cover types and topography. Am J Trop Med Hyg 2005;73(1):157–65.

18. Atieli H, Zhou G, Lee M-C, Kweka E, Afrane Y, Mwanzo I, Githeko A, Yan G. Topography as a modifier of breeding habitats and concurrent vulnerability to malaria risk in the western Kenya highlands. Parasite Vectors 2011;4(1):241.

19. Githeko A, Ayisi J, Odada P, Atieli F, Ndenga B, Githure J, Yan G. Topography and malaria transmission heterogeneity in western Kenya highlands: prospects for focal vector control. Malar J 2006;5(1):107.

20. WHO. World Malaria Report. In. Geneva: WHO; 2009.

21. Ochomo E, Bayoh N, Walker E, Abongo B, Ombok M, Ouma C, Githeko A, Vulule J, Yan G, Gimnig J. The efficacy of long-lasting nets with declining physical integrity may be compromised in areas with high levels of pyrethroid resistance. Malar J 2013;12(1): 368.

22. Georghiou GP, Taylor C. Factors influencing the evolution of résistance. In: Glass EH. (Ed.) Pesticide Résistance: Strategies and Tactics for Management. Washington, DC: National Academy of Sciences, 1986: pp. 157–169.

23. Tabashnik BE. Implications of gene amplification for evolution and management of insecticide resistance. J Econ Entomol 1990;83(4):1170–6.

24. Koenker H, Kilian A. Recalculating the Net Use Gap: A Multi-Country Comparison of ITN Use versus ITN Access. PLoS ONE 2014;9(5):e97496.

25. Magesa S, Lengeler C, deSavigny D, Miller J, Njau R, Kramer K, Kitua A, Mwita A. Creating an "enabling environment" for taking insecticide treated nets to national scale: the Tanzanian experience. Malar J 2005;4(1):34.

26. Zhou G, Afrane Y, Dixit A, Atieli H, Lee M-C, Wanjala C, Beilhe L, Githeko A, Yan G. Modest additive effects of integrated vector control measures on malaria prevalence and transmission in western Kenya. Malar J 2013;12(1):256.

27. Kolaczinski J, Kolaczinski K, Kyabayinze D, Strachan D, Temperley M, Wijayanandana N, Kilian A. Costs and effects of two public sector delivery channels for long-lasting insecticidal nets in Uganda. Malar J 2010;9(1):102.

28. Jima D, Getachew A, Bilak H, Steketee R, Emerson P, Graves P, Gebre T, Reithinger R, Hwang J, EMISWG. Malaria indicator survey 2007, Ethiopia: coverage and use of major malaria prevention and control interventions. Malar J 2010;9(1):58.

29. Hakizimana E, Cyubahiro B, Rukundo A, Kabayiza A, Mutabazi A, Beach R, Patel R, Tongren J, Karema C. Monitoring long-lasting insecticidal net (LLIN) durability to validate net serviceable life assumptions, in Rwanda. Malar J 2014;13(1):344.

30. Ogoma SB, Lweitoijera DW, Ngonyani H, Furer B, Russell TL, Mukabana WR, Killeen GF, Moore SJ. Screening mosquito house entry points as a potential method for integrated control of endophagic filariasis, arbovirus and malaria vectors. PLoS Neg Trop Dis 2010;4(8):e773.

31. Liu JX, Bousema T, Zelman B, Gesase S, Hashim R, Maxwell C, Chandramohan D, Gosling R. Is Housing Quality Associated with Malaria Incidence among Young Children and Mosquito Vector Numbers? Evidence from Korogwe, Tanzania. PLoS ONE 2014;9(2):e87358.

32. Atieli H, Menya D, Githeko A, Scott T. House design modifications reduce indoor resting malaria vector densities in rice irrigation scheme area in western Kenya. Malar J 2009;8(1):108.

33. Massebo F, Lindtjorn B. The effect of screening doors and windows on indoor density of Anopheles arabiensis in south-west Ethiopia: a randomized trial. Malar J 2013;12(1): 319.

34. Wanzirah H, Tusting LS, Arinaitwe E, Katureebe A, Maxwell K, Rek J, Bottomley C, Staedke SG, Kamya M, Dorsey G, Lindsay SW. Mind the gap: house structure and the risk of malaria in Uganda. PLoS ONE 2015;10(1):e0117396.

35. Ondiaka S, Masinde E, Koenraadt C, Takken W, Mukabana W. Effects of fungal infection on feeding and survival of Anopheles gambiae (Diptera: Culicidae) on plant sugars. Parasite Vectors 2015;8(1):35.

36. Mnyone L, Lyimo I, Lwetoijera D, Mpingwa M, Nchimbi N, Hancock P, Russell T, Kirby M, Takken W, Koenraadt C. Exploiting the behaviour of wild malaria vectors to achieve high infection with fungal biocontrol agents. Malar J 2012;11(1):87.

37. Ferguson H, Ng'habi K, Walder T, Kadungula D, Moore S, Lyimo I, Russell T, Urassa H, Mshinda H, Killeen G, Knols B. Establishment of a large semi-field system for experimental study of African malaria vector ecology and control in Tanzania. Malar J 2008;7(1):158.

38. Olanga E, Okal M, Mbadi P, Kokwaro E, Mukabana W. Attraction of Anopheles gambiae to odour baits augmented with heat and moisture. Malar J 2010;9(1):6.

39. Nyasembe VO, Tchouassi DP, Kirwa HK, Foster WA, Teal PEA, Borgemeister C, Torto B. Development and assessment of plant-based synthetic odor baits for surveillance and control of malaria vectors. PLoS ONE 2014;9(2):e89818.

40. Tabuti JRS. Herbal medicines used in the treatment of malaria in Budiope county, Uganda. J Ethnopharmacol 2008;116(1):33–42.

41. Kweka E, Mosha F, Lowassa A, Mahande A, Kitau J, Matowo J, Mahande M, Massenga C, Tenu F, Feston E, Lyatuu E, Mboya M, Mndeme R, Chuwa G, Temu

E. Ethnobotanical study of some of mosquito repellent plants in north-eastern Tanzania. Malar J 2008;7(1): 152.

42. Kweka E, Mosha F, Lowassa A, Mahande A, Mahande M, Massenga C, Tenu F, Lyatuu E, Mboya M, Temu E. Longitudinal evaluation of Ocimum and other plants effects on the feeding behavioral response of mosquitoes (Diptera: Culicidae) in the field in Tanzania. Parasite Vectors 2008;1(1):42.
43. Kweka E, Munga S, Mahande A, Msangi S, Mazigo H, Adrias A, Matias J. Protective efficacy of menthol propylene glycol carbonate compared to N, N-diethyl-methylben - zamide against mosquito bites in Northern Tanzania. Parasite Vectors 2012;5(1):189.
44. Seyoum A, Killeen GF, Kabiru EW, Knols BGJ, Hassanali A. Field efficacy of thermally expelled or live potted repellent plants against African malaria vectors in western Kenya. Trop Med Int Health 2003;8(11):1005–11.
45. Seyoum A, Pålsson K, Kung'a S, Kabiru EW, Lwande W, Killeen GF, Hassanali A, Knots BGJ. Traditional use of mosquito-repellent plants in western Kenya and their evaluation in semi-field experimental huts against Anopheles gambiae: ethnobotanical studies and application by thermal expulsion and direct burning. Trans Royal Soc Trop Med Hyg 2002;96(3):225–31.
46. Dekker T, Ignell R, Ghebru M, Glinwood R, Hopkins R. Identification of mosquito repellent odours from Ocimum forskolei. Parasite Vectors 2011;4(1):183.
47. Karunamoorthi K, Hailu T. Insect repellent plants traditional usage practices in the Ethiopian malaria epidemic-prone setting: an ethnobotanical survey. J Ethnobiol Ethnomed 2014;10(1):22.
48. Dube F, Tadesse K, Birgersson G, Seyoum E, Tekie H, Ignell R, Hill S. Fresh, dried or smoked? repellent properties of volatiles emitted from ethnomedicinal plant leaves against malaria and yellow fever vectors in Ethiopia. Malaria J 2011;10(1):375.
49. Brown AW. Insecticide resistance in mosquitoes: a pragmatic review. J Am Mosq Control Assoc 1986;2(2):123–40.
50. Hemingway J. Insecticide resistance in malaria vectors: a new approach to an old subject. Parassitologia 1999;41(1–3):315–8.
51. Okara R, Sinka M, Minakawa N, Mbogo C, Hay S, Snow R. Distribution of the main malaria vectors in Kenya. Malar J 2010;9(1):69.
52. Kamau L, Vulule J. Status of insecticide susceptibility in Anopheles arabiensis from Mwea rice irrigation scheme, Central Kenya. Malar J 2006;5(1):46.
53. Mulamba C, Riveron JM, Ibrahim SS, Irving H, Barnes KG, Mukwaya LG, Birungi J, Wondji CS. Widespread pyrethroid and DDT resistance in the major malaria

vector Anopheles funestus in East Africa is driven by metabolic resistance mechanisms. PLoS ONE 2014;9(10):e110058.

54. Stump AD, Atieli FK, Vulule JM, Besansky NJ. Dynamics of the pyrethroid knockdown resistance allele in western Kenyan populations of Anopheles gambiae in response to insecticide-treated bed net trials. Am J Trop Med Hyg 2004;70(6):591–6.

55. Mawejje HD, Wilding CS, Rippon EJ, Hughes A, Weetman D, Donnelly MJ. Insecticide resistance monitoring of field-collected Anopheles gambiae s.l. populations from Jinja, eastern Uganda, identifies high levels of pyrethroid resistance. Med Vet Entomol 2013;27(3):276–83.

56. Ojuka P, Boum Y, Denoeud-Ndam L, Nabasumba C, Muller Y, Okia M, Mwanga-Amumpaire J, De Beaudrap P, Protopopoff N, Etard J-F. Early biting and insecticide resistance in the malaria vector Anopheles might compromise the effectiveness of vector control intervention in Southwestern Uganda. Malar J 2015;14(1):148.

57. Ramphul U, Boase T, Bass C, Okedi LM, Donnelly MJ, Müller P. Insecticide resistance and its association with target-site mutations in natural populations of Anopheles gambiae from eastern Uganda. Trans R Soc Trop Med Hyg 2009;103(11):1121–6.

58. Verhaeghen K, Bortel WV, Roelants P, Okello PE, Talisuna A, Coosemans M. Spatio- temporal patterns in kdr frequency in permethrin and DDT resistant Anopheles gambiae s.s. from Uganda. Am J Trop Med Hyg 2010;82(4):566–73.

59. Morgan JC, Irving H, Okedi LM, Steven A, Wondji CS. Pyrethroid resistance in an Anopheles funestus population from Uganda. PLoS ONE 2010;5(7):e11872.

60. John R, Ephraim T, Andrew A. Reduced susceptibility to pyrethroid insecticide treated nets by the malaria vector Anopheles gambiae s.l. in western Uganda. Malar J 2008;7(1): 92.

61. Mulamba C, Irving H, Riveron J, Mukwaya L, Birungi J, Wondji C. Contrasting Plasmodium infection rates and insecticide susceptibility profiles between the sympatric sibling species Anopheles parensis and Anopheles funestus s.s: a potential challenge for malaria vector control in Uganda. Parasite Vectors 2014;7(1):71.

62. Verhaeghen K, Van Bortel W, Roelants P, Backeljau T, Coosemans M. Detection of the East and West African kdr mutation in Anopheles gambiae and Anopheles arabiensis from Uganda using a new assay based on FRET/Melt Curve analysis. Malar J 2006;5(1):16.

63. Melville AR, Wilson DB, et al. Malaria in Abyssinia. East Afr Med J 1945;22:285–94.

64. Abeku TA, van Oortmarssen GJ, Borsboom G, de Vlas SJ, Habbema JDF. Spatial and temporal variations of malaria epidemic risk in Ethiopia: factors involved and implications. Acta Trop 2003;87(3):331–40.

65. White GB. Malaria vector ecology and genetics. Br Med Bull 1982;38(2):207–12.

66. Hunt RH, Coetzee M, Fettene M. The Anopheles gambiae complex: a new species from Ethiopia. Trans Royal Soc Trop Med Hyg 1998;92(2):231–5.

67. Fettene M, Temu EA. Species-specific primer for identification of Anopheles quadriannulatus sp. B (Diptera: Culicidae) from Ethiopia using a multiplex polymerase chain reaction assay. J Med Entomol 2003;40(1):112–5.

68. Ribeiro JM, Seulu F, Abose T, Kidane G, Teklehaimanot A. Temporal and spatial distribution of anopheline mosquitos in an Ethiopian village: implications for malaria control strategies. Bull World Health Org 1996;74(3):299–305.

69. Balkew M, Gebre-Michael T, Hailu A. Insecticide susceptibility level of Anopheles arabiensis in two agrodevelopment localities in eastern Ethiopia. Parassitologia 2003;45(1):1–3.

70. Yewhalaw D, Wassie F, Steurbaut W, Spanoghe P, Van Bortel W, Denis L, Tessema DA, Getachew Y, Coosemans M, Duchateau L, Speybroeck N. Multiple insecticide resistance: an impediment to insecticide-based malaria vector control program. PLoS ONE 2011;6(1):e16066.

71. Abate A, Hadis M. Susceptibility of Anopheles gambiae s.l. to DDT, malathion, permethrin and deltamethrin in Ethiopia. Trop Med Int Health 2011;16(4):486–91.

72. Balkew M, Ibrahim M, Koekemoer L, Brooke B, Engers H, Aseffa A, Gebre-Michael T, Elhassen I. Insecticide resistance in Anopheles arabiensis (Diptera: Culicidae) from villages in central, northern and south west Ethiopia and detection of kdr mutation. Parasite Vectors 2010;3(1):40.

73. Yewhalaw D, Asale A, Tushune K, Getachew Y, Duchateau L, Speybroeck N. Bio- efficacy of selected long-lasting insecticidal nets against pyrethroid resistant Anopheles arabiensis from South-Western Ethiopia. Parasite Vectors 2012;5(1):159.

74. Kabula B, Tungu P, Matowo J, Kitau J, Mweya C, Emidi B, Masue D, Sindato C, Malima R, Minja J, Msangi S, Njau R, Mosha F, Magesa S, Kisinza W. Susceptibility status of malaria vectors to insecticides commonly used for malaria control in Tanzania. Trop Med Int Health 2012;17(6):742–50.

75. Kweka EJ, Mahande AM, Nkya WM, Assenga C, Lyatuu EE, Nyale E, Mosha FW, Mwakalinga SB, Temu EA. Vector species composition and malaria infectivity rates in Mkuzi, Muheza District, north-eastern Tanzania. Tanzania J Health Res 2008;10(1):46– 9.

76. Matowo J, Kulkarni M, Mosha F, Oxborough R, Kitau J, Tenu F, Rowland M. Biochemical basis of permethrin resistance in Anopheles arabiensis from Lower Moshi, north- eastern Tanzania. Malar J 2010;9(1):193.

77. Kabula B, Kisinza W, Tungu P, Ndege C, Batengana B, Kollo D, Malima R, Kafuko J, Mohamed M, Magesa S. Co-occurrence and distribution of East (L1014S) and West (L1014F) African knock-down resistance in Anopheles gambiae sensu lato population of Tanzania. Trop Med Int Health 2014;19(3):331–41.

78. Kabula B, Tungu P, Malima R, Rowland M, Minja J, Wililo R, Ramsan M, McElroy PD, Kafuko J, Kulkarni M, Protopopoff N, Magesa S, Mosha F, Kisinza W. Distribution and spread of pyrethroid and DDT resistance among the Anopheles gambiae complex in Tanzania. Med Vet Entomol 2014;28(3):244–52.

79. Okumu F, Chipwaza B, Madumla E, Mbeyela E, Lingamba G, Moore J, Ntamatungro A, Kavishe D, Moore S. Implications of bio-efficacy and persistence of insecticides when indoor residual spraying and long-lasting insecticide nets are combined for malaria prevention. Malar J 2012;11(1):378.

80. Haji K, Khatib B, Smith S, Ali A, Devine G, Coetzee M, Majambere S. Challenges for malaria elimination in Zanzibar: pyrethroid resistance in malaria vectors and poor performance of long-lasting insecticide nets. Parasite Vectors 2013;6(1):82.

81. Protopopoff N, Matowo J, Malima R, Kavishe R, Kaaya R, Wright A, West P, Kleinsch - midt I, Kisinza W, Mosha F, Rowland M. High level of resistance in the mosquito Anopheles gambiae to pyrethroid insecticides and reduced susceptibility to bendiocarb in north-western Tanzania. Malar J 2013;12(1):149.

82. Oxborough RM, Mosha FW, Matowo J, Mndeme R, Feston E, Hemingway J, Rowland M. Mosquitoes and bednets: testing the spatial positioning of insecticide on nets and the rationale behind combination insecticide treatments. Ann Trop Med Parasitol 2008;102(8):717–27.

83. Oxborough R, Kitau J, Jones R, Feston E, Matowo J, Mosha F, Rowland M. Long-lasting control of Anopheles arabiensis by a single spray application of micro-encapsulated pirimiphos-methyl (Actellic(R) 300 CS). Malar J 2014;13(1):37.

84. Kulkarni M, Rowland M, Alifrangis M, Mosha F, Matowo J, Malima R, Peter J, Kweka E, Lyimo I, Magesa S, Salanti A, Rau M, Drakeley C. Occurrence of the leucine-to- phenylalanine knockdown resistance (kdr) mutation in Anopheles

arabiensis populations in Tanzania, detected by a simplified high-throughput SSOP-ELISA method. Malar J 2006;5(1):56.

85. Mahande AM, Dusfour I, Matias JR, Kweka EJ. Knockdown Resistance, rdl Alleles, and the Annual Entomological Inoculation Rate of Wild Mosquito Populations from Lower Moshi, Northern Tanzania. J Global Infec Dis 2012;4(2):114–9.

86. Protopopoff N, Van Bortel W, Marcotty T, Van Herp M, Maes P, Baza D, D'Alessandro U, Coosemans M. Spatial targeted vector control in the highlands of Burundi and its impact on malaria transmission. Malar J 2007;6:158.

87. Protopopoff N, Van Bortel W, Marcotty T, Van Herp M, Maes P, Baza D, D'Alessandro U, Coosemans M. Spatial targeted vector control is able to reduce malaria prevalence in the highlands of Burundi. Am J Trop Med Hyg 2008;79(1):12–8.

88. Yadouleton A, Asidi A, Djouaka R, Braima J, Agossou C, Akogbeto M. Development of vegetable farming: a cause of the emergence of insecticide resistance in populations of Anopheles gambiae in urban areas of Benin. Malar J 2009;8(1):103.

89. N'Guessan R, Corbel V, Akogbeto M, Rowland M. Reduced efficacy of insecticide- treated nets and indoor residual spraying for malaria control in pyrethroid resistance area, Benin. Emerg Infec Dis 2007;13(2):199–206.

90. Yadouleton AW, Padonou G, Asidi A, Moiroux N, Bio-Banganna S, Corbel V, N'Guessan R, Gbenou D, Yacoubou I, Gazard K, Akogbeto MC. Insecticide resistance status in Anopheles gambiae in southern Benin. Malar J 2010;9:83.

91. Wondji CS, Dabire RK, Tukur Z, Irving H, Djouaka R, Morgan JC. Identification and distribution of a GABA receptor mutation conferring dieldrin resistance in the malaria vector Anopheles funestus in Africa. Insect Biochem Mol Biol 2011;41(7):484–91.

92. Nkya T, Poupardin R, Laporte F, Akhouayri I, Mosha F, Magesa S, Kisinza W, David J-P. Impact of agriculture on the selection of insecticide resistance in the malaria vector Anopheles gambiae: a multigenerational study in controlled conditions. Parasite Vectors 2014;7(1):480.

93. Awolola S, Adeogun A, Olojede J, Oduola A, Oyewole I, Amajoh C. Impact of PermaNet 3.0 on entomological indices in an area of pyrethroid resistant Anopheles gambiae in south-western Nigeria. Parasite Vectors 2014;7(1):236.

94. Bloomquist JR. Ion channels as targets for insecticides. Annu Rev Entomol 1996;41(1): 163–90.

95. IRAC. Prevention and management of insecticide resistance in vectors of public health importance. In: Resistance Management for Sustainable Agriculture

and Improved Public Health, Second Edition: Insecticide Resistance Action Commitee; 2010: 72.

96. Weetman D, Donnelly MJ. Evolution of insecticide resistance diagnostics in malaria vectors. Trans R Soc Trop Med Hyg 2015;109(5):291–3.

97. Ngufor C, Tungu P, Malima R, Kirby M, Kisinza W, Rowland M. Insecticide-treated net wall hangings for malaria vector control: an experimental hut study in north-eastern Tanzania. Malar J 2014;13(1):366.

98. Kitau J, Oxborough R, Matowo J, Mosha F, Magesa S, Rowland M. Indoor residual spraying with microencapsulated DEET repellent (N, N-diethyl-m-toluamide) for control of Anopheles arabiensis and Culex quinquefasciatus. Parasite Vectors 2014;7(1):446.

99. Curtis CF, Mnzava AEP. Comparison of house spraying and insecticide-treated nets for malaria control. Bull World Health Org 2000;78:1389–400.

100. Mosha FW, Lyimo IN, Oxborough RM, Malima R, Tenu F, Matowo J, Feston E, Mndeme R, Magesa SM, Rowland M. Experimental hut evaluation of the pyrrole insecticide chlorfenapyr on bed nets for the control of Anopheles arabiensis and Culex quinquefasciatus. Trop Med Int Health 2008;13(5):644–52.

101. Deressa W, Yihdego Y, Kebede Z, Batisso E, Tekalegne A, Dagne G. Effect of combining mosquito repellent and insecticide treated net on malaria prevalence in Southern Ethiopia: a cluster-randomised trial. Parasite Vectors 2014;7(1):132.

102. Mng'ong'o FC, Sambali JJ, Sabas E, Rubanga J, Magoma J, Ntamatungiro AJ, Turner EL, Nyogea D, Ensink JHJ, Moore SJ. Repellent plants provide affordable natural screening to prevent mosquito house entry in tropical rural settings—results from a pilot efficacy study. PLoS ONE 2011;6(10):e25927.

103. Kitau J, Oxborough RM, Tungu PK, Matowo J, Malima RC, Magesa SM, Bruce J, Mosha FW, Rowland MW. Species shifts in the Anopheles gambiae complex: do LLINs successfully control Anopheles arabiensis? PLoS ONE 2012;7(3):e31481.

104. Bayoh MN, Mathias D, Odiere M, Mutuku F, Kamau L, Gimnig J, Vulule J, Hawley W, Hamel M, Walker E. Anopheles gambiae: historical population decline associated with regional distribution of insecticide-treated bed nets in western Nyanza Province, Kenya. Malar J 2010;9(1):62.

105. Kawada H, Ohashi K, Dida G, Sonye G, Njenga S, Mwandawiro C, Minakawa N. Insecticidal and repellent activities of pyrethroids to the three major pyrethroid- resistant malaria vectors in western Kenya. Parasite Vectors 2014;7(1):208.

106. Russell T, Govella N, Azizi S, Drakeley C, Kachur SP, Killeen G. Increased proportions of outdoor feeding among residual malaria vector populations following increased use of insecticide-treated nets in rural Tanzania. Malar J 2011;10(1):80.

107. Okiro E, Alegana V, Noor A, Mutheu J, Juma E, Snow R. Malaria paediatric hospitalization between 1999 and 2008 across Kenya. BMC Med 2009;7(1):75.

108. Okiro E, Alegana V, Noor A, Snow R. Changing malaria intervention coverage, transmission and hospitalization in Kenya. Malar J 2010;9(1):285.

109. Ishengoma D, Mmbando B, Segeja M, Alifrangis M, Lemnge M, Bygbjerg I. Declining burden of malaria over two decades in a rural community of Muheza district, north- eastern Tanzania. Malar J 2013;12(1):338.

110. Mmbando B, Vestergaard L, Kitua A, Lemnge M, Theander T, Lusingu J. A progressive declining in the burden of malaria in north-eastern Tanzania. Malar J 2010;9(1):216.

111. De Beaudrap P, Nabasumba C, Grandesso F, Turyakira E, Schramm B, Boum Y, Etard J-F. Heterogeneous decrease in malaria prevalence in children over a six-year period in south-western Uganda. Malar J 2011;10(1):132.

112. O'Meara WP, Mangeni JN, Steketee R, Greenwood B. Changes in the burden of malaria in sub-Saharan Africa. Lancet Infec Dis 2010;10(8):545–55.

113. Okumu F, Kiware S, Moore S, Killeen G. Mathematical evaluation of community level impact of combining bed nets and indoor residual spraying upon malaria transmission in areas where the main vectors are Anopheles arabiensis mosquitoes. Parasite Vectors 2013;6(1):17.

114. Okia M, Ndyomugyenyi R, Kirunda J, Byaruhanga A, Adibaku S, Lwamafa D, Kironde F. Bioefficacy of long-lasting insecticidal nets against pyrethroid-resistant populations of Anopheles gambiae s.s. from different malaria transmission zones in Uganda. Parasite Vectors 2013;6(1):130.

115. Asale A, Getachew Y, Hailesilassie W, Speybroeck N, Duchateau L, Yewhalaw D. Evaluation of the efficacy of DDT indoor residual spraying and long-lasting insecticidal nets against insecticide resistant populations of Anopheles arabiensis Patton (Diptera: Culicidae) from Ethiopia using experimental huts. Parasite Vectors 2014;7(1):131.

116. Ranson H, Jensen B, Vulule JM, Wang X, Hemingway J, Collins FH. Identification of a point mutation in the voltage-gated sodium channel gene of Kenyan Anopheles gambiae associated with resistance to DDT and pyrethroids. Insect Mol Biol 2000;9(5): 491–7.

117. Thomsen EK, Strode C, Hemmings K, Hughes AJ, Chanda E, Musapa M, Kamuliwo M, Phiri FN, Muzia L, Chanda J, Kandyata A, Chirwa B, Poer K, Hemingway J, Wondji CS, Ranson H, Coleman M. Underpinning sustainable vector control through informed insecticide resistance management. PLoS ONE 2014;9(6):e99822.

118. Mani GS. Evolution of resistance in the presence of two insecticides. Genetics 1985;109(4):761–83.

119. Penilla PR, RodrÍGuez AD, Hemingway J, Torres JL, Arredondo-JimÉNez JI, RodrÍ Guez MH. Resistance management strategies in malaria vector mosquito control. Baseline data for a large-scale field trial against Anopheles albimanus in Mexico. Med Vet Entomol 1998;12(3):217–33.

120. Killeen G, Okumu F, N'Guessan R, Coosemans M, Adeogun A, Awolola S, Etang J, Dabire R, Corbel V. The importance of considering community-level effects when selecting insecticidal malaria vector products. Parasite Vectors 2011;4(1):160.

121. Tungu P, Magesa S, Maxwell C, Malima R, Masue D, Sudi W, Myamba J, Pigeon O, Rowland M. Evaluation of PermaNet 3.0 a deltamethrin-PBO combination net against Anopheles gambiae and pyrethroid resistant Culex quinquefasciatus mosquitoes: an experimental hut trial in Tanzania. Malar J 2010;9(1):21.

122. Chareonviriyaphap T, Bangs M, Suwonkerd W, Kongmee M, Corbel V, Ngoen-Klan R. Review of insecticide resistance and behavioral avoidance of vectors of human diseases in Thailand. Parasite Vectors 2013;6(1):280.

123. WHO. Pesticides and their application for the control of vectors and pests of public health importance. In: WHO/CDS/NTD/WHOPES/GCDPP/2006.1, sixth Edition edn. Geneva: WHO; 2006: pp. 1–125.

124. Fillinger U, Ndenga B, Githeko A, Lindsay SW. Integrated malaria vector control with microbial larvicides and insecticide-treated nets in western Kenya: a controlled trial. Bull World Health Org 2009;87:655–65.

125. Sezi C. The phenomenon of diminishing -returns in the use of bed nets and indoor house spraying and the emerging place of antimalarial medicines in the control of malaria in Uganda. Afr Health Sci 2014;14:100–10.

126. Bekele D, Belyhun Y, Petros B, Deressa W. Assessment of the effect of insecticide-treated nets and indoor residual spraying for malaria control in three rural kebeles of Adami Tulu District, South Central Ethiopia. Malar J 2012;11(1):127.

127. Toe KH, Jones CM, N'Fale S, Ismail HM, Dabire RK, Ranson H. Increased pyrethroid resistance in malaria vectors and decreased bed net effectiveness, Burkina Faso. Emerg Infect Dis 2014;20(10):1691–6.

128. Mnzava A, Knox T, Temu E, Trett A, Fornadel C, Hemingway J, Renshaw M. Implementation of the global plan for insecticide resistance management in malaria vectors: progress, challenges and the way forward. Malar J 2015;14(1):173.

129. Bauch J, Gu J, Msellem M, Martensson A, Ali A, Gosling R, Baltzell K. Perception of malaria risk in a setting of reduced malaria transmission: a qualitative study in Zanzibar. Malar J 2013;12(1):75.

130. Kawada H, Dida GO, Ohashi K, Komagata O, Kasai S, Tomita T, Sonye G, Maekawa Y, Mwatele C, Njenga SM, Mwandawiro C, Minakawa N, Takagi M. Multimodal pyrethroid resistance in malaria vectors, Anopheles gambiae s.s., Anopheles arabiensis, and Anopheles funestus s.s. in Western Kenya. PLoS ONE 2011;6(8):e22574.

131. Kawada H, Ohashi K, Dida G, Sonye G, Njenga S, Mwandawiro C, Minakawa N. Preventive effect of permethrin-impregnated long-lasting insecticidal nets on the blood feeding of three major pyrethroid-resistant malaria vectors in western Kenya. Parasite Vectors 2014;7(1):383.

132. Chen H, Githeko AK, Githure JI, Mutunga J, Zhou G, Yan G. Monooxygenase levels and knockdown resistance (kdr) allele frequencies in Anopheles gambiae and Anopheles arabiensis in Kenya. J Med Entomol 2008;45(2):242–50.

133. Protopopoff N, Verhaeghen K, Van Bortel W, Roelants P, Marcotty T, Baza D, D'Ales-sandro U, Coosemans M. A significant increase in kdr in Anopheles gambiae is associated with an intensive vector control intervention in Burundi highlands. Trop Med Int Health 2008;13(12):1479–87.

134. Yewhalaw D, Bortel WV, Denis L, Coosemans M, Duchateau L, Speybroeck N. First evidence of high knockdown resistance frequency in Anopheles arabiensis (Diptera: Culicidae) from Ethiopia. Am J Trop Med Hyg 2010;83(1):122–5.

135. Nkya T, Akhouayri I, Poupardin R, Batengana B, Mosha F, Magesa S, Kisinza W, David J-P. Insecticide resistance mechanisms associated with different environments in the malaria vector Anopheles gambiae: a case study in Tanzania. Malar J 2014;13(1):28.

136. Jones CM, Toé HK, Sanou A, Namountougou M, Hughes A, Diabaté A, Dabiré R, Simard F, Ranson H. Additional selection for insecticide resistance in urban malaria vectors: DDT resistance in Anopheles arabiensis from Bobo-Dioulasso, Burkina Faso. PLoS ONE 2012;7(9):e45995.

137. USAID/PMI. President's Malaria Initiative. Rwanda, Malaria Operational Plan FY 2014, p. 29.

11 Chapter

EMERGING INSECT-TRANSMITTED DISEASES OF AGRICULTURAL, MEDICAL, AND VETERINARY SIGNIFICANCE

1. INTRODUCTION: AN OVERVIEW

In 1962, Rachel Carson's book Silent Spring chronicled the harmful environmental impacts of the uncontrolled application of synthetic pesticides. The book asserted that DDT and other pesticides have been demonstrated to induce cancer and that its application in agriculture poses a danger to wildlife, especially avian species. She directly accused the chemical business of disseminating deceptive information and public officials of uncritically adopting corporate assertions regarding effects. Notwithstanding the vigorous resistance from chemical corporations, the book significantly influenced the American populace, marking a pivotal moment for the environmental movement and prompting a shift in national pesticide policy. The problem remains highly relevant, given the recent discourse over the impact of neonicotinoids on honey bees and avian species. [5].

In 1972, DDT was banned on the agricultural uses in the US, and soon after in EU. Before DDT was banned, more than 600,000 tonnes were applied in the US. The environmental movement led to the creation of the US EPA (Environmental Protection Agency). The first resistance episode concerned DDT in 1914 [6]. In May 2015, US President Obama, considering the final report of the Commission, stated the primary importance of antibiotic multiresistance, leading to the ban of the use of antibiotics in agriculture and farm practices. However, deep divide exists between American and European regulation of pesticides and other chemicals, many chemicals that are banned or strictly regulated in EU are permitted in the US.

Accordingly to the current definitions, resistance can be defined as "a heritable change in the sensitivity of a pest population that is reflected in the repeated failure of a product to achieve the expected level of control when used according to the label recommendation for that pest species" [7]. In this chapter, we will consider resistance as "the inherited ability of an organism to become tolerant to a dosage of the chemical that would be lethal to a definite species." Evidence for pesticide resistance in arthropods of agricultural and medical importance is an emerging threat. It is possible that in the next 20–30 years, all the synthetic pesticides now employed for pest control will lose their efficacy. Research on newer and safer control tools may be helpful in future scenarios for mankind, dramatically involving feed and food.

Synthetic insecticides are usually utilized to reduce damages caused by insects that destroy crops or transmit diseases. To be effective, an insecticide should be lethal to the majority of the individuals in a normal wild population. However, the insecticide can lose its efficacy, thus many pest populations developed resistance to the toxic effects [8]. This key point is only a further example of the consequences of human tendency to amplify the natural resources beyond any limit in order to obtain the maximum of the effects and not considering the consequences. The problem is inherent: the resistance is related to a massive and persistent use of pesticides, exactly like for antibiotics in microorganisms. Many species have resistant populations, which can resist to one or many treatments [9]. In the moment of the insecticide use, some individuals result resistant. Sensitive insects exposed to the insecticide will die, except the few resistant ones, which can continue to develop and proliferate. Continuing the use, they will be favored. More use of pesticide fuels the dominance of resistant populations. The consequence of the mechanism is that, in the right time, insecticides, once effective, are not sufficient in controlling insects [10]. Nowadays, the above definitions risk to be insufficient to describe the complex system that generates the resistance phenomenon to insecticides or the absence of any real efficacy in control of insect pests. The solution can be considered simply as "find the best weapon to kill the bacterium and solve the problem," as so far mainly considered. As in the general medicine, where the "chemical magic bullet" was considered the central solution for any disease, this approach is nowadays in crisis because the physiologic aspects are more complex and complicated by interactions at several levels. Resistance is the consequence of a series of events.

This chapter attempts to overcome the paradigm substance → replay of organism → resistance, in favor of a more complex approach, leading to integrated pest management. A focus on two current vector-borne emergencies was provided.

2. A NEW SITUATION

So far, much attention of research and public concern was focused on vector-borne human diseases in order to eradicate their presence and to save as many lives as possible [11]. However, if the resistance will affect our life supports, the surviving struggle will be in balance due to two key factors: resistance to many pesticides and/or the impossibility of using current effective substances because of the consequences of the effects on environment. Even pyrethroids are now considered dangerous for nontarget animals since they can impair memory and movement [12]. Currently, we have to face new epidemic emergencies due to several factors (including climate change) concerning crops and livestock. Insects are vectors of important diseases involving nonhuman targets, causing important effects on plants and animals of strategic economic relevance. Recently, some of such diseases are of increasing concern to the general population, attracting a level of attention never experimented before and generating great alarms for the consequences of their rapid diffusion. The economic negative effects are enormous, and the damages on the local economic system are dramatic. Agricultural production must increases in relation to the needs of world population. However, losses due to arthropod pests account now for around 20–30% of the production [13]. Insect damages are important in the field and in the stored products. Agricultural production resorts to the use of a large quantity of insecticides to raise production and preservation of foodstuff. According to the increasing needs, the use of insecticides has increased even more than necessary, although it was demonstrated that the excessive and inappropriate use of synthetic pesticides is frequently even counterproductive. Effects are not limited to the treated field, involving undesirable consequences on public health and environment. Current pesticide pollution of Adige valley in Northern Italy, due to the continuous heavy treatments of apple monoculture, is a clear example. The introduction of OGM did not afford solutions, inducing additional problems to farmers.

Most of insecticides are usually employed to reduce the damages caused by insects that destroy crops or transmit diseases. Up to now, agricultural pests account for 59% of the resistant insect species, while veterinary pests account for 41%. Antibiotics in the US and the UK mainly utilized to treat livestock are considered immunostimulants, similarly, most insecticides are used in agricultural practice to improve the production and preservation of foodstuff. In both cases, the use is now widespread, excessive, and inappropriate.

Recently, some of such diseases have rapidly gained media's attention, generating great alarm for the consequences of their diffusion. The economic negative effects are enormous, and the damages on the local living system are dramatic. Several epidemic emergencies are in act, and the emergency is going to be converted into a normal trend, as a consequence of the permanence of several factors, including in first place the climate changes [14].

On the basis of novel knowledge, some new approaches are emerging, changing the aspect of insect control. Integrated pest management is an important approach, developed in the last years to control disease vectors and limit economic agricultural damages, improving crop yield with minimum cost. Main goals are (a) to increase basic knowledge of biology of the insect pest and its relationship with other organisms sharing the same ecological niche [15]; (b) to reduce pesticide application and quantity developing biological controls, farming practices, farmers collaboration, and mechanical and physical controls; and (c) to build new models of integrated managements on the basis of laboratory and field experiments, including the research of new active compounds. Novel pesticides to be suitable must be low cost, eco- friendly, from renewable raw material, nontoxic to nontarget organisms, of rapid degradation, and no accumulation in the environment.

However, resistance is only the last consequence of a series of events. Most of this paper will be dedicated to the deep knowledge of this sequence, being considered the key to struggle resistance. Insect-borne diseases are the result of complex multiorganism interactions. The network of several different collaborating organisms is on the basis of diffusion, effectiveness, and metabolism of insect vectors, including the resistance phenomenon. The integrated network acts like a "superorganism," integrating functions of all the different types of involved organisms. Disease is the result of a brave and useful collaboration between organisms totally different, from bacteria to insect, giving rise to an integrated system that is the key to survive and proliferate. It is the example that we must learn, asking for several levels of eco-friendly interactions. The consequences that we consider as negative are only the collateral effects of the competitive struggle.

In these years, the Mediterranean Sea was an incubation sap of several massive migrations of organisms, mainly due to climate changes and commercial routes, which radically modified previous equilibrium. Migrations start from distant sites but are able to spread in a large area until they find the right conditions to set up and rapidly become dominant.

Therefore, insects, like microorganisms, are particularly able to change their genome. It is a problem of survival. In some cases, the change generates an organism more aggressive and virulent. Previous treatments are usually not useful, in particular when they are the cause of the genetic change, like the exaggerated use of pesticides and/or climate changes.

In Italy, at least three cases are focusing on scientific, social, and policy attention, causing strong alert for the consequences of their anomalous increasing speed of spreading. The first one concerned Aedes albopictus (Skuse), commonly known as the Asian tiger mosquito [16–18]. This species is currently retained the most invasive mosquito species in the world since it is able to rapidly adapt to different anthropogenic environments, thanks to its ecological and physio□ logical plasticity [19]. Recently, the Asian tiger mosquito has invaded many countries, spreading rapidly to Europe, North and South America, the Caribbean, Africa, and the Middle East [20,21]. A. albopictus is both a nuisance and a disease vector. Its medical importance is mainly due to the aggressive daytime human-biting behavior and to its ability to transmit many diseases. It works as a vector for many viruses, including dengue, yellow fever, West Nile, Japanese encephalitis, St. Louis, encephalitis virus (Flaviridae, genus Flavivirus); chikungunya, Eastern equine encephalitis, Venezuelan equine encephalitis, Western equine encephalitis, Ross River, Sindbis, Mayaro, Getah (Togaviridae, genus Alphavirus); Potosi, San Angelo, La Crosse, Jamestown Canyon (Bunyaviridae, genus Bunyavirus); Rift Valley fever (Bunyaviridae, genus Phlebovirus), and Orungo virus (Reoviridae, genus Orbivirus). A. albopictus is also the vector of different filariasis, such as Dirofilaria immitis Leidy, Dirofilaria repens Railliet and Henry, and Setaria labiatopapillosa Perroncito [19]. Although the introduction of Tiger mosquito was casual, probably due to the commerce of old tires, the permanence is clearly due to the climate change with the rising of temperature.

Two recent exceptional cases of overflowing insect-borne diseases not directly involving human health are reported to evidence of the difficulties of fighting new insect emergencies. Both emergencies are the results of a complex multiorganism interaction. The network of several different organisms is on the basis of the mechanism of survival and diffusion, conditioning effectiveness and metabolism of insect vectors, including the resistance phenomenon. The network acts like a "superorganism," integrating functions of all the different types of involved organisms, asking for several levels of eco-friendly actions. Integrated methods are urgently needed for control of these pests. On the basis of novel knowledge, some new approaches are emerging, changing the aspect of insect control.

Both cases present several, not casual, similarities and therefore can be considered as para-digms of next future or actual situations. The first case is the expanding relevance of bluetongue disease, vectored by Culicoides sp., so far concerning Southern Europe (Sardinia in particular) and going to be present in other countries. The second case is the alarm concerning the olive trees epidemic disease, probably due to Xylella fastidiosa, which may lead to the disappearance of extensive areas cultivated with olive trees in Southern Italy. Actually, no useful control tools have been reported.

3. XYLELLA FASTIDIOSA: A THREAT FOR OLIVE TREES

In 2013, one of the most beautiful part of South Italia, the Salento Peninsula, well known for the production of the olive oil and wines, was alerted by a dramatic phenomenon, never reported since human memory. The centenary olive trees that are the hearth of the monumental natural architecture of the region started to die, as Golias killed by invisible Davids. The sentence was that the responsible bacterium, X. fastidiosa, causes so-called Pierce's disease. So far, Pierce's disease (PD) was mainly known as a deadly disease of grapevines [22]. It is caused by the bacterium X. fastidiosa, which is spread by xylem feeding leafhoppers, known as sharpshooters.

PD is known to be prevalent within the US from Florida to California and outside the US in Central and South America (Table 1).

Figure 1. The effect of Xylella fastidiosa on olive trees in Apulia (Photo: Marcello Nicoletti, October 2014).

Table 1. Chronology of Xylella fastidiosa outbreaks

1870	Reports in California of grape wine "mysterious disease"
1890	The disease practically disappeared
1892	Newton B. Pierce reports on the disease on grapes in California.
1920	New epidemic diseases in California, apparently not linked to the previous episode
1920	Alfa-alfa disease (AD), no other cases reported
1930	Hewitt names the rediscovered grape wine disease as Pierce disaster (PD)
1930	Reports on peach disease
1933	PD spreads in South US
1940	Major epidemic disaster; vectors come from alpha-alpha (AD) through the some "virus"; they are xylem sap- feeders and disease is xylem-limited
1970	Almonds and oaks also affected; symptomless plant host discovered
2011	First cases of dehydrated olive trees near Lecce town in Puglia
2012–2014	Most of the peninsula of Puglia region, named Salento, evidences the presence of Xylella in the centenary olive trees.
2014	The disease spreads interesting more than 9000 hectares.
2015	The eradication campaign, in accordance with the EU protocols, starts with the destruction of dozens of trees, against the population concern. The regional court accepts the considerations against the eradication, but the campaign goes on.

X. fastidiosa works blocking the xylem, which conducts the water around the plant. Symptoms include chlorosis and scorching of leaves, and entire vines will die after 1–5 years. Pierce's disease is less prevalent where winter temperatures are cold, that is, at high altitudes and in inland northern areas [23–26]. The anomalous recent diffusion of the bacterium X. fastidiosa is causing a great alert and enormous damage. The disease risks to menace the surviving of olive trees, at least in Southern Italy. Starting from a little area in Gallipoli, near the town Lecce, most of the olive trees of a great part of Apulia region in the South of Italy were totally destroyed during the last 2 years. Therefore, the diffusion was very rapid, epidemic, and devastating. However, the disease is known by long time and so far considered mainly affecting grape wine and sporadically other plants, like oleander, almond, cherry tree, Polygala myrtiflora, and Spartium junceum. Concerning olive trees, so far it was considered one of the hundred diseases affecting the species, without any report of epidemic virulence.

In October 2013, the bacterium was found to be infecting olive trees in the region of Apulia in southern Italy. The disease was causing a rapid decline in olive plantations, and by April 2015, it was affecting the whole Province of Lecce and other zones of Apulia, focused in the Salento Peninsula. Almond and oleander plants in the region have also tested positive for the pathogen. The disease has been called Olive Quick Decline Syndrome (OQDS). The disease causes withering and desiccation of terminal shoots, distributed randomly at first but which then expands to the rest of the canopy. This results in the collapse and death of the trees (Fig. 1). In the affected groves, all of the plants show symptoms. By the beginning of 2015, it had infected up to a million trees in the southern region of Apulia [25–30]. The epidemic damage affected thousands of centenary olive trees, completely dehydrated by the disease. After the eradication, the treated areas appeared totally desertified. Beside the great economic damage, loss of the olive trees means a tremendous cultural and environmental impact for a territory, where they are the symbol of region's identity. Centenary olive trees are the living monuments of the Apulia. Furthermore, there are convincing hypotheses that the epidemic infection will propagate rapidly to the neighboring regions and later in several parts of the Mediterranean area. Therefore, interested countries, like France and Greece, are asking for a rapid control of the disease before further pandemic diffusion.

In conclusion, there are two main hypotheses about future scenarios: (a) a natural stress- induced dieback: this is consistent with widespread groves of various ages all suffering to different degrees and slowly declining rather than a virulent point infection that can be seen to spread. In other words, the disease is due to a "normal" increasing of virulence coupled with the "stress effect," derived from climate change and agricultural loss, that will affect mainly the old trees, causing a turnover in favor of the new stronger generations; (b) a modified, more virulent pathogen appeared, and plants defenses will be not able to face the new challenge, with devastating consequences.

So far, the only real performed action was the application of the UE protocol consisting in the eradication of any olive tree and creating a defensive line of 2 km in extension in the northern part of Salento Peninsula, where any plant must be eradicated, in order to isolate the disease. This measure should stop the diffusion of X. fastidiosa? The bacterium relies to insect vectors. Known vectors of X. fastidiosa are xylem-sap feeder insects belonging to the families Cicadellidae, Aphrorophoridae, Cercopidae, and Cicadidae within the Cicadomorpha. Among them, the meadow spittlebug, Philaenus spumarius, is one of the most abundant field insect in that region, although other species are probably involved. The spittlebug xylem-sap feeding possesses a piercing-sucking beak, named rostrum. Introducing the rostrum into the tree for feeding, the insect causes the infection and the bacterium propagation close the xylem vessels, causing the dehydration of the plant. It is very difficult that the EU protective approach will give significant effects. The vectors are not good flyers; they usually move by jumping, but they can be efficiently transported by the wind and human of animal occasional transportation, travelling for many kilometers in only one day. The block of plant import from America is in act. Also, in this case, the total control is quite impossible.

The bacterium cannot be controlled by the use of antibiotics, either because they are banned in agriculture in EU or because they are highly costly and complicated. The insect vector could be controlled by adequate insecticide. Insecticides usually select as target the adults, but the larval stage is the best situation to act on the insects, before they are able to move and fly away.

In any case, to obtain any real result, we must learn the Nature's lesson. Olive tree disease evidences three main actors: the bacterium, the vector and the plant, and probably a symbiotic fungus. They work together, acting like a "superorganism." It is a very complex system, but in some way also very efficient. The only way to face the X. fastidiosa challenge is an integrated pest management. It is necessary to operate considering together the several involved aspects: a treatment of soil to sustain the plant; an insecticidal agent to control the insect; a natural, low cost, and eco-friendly antibiotic to be inserted inside the plant.

A key step is the reply to the following question: How did X. fastidiosa become so dangerous in the last 2 years? We know the presence of bacterium in Italy from at least 30 years, and so far it was considered just one of the several diseases involving olive oil. Something happened during the last years changing completely the equilibrium between the microorganism and the host. There are several hypotheses about the causes of the change and a consequent relevant debate. The local official institutions have accepted the idea that some infected plants of oleander imported from Costa Rica were the epidemic start. Everything could be, but it is strange that only a single little point of Apulia was affected by the only infected imported plant.

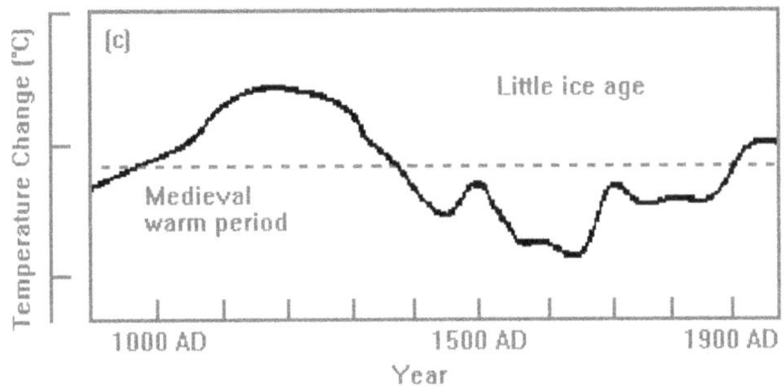

Figure 2. Temperature changes over the last centuries.

The second hypothesis is that a change in the X. fastidiosa genoma occurred, giving rise to more aggressive and dangerous strains. In this case, there are two possibilities: the change is derived from some experiment or a biological cause. Due to climate change (Figure 2), some virulent strains from hotter countries could be able to survive and proliferate against local populations. Other causes, like the absence of the traditional treatment of the soil and the trees due the urbanization of the population, could have interfered. The phenomenon is quite well known and possible, as explained in the next part of this chapter.

4. THE CLIMATE HYPOTHESIS

There have been several marked changes in climatic conditions in Europe during the last century. In the 20th century two main periods of warming have occurred in Europe (Fig. 2). The second was the warmest decade (the 1976–2000) on record, and it is still in act. The evidence was an increasing of approximately 1.2°C over the past 100 years, which means twice the average global rate. Turnover of hot and glacial periods is a normal trend for our planet, mainly due to the CO2 the quantity of this gas in the atmosphere. We are now in a warming period, and the trend is accelerated by the emission due to human activities [30–34]. Warning consequences are higher nighttime temperature, with limited difference between day and night, and few frost days in winter, associated with milder temperature in all winter period, longer dry periods, and peaks of temperature. In particular, temperature increases were most marked in both Central Europe (Italy, Corsica, and Balearic Islands) and Eastern Europe (western Bulgaria, northern Greek, Albania, Macedonia, Bosnia, Montenegro, and Croatia). On the contrary, central Iberia and the region around the border between Morocco and Algeria have cooled. Simplifying the model tendency, Europe is getting warm and North Africa is cooling. This resulted in changes in precipitation dividing Europe in two parts: the number of wet days have increased in the North Europe and decreased on the South, increasing the already presence tendency to desertification of several regions. North Europe, including the UK, northern Iberia, and Scandinavia, is becoming wetter, whereas southern Iberia, France, Germany, and Italy are becoming drier.

Vector-borne pathogens are particularly sensitive to climate, a fact that has led to widespread and continued speculations that anthropogenic climate change will increase the incidence and intensity of their transmission. Other nonclimatic abiotic and biotic factors can also affect disease distribution. Diffusion can be very rapid and effective. Adult insects are usually not strong fliers, but they can be passively dispersed by the wind, possibly up to several kilometers in a single night, especially over the sea. Thus, natural barriers cannot be considered an efficient control of the diffusion. Otherwise, they can travel utilizing ancient transportations, like other animals (street ruminants), or new unexpected ones, i.e., inside old tires as happened for A. albopictus.

There is an urgent need for ecologically sound, equitable, and ethical pest management, based on control agents that are pest-specific, nontoxic to humans and other biota, biodegradable, less prone to pest resistance and resurgence, and relatively less expensive. The last aspect is fundamental for a large-scale use in emerging countries.

5. BLUETONGUE SPREAD IN EUROPE

Bluetongue is a devastating disease of ruminants, mostly restricted to certain breeds of sheep, particularly fine wool and mutton breeds common in Europe [35] (Table 2). Until 1998, bluetongue has made only sporadic incursions in Europe, until six strains of the blue tongue virus (BTV), from the Middle East, were transferred to Europe, through two main pathways. One spread northward involving Greece and Balkans. The second one interested the North Africa, and from Tunisia/Algeria landed to Sardinia, Sicily, Corsica, and Balearic islands. Clearly, these pathways are coincident with the traditional livestock trade routes, such ruminant street. The same routes were used in the last years by human migrants to reach the Italian peninsula, i.e., across the Adriatic Sea from Albania and across the Sicily channel from North Africa [36, 37].

Table 2. Chronology of blue tongue virus (BTV) spread in Europe

1969	First isolation in Greece of strain BTV-9 and BTV-1
1998	Isolation in Tunisia of the strain BTV-2, endemic of sub-Saharan Africa, and belonging to strains from South Africa, Nigeria, Sudan, US.
2000–2001	Isolation in Greece and Turkey of the European strain BTV-1, similar to viruses that have been isolated in India. First outbreak in Bulgaria.
2003	A new strain BTV-4 type, different from that of Greece and Turkey, is isolated in Corsica, Sardinia and Balearics.

2004	Detection of strain BTV-4 in Sicily and France.
2005–2014	Distinct strains are still entering in Europe, affecting at least 12 countries and more than 800 km further north than before.
2014	In Sardinia, the BT disease caused the deaths of 13,000 sheep and damages for 42 million of euro. 5772 infection sites were detected.

Since its arrival, BTV has caused the deaths of more than one million sheep, and the loss of trade in animals and animal products, with an estimated damage of US$125 million in the US alone [38]. Sardinia was in particular affected, being the economy largely based on sheep, producing very appreciated like cheese (pecorino) and fine wool. The widespread use of a vaccine, although effective, caused a series of problems for its distribution and episodes of corruption.

6. NATURAL PRODUCTS HELP

Natural products are mainly derived from plants as the result of coevolution between organisms and environment. For this reason, they are used for centuries in popular and traditional medicines, as well as often as spices and insecticides. Unlike modern pharmacology and drug development based on single chemical entity, natural product preparations are multi-ingredient, derived from the historical references and empirical experiences. A single herbal drug contains at least hundred of compounds making a complex matrix, named phytocomplex in which not the single active constituent is considered the only responsible for the overall efficacy. The phytocomplex utilization is not a philosophy because many data afford the validity of this approach and others can be obtained using the modern pharmacological devices.

The study produced in 2010 by MIT and the Broad Institute of Harvard University (US) is a clear step in the direction of a scientific validation of natural products [39]. The argument strictly relates to the past and future role of natural products and the endless debate about their efficacy, often resulting into a fighting contrast between natural products supporters vs. synthetic drugs defenders. The key argument was to understand what is going on between the two main levels of the metabolism, involving the functional connection between genes and genes products and between targets and genes. The MIT researchers decided to commit the argument to the neutral judgment of artificial intelligence. The computational work was based on the comparison of cumulative connectivity distribution of small molecules, natural or synthetic, grouped according to connectivity associated with the target, assuming that proteins form biological networks. The result is simple: natural products target the proteins with a high number of protein–protein functional interactions (higher network connectivity), whereas the synthetic ones act on limited protein network: "We observe that approved drug targets that are not also natural product target exhibit a connection distribution much closer to the case for human disease genes that natural product targets, which remain the most highly connected targets."

Natural products tend to target proteins more essential and general to an organism than other groups of small-molecule targets, like those related to disease genes. They therefore work as a nonspecific basic defense against predators or pathogens acting on more highly connected proteins, interrupting essential protein activity of the environmental competitor or invader. However, natural products are not only defense and toxic substances. On the contrary, the story of plant evolution and the experience evidence the progressive production of positive substances produced in favor of a collaboration with the animals present in the same habitat. They may be tailored for a positive or negative influence in physiologic activities and basic metabolism. These argumentations are in favor of the potential use of natural products as insecticides.

7. THE NEEM OPPORTUNITY

The tree Azadirachta indica A. Juss (sin. Melia azadirachta) is commonly known as neem or nimba, margosa or Indian neem, Indian lilac, the last one to distinguish from the similar species Melia azedaracht L., named Melia o Persian lilac (Fig. 3). Several exceptional terms were used to describe the importance and the value of neem, i.e., "the marvellous tree, the tree of XXI century, the divine tree, India's tree of life, Nature's drugstore, Panacea for all diseases, a tree for solving global problems" [40–45]. In 1989, WHO/UNEP considered the neem tree as one of the most promising tree of the 21st century. In 1992, the US National Academy of Sciences published a report having the significant title "Neem—A Tree for Solving Global Problems" [45]. The medicinal use of neem is strongly eradicated into the Indian tradition. All its parts are largely used for many illnesses, and in Indian rural areas, the plant is called "the village pharmacy." Neem is considered a natural exemplar insecticide.

Figure 3. The neem tree (Azadirachta indica A. Juss)

The neem tree pertains to the Meliaceae (Mahogany family). It is a fast-growing evergreen tree, native of Indian subcontinent and distributed in the Tropical and Subtropical areas. Neem [46– 48], owing to its ability of growing so easily and surviving on dry, nutrient-lean soil, is now cultivated in tropical and subtropical countries, including South Asia, West Africa, central (i.e. Cuba) and South America, and Australia. Flowering occurs from January to May. Flowers are fragrant, beautiful, and abundant. Fruits ripening from June through August are green ellipsoidal drupes containing one seed. A single mature tree may produce annually 5–8 kg of seeds.

The main product of neem is the oil obtained by expressing the kernels containing the seeds (Figs. 4 and 5). Neem seed oil (NSO) is obtained by different extraction methods. Most of the NSO is produced in India by familiar little producers using very simple machines, but many other countries are now producing NSOs. Therefore, considering also the possible different geographical origin of the raw material, combined pre- and postharvesting factors can result in great differences in constituents present in marketed NSOs, as already reported [49].

The chemistry of neem is very complicated and still far to be completed, despite the great number of dedicated researches. More than 300 compounds have been characterized from the seeds. One-third of them are nortriterpenoids, which means triterpenoid lacking some carbon atoms. Partial loss of lateral chain is combined by a complicated rearranging of the remaining part, giving rise to different polyclicic molecular skeletons full of oxygenated functional groups, partially acylated.

Figure 4. The kernels, containing the seeds, that are used as raw material for the production of the oil, mainly produced in India by little and simple producers (figure on the right).

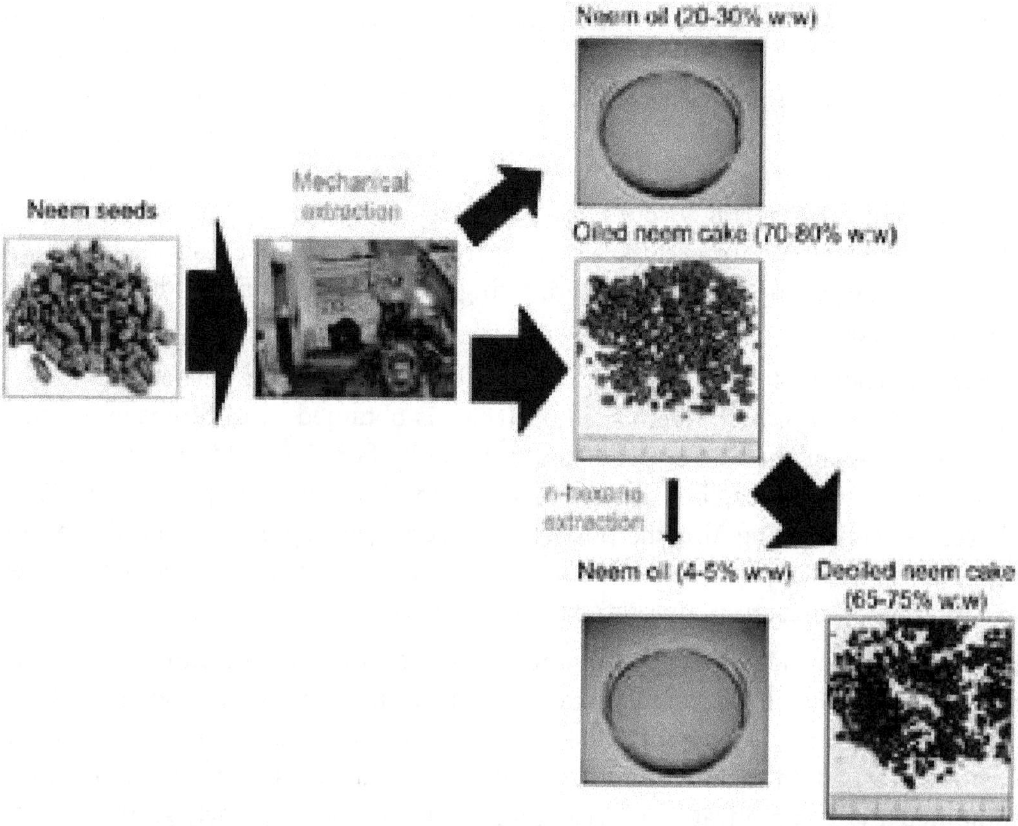

Figure 5. The process of neem oil production and neem cake (modified from Benelli et al., 2015).

Among major nor-triterpenes are limonoids, azadirachtin A and B, nimbin, nimbidin, salannin, and nimbolides (Fig. 6) [50]. Unlike chemical insecticides, neem compounds work on the insect's hormonal system, not on the digestive or nervous system, and therefore do not lead to development of resistance in future generations. The limonoids present in neem make it a harmless and effective insecticides, pesticide, nematicide, fungicide, etc. The most significant limonoids found in neem with proven ability to block insect growth are azadirachtin, salanin, meliantriol, and nimbin. Azadirachtin is currently considered as neem main agent for controlling insects. It appears to cause 90% of the effect on most pests. It does not kill insects – at least not immediately – instead both repels and disrupts their growth and reproduction. Research over the past years has shown that it is the most potent growth regulator and feeding deterrent ever assayed. It can repel or reduce the feeding of many species of pest insects as well as some nematodes. In fact, it is so potent a deterrent that a mere trace of its presence prevents some insects from even touching plants.

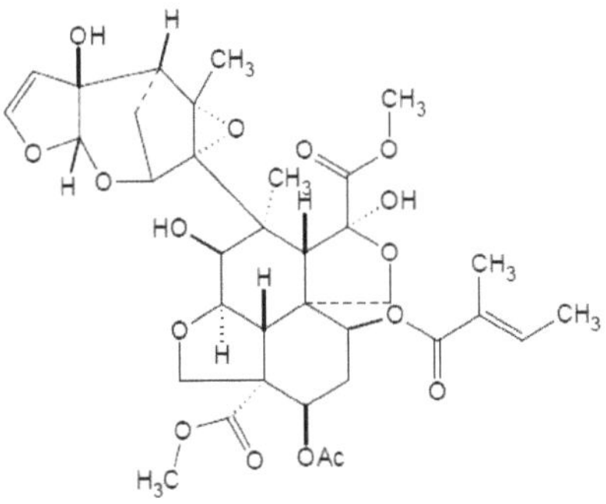

azadirachtin A

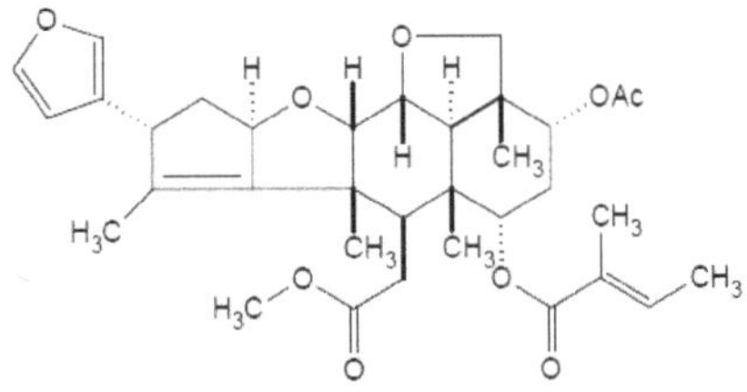

salannin

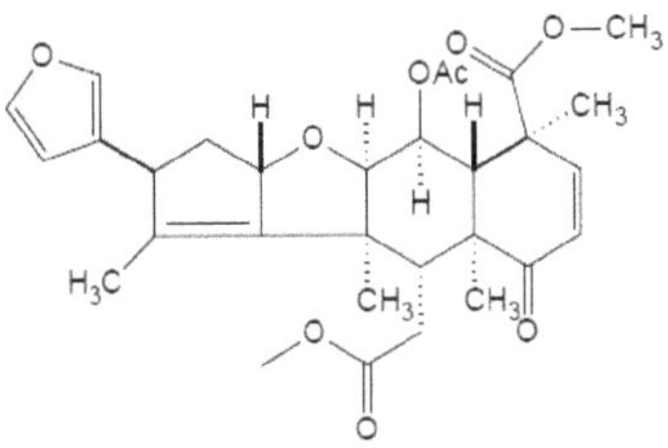

nimbin

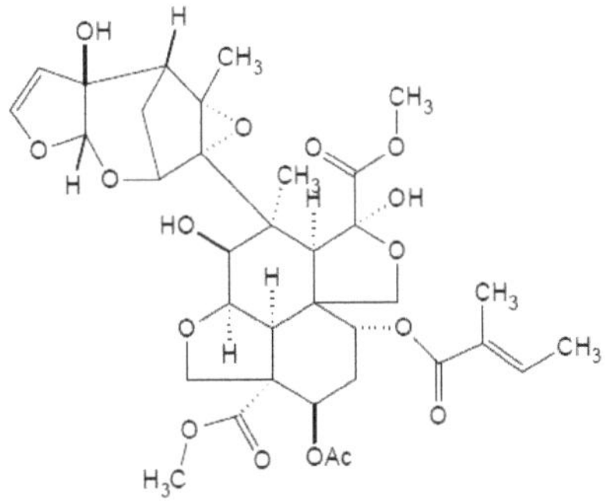

azadirachtin B

Figure 6. The main limonoids of neem.

Many formulations deriving from neem seeds show antifeedancy, fecundity suppression, ovicidal and larvicidal activity, insect growth regulation, and/ or repellence against insect pests, even at low dosages [51–56], including ticks, house dust mites, cockroaches, raptor bugs, cat fleas, bed bugs, Sarcoptes scabiei mites infesting dogs, poultry mites, and beetle larvae parasitizing the plumage of poultry. The insecticidal properties, environmental safety, and public acceptability of neem and its products have been certified by the US EPA [57] and have led to its adoption into some control programs against Diptera pests [58]. Noticeably, emulsified formulations of A. indica oil showed an excellent larvicidal potential against different mosquito genera, including Aedes, Anopheles, and Culex, also under field conditions.

Action mechanisms include repellence, feeding and oviposition deterrence, but hormonal effects are the key of the inversion of control strategy, changing the target from the adult everywhere dispersed to the locally maintained larvae, through growth inhibition, mating disruption, chemo-sterilization, etc. In fact, hormones are necessary for to complete the process of metamorphosis as the insects pass from larva to pupa to adult. In any case, if the larva manages to enter the pupal stage, the adult emerging from the pupa is 100% malformed, absolutely sterile without any capacity for reproduction. The insect populations decline drastically as they become unable to reproduce. However, also antifeedant and deterrent activities are important to defend crops. The ideal plant-derived product, including insecticide, should be eco-friendly, sustainable, low cost, and target specific, leaving unaffected the beneficial ones. Neem products do not leave any residue on the field, being biodegradable by the action of sunlight. Azadirachtin in open space after dissipation has a half-time of about 20 h. The degradation slowly occurs also when neem products are stored under appropriate conditions [59–60]. Neem, at usual concentrations, is harmless to nontarget and beneficial organisms like pollinators, honeybees, mammals, and other vertebrates [61]. The absence of toxicity is largely evidenced by the millenary use in Indian traditional medicine, as well as by the EPA report and the large use during the last 30 years, including products for pet care.

Neem cake is the residue that is left over when the kernel is crushed from neem kernels containing seeds and the remaining is pressed to obtain the oil (Fig. 5). In fact, although the overall marketed name is seed neem oil, not only the seeds are utilized. Neem cake looks more like flour than cake, with differences in color and size of particles. Two products are therefore in the market: neem oil cake obtained by cold pressure, with 6% of the oil still residue, and neem cake deoiled, with a residue up to 1.5% [62].

Actually, it is not approved as pesticide, and mainly it is highly appreciated as organic fertilized. Neem cake acts as a natural fertilizer with pesticide properties, protecting crops from nematodes, soil grubs, and white ants.

India alone has an annual potential of 80,000 tons of oil and 330,000 tons of neem cake from 14 million plants that grow naturally. To this potentiality, the high number of cultivations actually occurring in many parts of the world must be added. This situation evidences the neem's high sustainability and possibility to have an increasing production of low cost products to be utilized in many fields, not only insecticides, from medicine to the cosmetic one. The importance of the neem future is strictly related to this wild range of utilizations, which are strictly linked to the new market of plant natural products.

8. THE NEEM CAKE ALTERNATIVE

Despite the evidence of efficacy, several factors limit the massive use of neem oil in control of insect vectors. Limits include high cost, photosensibility, and persistence in the soil. A network of several Italian universities and research institutions decided to investigate the larvicidal activity of the neem cake as an alternative. In fact, neem cake is a low cost by-product of neem oil production (Fig. 7).

Figure 7. Nem cake ready for exportation.

The first step of the network was analytical. Several neem cakes from different producers and importers were analyzed by high-performance liquid chromatography (HPLC), evidencing still the presence of neem limonoids, but with different pattern in comparison with neem oil. Percentage was low but very different in each sample and salannin was the prevalent nortriterpene (3750 ppm of azadirachtin A+B, 7980 ppm of salannin, and 1850 ppm of nimbin) [63– 65]. The high-performance thin layer chromatography (HPTLC) analyses, performed in the laboratories of Environmental Biology at the Sapienza University of Rome, allowed to evidence a great complexity of the neem cake extract, showing at least more than 30 secondary metabolites spread in all range of polarity (Fig. 8). On the basis of the information obtained in the chromatographic analyses, a neem cake product was selected and used for the activity tests, realized at the ENEA laboratories. Laboratory essays evidenced a significant activity of neem cake n-hexane and ethylacetate extract against A. albopictus mosquito larvae [66].

In the same time, another group of the network, at the University of Sassari in Sardinia, worked on Blue Tongue disease. Culicoides species are vectors of BTV [67]. These insects breed in mist microhabitats, like small pools, irrigation channels, beverage sites, and drainage pipes. Culicoides imicola is the main vector, representing about 10% of all emerged Culicoides adults. In the laboratory essays, larvae of C. imicola resulted highly sensitive to the commercial neem cake. The larval mortality in water after 7 days gave a lethal concentration value (LC50) of 0.37 g/l. In order to define the chemical nature of active constituents, a neem cake methanol extract was separated by different solvents. Fractions of increasing polarity were assayed on Culicoides larvae. The most active resulted the ethyl acetate fraction, containing 1 ppm of azadirachtin, 1.5 ppm salannin, and 0.3 ppm of nimbin. The fraction was more toxic than a commercial formulation at the same azadirachtin concentration.

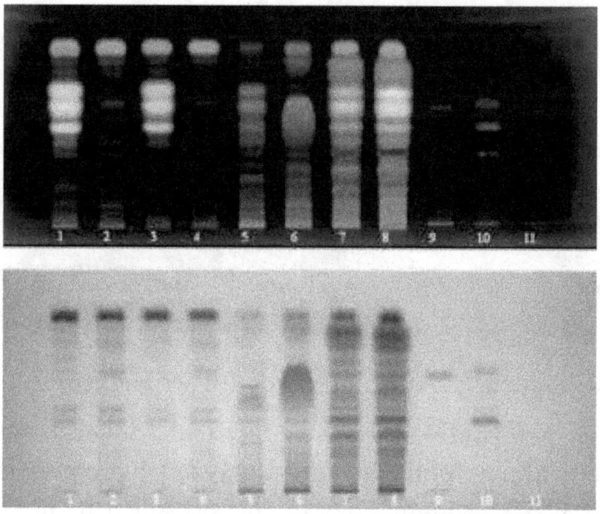

Figure 8. HPTLC analysis of different neem products. Mobile phase: toluene, ethyl acetate (4:6 v/v). Derivatization: Anisaldehyde. Plate on the top, visualization: UV366 nm. Plate on the bottom visualization: white light. Tracks: (1) neem oil marketed in Italy extracted with ethyl acetate, (2) neem oil marketed in India extracted with ethyl acetate, (3) neem oil marketed in Italy, (4) neem oil marketed in India, (5 and 6) neem cakes extracted with ethyl acetate, (7 and 8) neem cakes of tracks, (5) defatted and concentrated (track 8 more concentrated), (9) nimbin, (10) salannin, and (11) azadirachtin A.

Strategy in field trials was based on the deposit of neem cake in the typical larval sites of Culicoides, and again the product was found to be very effective. A treatment with neem cake at dose of 100 g/m was applied in a larval breeding site of Culicoides located in a riverside of a pond margin of a livestock farm in Sardinia, Italy. The treatment with the neem cake resulted in a significant reduction in Culicoides emergence until 28 days.

Finally, activity tests are now in progress at the ENEA laboratories to measure the larvicidal toxicity against P. spumarius, as main vector of X. fastidiosa. First experiments were positive but limited to the laboratory conditions. Field experiments are urgently needed.

Although the mentioned results need confirmation and utilization in larger scale, neem cake is a promising material for the development of newer products useful in the control of vectors of insect-borne diseases at the larval stage.

9. CONCISE SUMMARY

The above examples pertain to Southern Italy; nevertheless, circumstances in other regions of the world are likely somewhat analogous. Chun-Xiao et al. documented the correlation between pesticide resistance and genomic alterations in Aedes aegypti in Southern China [68]. The correlation has been identified for numerous pesticides, however the mechanisms of resistance remain incompletely elucidated. Knockdown resistance to both pyrethroids and DDT has been associated with amino acid alteration in the sodium channel. The paper attributes the causes of resistance not only to the extensive recent use of pyrethroids and various factors, primarily climate change, but also to the rapid development of tourism, transportation, and increasing urbanization, which may enhance A. aegypti breeding sites. Consequently, the emergence of resistance should be regarded as a complex multifactorial event. The intricate solution must have a multitreatment involving a minimum of three stages.

The soil must be considered not only to sustain the plant with adequate fertilizer able to give the necessary elements but also to positively change the biome living underground. Roots must be considered not only as the corm part necessary for collection of water and minerals but also as a part of the plant integrated to the underground habitant, including the living system. Insect vectors must be controlled possibly at the larval stage, when the insects are not able to move and need mist conditions to survive. Insecticides should be environmentally sustainable and economically viable, aimed at safeguarding pollinators and beneficial insects. Natural compounds may serve as the foundation for the development of next-generation antibiotics, utilizing diverse mechanisms and suitable for large-scale application without significant environmental harm. This multidisciplinary approach underscores the necessity for enhanced collaboration among pharmacologists, chemists, parasitologists, entomologists, and behavioral ecologists.

REFERENCES

1. Carlson R. Silent Spring. Farwcett World Library. USA. (1962).
2. Caspar A. Hallmann, Ruud P. B. Foppen, Chris A. M. van Turnhout, Hans de Kroon, Eelke Jongejans (9 July 2014). Declines in insectivorous birds are associated with high neonicotinoid concentrations. Nature. Retrieved 14 July 2014. doi: 10.1038/ nature13531.
3. J. P. van der Sluijs, V. Amaral-Rogers, L. P. Belzunces, M. F. I. J. Bijleveld van Lex mond, J-M. Bonmatin, M. Chagnon, C. A. Downs, L. Furlan, D. W. Gibbons, C. Giorio, V. Girolami, D. Goulson, D. P. Kreutzweiser, C. Krupke, M. Liess, E. Long, M. McField, P. Mineau, E. A. D. Mitchell, C. A. Morrissey, D. A. Noome, L. Pisa1, J. Settele, N. Simon-Delso, J. D. Stark, A. Tapparo, H. Van Dyck, J. van Praagh, P. R. Whitehorn, M. Wiemers (10 October 2014). Conclusions of the worldwide integrated assessment on the risks of neonicotinoids and fipronil to biodiversity and ecosystem functioning. Environmental Science and Pollution Research 22: 3229. doi: 10.1007/ s11356-014-3229-5.

4. Neonicotinoid Pesticides and Honey Bees. Washington State University Extension Fact Sheet FS122E.

5. Gill, R. J., O. Ramos-Rodriguez, N. E. Raine (2012). Combined pesticide exposure severely affects individual and colony-level traits in bees. Nature 491(7422): 105–108. doi: 10.1038/nature11585.

6. Tadeusz Banaskiewicz (2010). Evolution of pesticide use. In: Contemporary Problems of Managements and Environmental Protection. Vol. 5. Influence of the Pesticide Dump on the Environment, p. 7–18.

7. Insecticide Resistance Action Committee. Resistance Management for Sustainable Agriculture and Improved Public Health. 2007. Retrieved December 2014.

8. J. Hemingway, H. Ranson (2000). Insecticide resistance in insect vectors of human disease. Annual Review of Entomology 45: 371–391.

9. D. W. Onstad (2008). Insect Resistance Management. Elsevier: Amsterdam.

10. S. U. Karaagac (2011). Insecticide Resistance. In: Advances in Integrated Pest Management, Farzana Perveen (ed.). InTech, Rijeka, Croatia, p.469–478.

11. G. Benelli (2015). Research in mosquito control: current challenges for a brighter future. Parasitology Research. doi: 10.1007/s00436-015-4586-9.

12. W. G. Ware, D. M. Whitacre (2004). The Pesticide Book. Meister Media Worldwide, Willoughby, Ohio.

13. Bayo F. Sanchez (2011). Ecological impact of insecticides. In: Advances in Integrated Pest Management, Farzana Perveen (ed.). InTech. Rijeka, Croatia, p. 61–90.

14. C. Gordon (2003). Role of environment stress in the physiological response to chemical intoxicants. Environmental Research 92: 1–7.

15. G. Benelli, K. M. Daane, A. Canale, C. Y. Niu, R. H. Messing, R. Vargas (2014). Sexual communication and related behaviours in Tephritidae—current knowledge and potential applications for integrated pest management. Journal of Pest Science 87, 385– 405.

16. R. G. Estrada-Franco, G. B. Craig (1995). Biology, Disease Relationship and Control of Aedes albopictus. Pan American Health Organization, Washington, DC. Technical Paper No. 42.

17. C. J. Mitchell (1995). Geographic spread of Aedes albopictus and potential for involvement in arbovirus cycles in the Mediterranean Basin. Journal of Vector Ecology 20: 44–58.

18. R. Romi (1995). History and updating of the spread of Aedes albopictus in Italy. Parassitologia 37: 99–103.

19. M. Q. Benedict, R. S. Levine, W. A. Hawley, L. P. Lounibos (2007). Spread of the tiger: global risk of invasion by the mosquito Aedes albopictus. Vector Borne and Zoonotic Disease 7: 76–85.

20. C. Caminade, J. M. Medlock, E. Ducheyne, K. M. McIntryre, S. Leach, M. Baylis, A. Morse (2012). Suitability of European climate for the Asian tiger mosquito Aedes albopictus: recent trends and future scenarios. Journal of the Royal Society Interface 9: 2708–2717.

21. C. Paupy, H. Delatte, L. Bagny, V. Corbel (2009). Fontenille. Aedes albopictus, an arbovirus vector: from the darkness to light. Microbes and Infection 11:1177–1185.

22. J. M. Wells, B. C. Raju, H. Y. Hung, W. G. Weisburg, L. M. Parl, D. Beemer (1987). Xylella fastidiosa gen. nov., sp. nov.: Gram-negative, xylem-limited, fastidious plant bacteria related to Xanthomonas spp. International Journal of Systematic Bacteriology 37(2): 136–143. doi:10.1099/00207713-37-2-136.

23. Minimizing the Spread of Disease in Italy's Famous Olive Trees. Our Environment at Berkeley (in English) (9 February 2015). University of California at Berkeley, Department of Environmental Science, Policy, and Management (ESPM). Retrieved 5 May 2015.

24. Public Intellectual Property Resource for Agriculture. PIPRA. Pierce's disease. Pierce's Disease website. "Major consequences" of olive disease spreads across EU. BBC News. 9 January 2015.

25. First report of Xylella fastidiosa in the EPPO region. European and Mediterranean Plant Protection Organization (EPPO).

26. Expert Says Eradication of New Olive Tree Disease in Europe Unlikely (29 March 2014). Olive Oil Times. Retrieved 1 March 2015.

27. Italy warns deadly olive tree bacteria could spread across Europe (27 February 2015). The Telegraph. Retrieved 1 March 2015.

28. S.E. Randolph (2004). Evidence that climate change has caused "emergence" of tick- borne disease in Europe. International Journal of Medical Microbiology 37, 5–15.

29. P. Reiter (1998). Global warning and vector disease in temperate regions and at high latitudes. Lancet 351, 839–849.

30. P. Reiter (2001). Climate change and mosquito-borne disease. Environmental Health Perspectives 109 (Suppl. 1): 141–161.

31. V. P. Pursue, P. S. Mellor, D. J. Rogers, A. R. Samuel, P. C. Mertens, M. Baylis (2005). Climate change and the recent emergence of bluetongue in Europe. Nature Review/ Microbiology 3: 171–181.

32. T. M. L. Wigkey, D. Schimel, eds. (2000). The Carbon Cycle. Cambridge, Cambridge University Press.

33. G. Cook (1992). Effect of global warning on the distribution of parasitic and other infectious diseases: a review. Journal of the Royal Society of Medicine 85: 688–691.

34. P. S. Melior, E. J. Wittiman (2002). Bluetongue virus in the Mediterranean basin, 1998–2001. Veterinary Journal 164: 20–37.

35. P. Calistri, V. Caporale (2003). Bluetongue in Italy: a brief description of the epidemiological situation and the control measures applied. Bulletin—Office International des Épizooties 15–17.

36. J. H. W. Slingenbergh, G. Hendrickx, G. R. W. Wingt (2002). Will the new livestock revolution succeed? Agriworld Vision 2: 31–331.

37. W. J. Tabachnick (1996). Culicoides variipennis and bluetongue virus epidemiology in the United States. Annual Review of Entomology 41: 23–43.

38. V. Dancik, K. P. Seile, D. W. Young, S. L. Schreiber, P. A. Clemons (2010). Distinct biological network properties between the targets on natural products and diseases genes. JACS Communications.

39. National Research Council (1992). Neem: a tree for solving global problems. Academic, Washington.

40. R. P. Singh, R. C. Saxena (1999). Azadirachta indica A. Juss. Oxford 7 IBH Publishing Co. Pvt. Ltd.

41. H. Schmutterer (1995). The Neem Tree, VCH. Verlagsgesellschaft mnH, Germany.

42. D. N. Tewari (1992). Monograph of Neem. International Book Distributors, Dehradun, India.

43. N. S. Randhawa, B. S. Parmar (eds.) Neem Research and Development, Society of Pesticide Science, India.

44. H. S. Puri (1999). Neem the Divine Tree, Harwod Academic Publishers.

45. M. Nicoletti, K. Murugan (2013). Neem the tree of 21st century. Pharmacologyonline 3, 115–121.

46. K. N. Kabra (2000). Development & Ecological Role of Neem in India, Neem Foundation, India.

47. R. T. Gahukar (1995). Neem in Plant Protection. Agri-Horticultural Publishing House.

48. R. P. Singh et al. (1996). Neem & Environment. Oxford & IBH Publishing Co. Pvt. Ltd.

49. F. R. Gallo, G. Multari, E. Federici, G. Palazzino, M. Nicoletti, V. Petitto (2012). The modern analytical determination of botanicals and similar novel natural products by the HPTLC fingerprint approach in studies in natural products chemistry. Atta-ur- Rahman (Ed.). Elsevier 37: 217–258.

50. M. Nicoletti, C. Toniolo (2013). Fingerprint HPTLC analysis of marketed products. Pharmacologyonline 3(6): 30.

51. A. J. Mordue, A. Blackwell (2003). Azadirachtin: an update. Journal of Insect Physiology 39: 903–924.

52. M. Rossi Forim, M. F. da Silva, F. das Gracas, J. B. Fernandes. Secondary metabolism as a measurement of efficacy of botanical extracts: the use of Azadirachta indica (Neem) as a model in insecticides—advances in integrated pest management. Farzana Perveen (Ed.). InTech. Rijeka, Croatia. 2011, p. 367–390.

53. C. B. Wandscheer, J. E. Duque, M. A. N. da Silva, Y. Fukuyama, J. L. Wohlke, J. Adelmann, J. D. Fontana (2004). Larvicidal action of ethanolic extracts from fruit endocarps of Melia azedarach and Azadirachta indica against the dengue mosquito Aedes aegypti. Toxicon 8:829–835.

54. R. P. Prakash Srivastava (2001). Neem and Pest Management. International Book distributing Co. T. Amorose (1995). Larvicidal efficacy of neem (Azadirachta indica) oil and defatted cake on Culex quinquefasciatus Say. Geobios 22: 169–173.[55]

55. M. S. Mulla, T. Su (1999). Activity and biological effects of neem products against arthropods of medicinal and veterinary importance. Journal of the American Mosquito Control Association 15: 133–152.

56. F. O. Okumu, B. G. J. Knols, U. Fillinger (2007). Larvicidal effects of a neem (Azadirachta indica) oil formulation on the malaria vector Anopheles gambiae. Malaria Journal 6: 63–67

57. Cold Pressed Neem Oil (025006) Fact Sheet. EPA. 2012. Cold Pressed Neem Oil (025006).

58. V. P. Sharma, R. C. Dhiman (1993). Neem oil as a sand fly (Diptera: Psychodidae) repellent. Journal of the American Mosquito Control Association 9:364–366.

59. T. Wei-Hong, S. Zhan-Qian (2006). Research on effect of four natural ultraviolet light absorbers on photostabilization of azadirachtin-A. Agricultural Sciences in China 5: 885–894.

60. S. J. Boeke, M. G. Boersma, G. M. Alink, J. J. van Loon, A. van Hnis, M. Dicke, I. M. Rietjens (2004). Safety evaluation of neem (Azadirachta indica) derived pesticides. Journal of Ethnopharmacology 94:25–41

61. M. Gandhi, R. Lal, A. Sarkaranarayaman, C. K. Baneyel, P. L. Sharma (1988). Acute toxicity of the oil from Azadirachta indica seed (neem oil). Journal of Ethnopharmacology 23:39–51 518.

62. G. Benelli, M. Nicoletti (2013). Shedding light on bioactivity of botanical by-products: neem cake compounds deter oviposition of the arbovirus vector Aedes albopictus (Diptera: Culicidae) in the field. Parasitology Research 110: 2013–202.

63. M. Nicoletti, M. Serafini, S. Mariani, T. Ciocchetti, K. Murugan (2012). Neem cake: chemical composition and larvicidal activity on Asian tiger mosquito. Parasitology Research 111 (1), 205–213.

64. S. Mariani, M. Nicoletti (2013). Antilarval activity of neem cake extracts against Aedes aldopictus. Pharmacologyonline 3:128–132.

65. M. Nicoletti, M. Serafini, A. Aliboni, A. D'Andrea, S. Mariani (2010). Toxic effects of neem cake extracts on Aedes albopictus larvae. Parasitology Research 107:89–84.

66. G. Benelli, K. Murugan, C. Panneerselvam, P. Madhiyazhagan, B. Conti, M. Nicoletti (2015). Old ingredients for a new recipe? Neem cake, a low-cost botanical by-product in the fight against mosquito-borne diseases. Parasitology Research 114:391–397.

67. C. Foxi, G. Delrio (2013). Efficacy of a neem cake for the control of Culicoides. Pharmacology Online 3: 110–114.

68. C.-X. Li, P. E. Kaufman, R.-D. Xue, M. H. Zhai, G. Wang, T. Yan (2015). Relationship between insecticide resistance and Kdr mutations in dengue vector Aedes aegypti in Southern China. Parasites Vectors 8(1): 325. doi: 10.1186/s13071-015-0933-7.

12 Chapter

RESISTANCE TO INSECTICIDES AND ASSOCIATED FITNESS COSTS

1. INTRODUCTION: AN OVERVIEW

1.1 Insecticides and mode of action

Insecticides are conventionally utilized in various human activities to eradicate or manage the population density of unwanted insects. Currently, despite the evident environmental consequences, the management of agricultural pests and disease vectors predominantly relies on the application of these compounds. Furthermore, in numerous instances, chemical substances serve as the primary method to disrupt pathogen transmission. Prior to the Second World War, the majority of insecticides were composed of inorganic compounds, along with a limited number of organic molecules, including nicotine, pyrethrin, and rotenone [1]. The contemporary epoch of organic insecticides commenced in the 1940s, an era referred to as the "pesticide revolution," during which DDT (dichlorodiphenyltrichloroethane) was utilized for the first time as an insecticide. [2].

Currently, there are 25 groups of insecticides and acaricides based on available evidence about their target sites and mode of action, according to the Insecticide Resistance Action Committee (IRAC) [3]. The World Health Organization Pesticide Scheme (WHOPES) promotes and coordinates the testing and evaluation of pesticides for public health purposes, since 1960. Its recommendations are generally adopted for national campaigns in several countries. The main insecticide classes used for vector control are: organochlorine (OC), organophosphates (OP), carbamates (CA), pyrethroids (PY), insect growth regulators (IGR), spinosyns (SP), and toxins derived from bacteria (Bacillus thuringiensis var. israelensis and Bacillus sphaericus) [4, 5]. The classes OC, OP, CA, PY, and SP include a broad range of compounds that act on the insect central nervous system and, thus, have an immediate effect.

The target site of OP and CA is the acetylcholinesterase (AChE), a conserved enzyme present in a wide variety of animals, including mammals, birds, reptiles, fish, and insects. This enzyme is responsible for the rapid hydrolytic degradation of the acetylcholine neurotransmitter at synapses, causing momentary interruption of the nerve impulse. The OP and CA insecticides bind in the AChE active site, compromising the acetylcholine hydrolysis and then accumulating the neurotransmitter at the synapses, causing repetitive nerve impulses.

The PY and OC (DDT and analogues) maintain the sodium channels in their opened conformation, generating a continuous influx of ions throughout the axons. Cyclodienes, another group of OC insecticides, act directly on the gamma-aminobutyric acid receptor (GABA), preventing the normal input of chlorine ions in the neurons, just after the nervous impulse. In all cases, regardless of the target site, OP, OC, and PY promote a continuous nerve impulse transmission that culminates in paralysis, convulsions, and death [6].

Unlike neurotoxic insecticides, IGRs do not induce an immediate death of the insects. However, they are toxic mainly against immature stages, affecting the moulting, metamorphosis processes, besides commitments in viability and reproduction of adults [7]. Based on the mode of action, the IGRs are classified into three major categories: (i) juvenile hormone mimics; (ii) ecdysone agonists; and (iii) chitin synthesis inhibitors [8].

Regarding bacterial toxins, Bacillus thuringiensis (Bs) var. israelensis (Bti) and B. sphaericus are the most commonly utilized as insecticides. Upon ingestion by larvae, the Bt toxins are activated by insect proteases and adhere to particular receptors in the midgut epithelium of the larvae. The ultimate consequence is osmotic stress that results in the breakdown of midgut membranes and, subsequently, death. [9]1.

2. INSECTICIDE RESISTANCE MECHANISMS

Insecticide resistance is considered the major challenge for control programs involving the use of chemicals. Up to 2014, populations of at least 590 species of insects were diagnosed as resistant to insecticides. Resistance to around 300 compounds, including the neurotoxics (OC, OP, CA, PY, and SP), the IGRs, and Bt toxins, has already been registered to one or more insect species [3].

Insecticide resistance has a genetic basis. Randomly arisen mutations can prompt several alterations in aspects of behavior, metabolism, and physiology of the insects, which may gain adaptive advantages in an insecticide-treated environment. Such alterations can be classified as: (i) behavioral changes; (ii) altered penetration (increased production of cuticular components that reduces intake of insecticide); (iii) target site modification; and (iv) metabolic resistance (detoxification enzymes and ABC transporters) [10]. Although evidenced, the two first aspects are less reported, whilst several studies have described and evaluated the target site and metabolic resistance mechanisms. These two, alone or combined, potentially induce a wide range of resistance levels to virtually all available insecticides [11].

Most insecticides target a single protein in the insect organism. The interaction between these molecules disrupts a normal biological process, leading to the toxicant effects. However, some mutations that induce structural alterations in the target protein can change the insecticide levels of toxicity. Moreover, most of these alterations are conserved among distinct insect orders. For instance, cyclodienes inhibit chloride ion transport by keeping the gama-aminobutyric acid (GABA) receptor in a close conformation [12]. The replacement of an alanine to a serine or glycine at the aminoacid position 302 (A302S/G) in the GABA, generally referred to as rdl mutations (resistance to dieldrin), confer resistance in several species, such as Drosophila melanogaster, Musca domestica, Hametobia irritans, Lucilia cuprina, Tribolium castaneum, Periplaneta americana, and Anopheles mosquitos [13].

The glycine-to-serine substitution (G119S)2 in the AChE (AChE-1, encoded by the ace-1 gene) confers resistance to OP and CA in Anopheles and Culex mosquitoes. Interestingly, this mutation was never found in Aedes mosquitoes, regardless of the intense use of OP against their populations. The most accepted hypothesis for this relies on the fact that the AChE-1 119 glycine is encoded by a GGA, differently from the GGC in other species. It means that in other mosquitoes a serine substitution (AGC) requires only one nucleotide change. By contrast, two concomitantly selected mutations would be necessary in Aedes mosquitoes, an unlikely situation referred to as codon constraint [14].

Similarly, several mutations associated with PY and DDT resistance are present in distinct insect orders: the kdr mutations, that impair the knockdown effect provoked by those insecticides. The most common kdr (knockdown resistance) mutation is a leucine-to-phenylalanine substitution in the 1014 codon3, although serine, histidine, cysteine, and tryptophan replacements are also found (reviews presented in Rinkevich et al., 2013 [15]). Several PY-resistant populations of major arthropod pests and disease vectors were found harboring kdr mutations. In this sense, for diagnostic purposes, different well-established tools for kdr genotyping have been implemented, specific for an increasing number of insect species. This allows a rapid and accurate access of the genetic background for PY resistance in natural populations [16].

The recent commercially introduced SP insecticides, which target the nicotinic acetylcholine receptors (nAChRs) [17], have been used for crop protection, animal health, and against human disease vectors. Three formulations of SP were approved by WHOPES for use in drinking water, increasing the chemical arsenal against mosquitoes [4]. However, resistance to this class of insecticides was already detected in a variety of insect species. A target-site point mutation (glycine-to-glutamate substitution G275E), for example, was identified in the nAChR of a Western flower thrips (Frankliniella occidentalis) in association with SP resistance [18]. Besides this single amino acid substitution, alternative splicing in the nAChR6 subunit seemed to be the mechanism selected in an SP-resistant population of the diamondback moth Plutella xylostella [19].

As exemplified above, mutations selected for resistance in the molecular targets of insecticides generally share homologous sites among different insects. These molecules are components of the nervous system, which are highly conserved among animals. Therefore, it is expected that few mutations can be maintained without impairing the essential physiological role of that molecule [20]. Target-site-resistant alleles are increasing in frequency and rapidly spreading, as well-recorded for malaria and dengue vectors. An interactive compilation of these data, organized in time and space scales, can be currently accessed on two distinct online platforms: IR Mapper (http://www.irmapper.com) and Popbio (https://www.vectorbase.org/ popbio/).

Detoxifying enzymes are naturally present in living organisms with a protective function against potential damages caused by xenobiotics and endogenous metabolites. In many cases, insecticide resistance occurs due to an increased activity of such enzymes, a mechanism known as metabolic resistance. In general, this mechanism is related with the intense use of insecticides. However, other toxic compounds, such as chemical pollutants and plant toxins can also select for metabolic resistance mechanisms in insect populations. In this sense, different xenobiotics present in the environment are probably related, at least in part, with a preadaptation for insecticide resistance in disease vector and agricultural pests [21, 22]. Basically, xenobiotics pass through a series of enzymatic steps that transform them in polar substances, soluble in water for an easier excretion [23]. The biotransformation is divided into three phases, with the participation of three main groups of enzymes. Phase I includes multiple function oxidases enzymes (MFO or P450) that carry out chemical modifications of a broad variety of xenobiotics. In phase II, glutathione S-transferases (GST) usually conduct conjugation reactions in the products resulting from the previous phase. The esterases (EST) can participate in both phase I and II, hydrolyzing ester bonds present in the xenobiotics. Finally, during phase III, the metabolites produced in the two first phases are actively exported out of the cells via ATP-binding cassette (ABC) transporters [24-27].

The metabolic resistance mechanisms are characterized by a gain in the ability for detoxifying molecules of insecticides, preventing them from reaching their targets. This acquisition can be selected by either an increase in the enzymatic activity over the insecticide (mutations that improve the detoxifying power) or an augment in the amount of copies of a specific enzyme (due to an increase in the transcription rate, for instance). Glutathione S-transferases, EST and MFO P450 enzymes are each comprised of tens of genes, composing supergene families, possibly resulting from duplication events along the evolutionary process, as well as independent gene duplications inside distinct species [28, 29]. Differently from target site mutations that can arise in homologous sites among different insect groups, several detoxifying genes are unique for some species and may be selected for insecticide resistance in a particular way.

The main questions that lie upon the molecular basis of insecticide resistance mechanisms are how many (and which) genes control the phenotype of resistance, how many mutations were selected within that gene(s), and if they are just spreading from one origin or appearing multiple times [30]. The advent of high-throughput screening molecular tools expands the searches for selected resistance mechanisms and their overall effects, toward beyond the target site mechanisms. Recent advances have revealed the complexity of metabolic systems enrolled in insecticide resistance at transcriptomic and genomic levels. Comparisons of the whole transcriptional profile between susceptible and resistant individuals generally indicate the participation of several genes in the physiological process of resistance [31-33]. In addition, genetic loci influencing the resistance can be physically mapped in the chromosomes through quantitative trait loci (QTL) approaches [34-36]. Likewise, a recent study identified several single nucleotide polymorphisms (SNPs), as well as an important and previously neglected copy number variation (CNV) related to insecticide resistance in Aedes aegypti, by combining genomic target enrichment with next-generation sequencing technologies [37].

3. EVOLUTION OF INSECTICIDE RESISTANCE

Insecticide resistance is an adaptive trait in which a set of genes are favorably selected to maintain the insect alive and able to reproduce under an environment exposed to pesticides. After being introduced, insecticides gradually eliminate the susceptible specimens, usually found at higher frequencies within populations. By contrast, harbors of resistant alleles, supposedly rare in the population, increase their frequencies along the time of continuous pesticide application. The importance of resistance alleles occurring prior to insecticide employment has been discussed since the 1950s with the works of Crown [38] and more recently incremented on ffrench-Constant's reviews [39, 40]. If resistance mechanisms hold elevated fitness cost in absence of insecticide (as discussed subsequently), the rareness of these alleles in nonexposed populations is then a direct assumption. In this case, the selection of resistance genes is a post adaptive response. On the other hand, pre adaptive selection of resistant alleles might have happened before the insecticide pressure, presumably if those alleles had another physiological role. Consequently, this type of resistance alleles would be less likely to carry a fitness cost [39].

The presence of insecticides in the environment is the basis for resistance selection. Operational factors, like formulation, dosage, frequency, and intensity of application, will determine the strength of that selection pressure. Likewise, environmental and intrinsic biological elements will determine the extension and velocity for the dispersion course of resistance alleles. The amount of resistance alleles and their initial frequency, as well as their dominance, penetrance, expressiveness, and interaction within the whole genetic background are the genetic components. In parallel, biological and ecological pieces in this scenario include the offspring size, generation turnover, mono or polygamy behaviors, together with degrees of mobility, isolation, and migration, mono or polyphagia, use of refuges, etc. [41]. Naturally, the knowledge of most of these aspects will optimize the design for more effective insect control strategies. Even considering all those parameters, insecticide application can play a strong selection pressure, able to change the profile of a population very quickly [42].

One parameter that probably has a large impact on the evolution of insecticide resistance is the side effects, usually negative, related to the resistance mechanisms. This is likely the main reason that explains the low frequency of resistance alleles in populations not exposed to chemicals. Therefore, the most common assumption is that when the use of insecticides is interrupted, the frequency of nonresistant specimens would tend to increase toward the establishment of the previous susceptibility levels of the population. This is especially what managers of campaigns against vector of pathogens anxiously look for, once the arsenal of insecticide compounds to this end is very restricted [4, 5].

The mode of insecticide application is crucial to the velocity of resistance evolution. Since the main goal of these control strategies is a prompt reduction of the targeted insect population, they often apply high dosages of insecticides, which combined with the indiscriminate use of the household or agriculture products, result in a strong selective pressure. Hence, even with a high impact on the fitness, some resistance alleles can spread among populations [43]. Besides physiological and reproductive hitched-hiked costs for resistance, a continuous pressure may favor the spread of mechanisms with lower side effects. An important factor resulting from the refining aspect of Natural Selection over the adaptation for resistance is the selection of "modifier genes", which neutralize or compensate deleterious effects [44]. The modifier genes can reduce drastic effects on the overall fitness previously induced by some resistance alleles, enhancing the adaptation to the environment with insecticides.

An emblematic example occurred in the in the Australian sheep blowfly L. cuprina, where a mutant allele for the carboxylesterase E3 is responsible for resistance to the OP diazinon, presenting, however, high disadvantage in environments free of insecticide. One of the effects on the overall fitness was a bilateral asymmetry in the resistant flies. With continuous use of insecticide over the resistant population, a modifier gene was subsequently selected, increasing the fitness and also neutralizing the negative effects over the asymmetry [45]. Later, it was verified that the candidate for that modifier was a gene with an important role in oogenesis, spermatogenesis, embryonic mesoderm formation, and eyes development. The authors hypothesized that the resistance allele had a broad pleiotropic effect causing developmental perturbations that affected bristles and wing development, presumably impelled by a role of the carboxylesterase E3 in cell adhesion. The selection of the modifier gene compensated these effects [46].

In Culex mosquitoes the ace-1R allele codes for the G119S mutant AChE resistant to OP, however, with 60% lower activity than the wild-type enzyme. Consequently, resistant individuals present a severe fitness cost, reflected with the decrease of the ace-1R allele frequency in the absence of insecticide, as observed in some Culex pipiens populations [47, 48]. The G119S mutation in Anopheles gambiae followed the same tendency [49]. The emergence of gene duplication in the ace-1 locus containing both resistant ace-1R and susceptible ace-1S alleles not only guaranteed resistance to OP but also diminished the resistance deleterious effects, once the physiological role of the enzyme was no longer compromised [50].

Another scenario of amelioration of resistance was richly described by Labbé et al. (2009) for a gradual replacement of resistant genes in a decade's time among populations of C. pipiens from Montpelier, Southern France. In that study, the authors found that the Ester1 allele (from Ester locus, enrolled with over production of EST) was selected for resistance to OP; however, it was later replaced by the Ester4 allele. This newer one conferred the same advantages over insecticides, nonetheless with lower pleiotropic effects and fitness cost. Interestingly, a third allele Ester2 with both higher advantage and fitness cost seemed to be replacing the previously selected Ester4. The hypothesis raised was that the first replacement (Ester1 to Ester4) occurred as a compensatory amelioration, since Ester4 is less costly and more "generalist". On the other hand, the Ester2 allele would be more "specialist" to insecticide-treated areas, conferring high resistance but with strong pleiotropic effects. The practices of insecticide use in different areas of Montpelier during that time certainly influenced the evolution of this Ester locus. If the intensity of treatment had decreased, Ester4 would have possibly been favored over the stronger resistant Ester2 allele, given the former's lower fitness cost [44].

Although a common class of insecticide can select the same mutation for resistance in different insects, its effects on fitness vary through the species or even among different populations of the same species. For instance, the A302S rdl mutation remained under high frequencies in natural populations and the resistance persisted despite the withdrawal of cyclodienes in the field for years, as reported to natural populations of Drosophila [51], the German cockroach [52], and to the mosquito A. gambiae [53]. On the other hand, a reduction in the rdl resistant allele without insecticide selection pressure was observed in natural populations of the horn fly H. irritans [13] and the Australian sheep blowfly L. cuprina from both field and laboratory caged strain [54]. In the same way, rdl mutant A. gambiae and Anopheles stephensi mosquitoes presented reduced fertility and fecundity [55].

One has to consider that the evaluation of the overall fitness effects of a given mutation is very challenging, once it is difficult to separate their own effects from those caused by other mechanisms possibly coselected for resistance. In these aforementioned rdl examples, the reduced fitness might be related to the A302S mutation itself, and/or to metabolic resistance mechanisms. Similarly, the persistence of the resistance allele in an environment free of dyeldrin might be explained by the rdl cross-resistance with other insecticide that had been continually applied, as well as by the selection of modifiers genes, as previously discussed.

4. EVALUATION OF FITNESS COST OF INSECTICIDE RESISTANCE

The main approach to investigate the fitness cost of resistance in field populations is to monitor the levels of resistance along the time in environments distinctly exposed to insecticides. Moreover, if the principal mechanism selected for resistance is known, the genotyping of resistance genes in place and time scales render important assumptions about their fitness cost. It is very difficult to access this kind of data from the field, however, since there are many variables occurring simultaneously.

For example, one population of A. gambiae from M'Bé, Côte D'Ivoire, used to be considered susceptible to most insecticides up to 2002, when a civil crisis broke and the monitoring was discontinued. Ten years later, a new study revealed important changes of the resistance mechanisms among A. gambiae populations from that locality. The main mechanisms that led them to become highly resistant to OC, PY, and CA were the L1014F kdr mutation and elevated activity of MFO and EST. The only well-known contexts that might explain this severe shift from susceptible to highly resistant were the pressure with deltamethrin-based products from rice paddles and the distribution of long-lasting PY impregnated nets (LLINs) since 2006 [56]. The alteration in the resistance profile over the time would suggest a low cost of the resistance alleles. However, little was known regarding the actual levels of insecticide pressure, migration from vicinity areas, and about the extent of the influence of surrounding environment. In this case, controlled laboratory assays could help to estimate the fitness costs of the selected resistance mechanisms.

For fitness studies in the laboratory, population cage experiments can evaluate the fluctuation of resistance itself and the selected mechanisms over successive generations, under an environment clearly free of insecticide and without interference of migration. In this matter, the cost of resistance can be measured according to the velocity that the resistance alleles decrease in confined lineages along the time. A laboratory lineage of A. aegypti resistant to PY due to the NaVR2 kdr mutation4 presented deleterious effects in a series of life-trait parameters. Population cage assays corroborated these negative costs, showing that the kdr allele severely decreased from 75% to almost zero along 15 generations [57]. Most of the studies have been making use of an opposite direction: populations from the field are confined and submitted to a selection pressure in the laboratory. In another example, also with A. aegypti, populations from distinct Mexican localities were pressured with the PY permethrin for at least five generations in the laboratory. All the lineages had an increase in the resistance levels, correlated with an augment in the frequency of the Val1016Ile kdr mutation and with a number of detoxifying genes differentially transcribed, generally distinct at each lineage. Interestingly, the lineages that reached the highest frequencies of the kdr mutation presented a lower number of altered detoxifying genes [58]. These results strongly suggested that this kdr mutation had a lower fitness cost compared to the metabolic resistance genes occurring at each genetic background.

The knowledge of alterations in physiological and reproductive aspects is generally achieved by comparing life-trait parameters between susceptible and resistant individuals. As the result of pleiotropic effects of an altered gene will depend on the whole genomic structure, it is important that susceptible and resistant groups have the most similar genetic background as possible. The parameters usually evaluated are larval developmental time, adult longevity, ability to avoid predators, fecundity, fertility, mating competitiveness, and reproductive potential. When treating of blood-sucking insects, probing, acceptance of blood meal, and amount of ingested blood can also be tested. Such studies demand well-controlled conditions and are generally highly laborious, so that most of them follow few parameters at a time. In addition, the knowledge of the biology of the species under investigation is a prerequisite for the definition of which aspects would be more informative.

Fitness studies in the laboratory necessarily have to consider a well-representative collection from the field, in order to contemplate most part of the whole amplitude of variable traits from the original population. An F1 offspring of this sampling may then be raised in the laboratory to sufficiently amplify the number of individuals to be tested, as well as to normalize the physiological condition among the different populations. A laboratory lineage control of susceptibility and vigor should also be raised in parallel, as an endogenous control of experimental conditions, whenever possible.

Selection pressure for insecticide resistance in the laboratory has the advantage of controlled strength of selection and environmental conditions, population size, and absence of migration. On the other hand, if a monogenic key-mechanism for resistance was under lower frequency in the field, it is likely that this gene is not present in the sampling that established the first generation in the laboratory. For this reason laboratory pressures tend to result in polygenic resistance, where several resistance traits of minor effects are selected, but with a larger response when emerged together in the same genetic background [30]. This could also explain the different patterns of selected mechanisms to the same class of insecticide, especially metabolic resistance, in a same species.

Another important issue to be aware of when evaluating fitness costs in the laboratory environment is that most of the studies have investigated the possible life-trait alterations under optimal conditions. The amount and quality of food, the composition of substrate (or water in case of aquatic insects), density of individuals along life cycle phases, and mainly temperature and humidity are usually controlled. By contrast, insects are continually exposed to a wide range of abiotic or biotic stresses in the field. Therefore, the physiological costs of resistance alleles are probably underestimated in laboratory optimal conditions [41, 59]. The evaluation of the fitness costs in resistant insects under stress conditions (in terms of nutrition, temperature, and larval density, for instance) can bring forth relevant data related to the evolution of resistance in the field. However, such investigations are still scarce [60-62].

5. POSSIBLE CHANGES ON DEVELOPMENT AND REPRODUCTION OF INSECTICIDE- RESISTANT INSECTS

As previously discussed, resistance genes may cause changes or even dysfunctions upon direct physiological process and indirect life history traits. The knowledge of the insecticide resistance costs and which parameters are altered are important to better design strategies of insect control, especially considering vectors of pathogens, once general developmental and reproductive life-traits are strongly associated to their vectorial capacity. In the following, we present some examples of resistance side effects in vector mosquitoes.

The longevity of insects is generally evaluated in fitness investigations as a key parameter of vector/parasite relationship. Decreased longevity has been detected in species resistant to different classes of insecticides. Both Culex pipiens pallens and A. aegypti selected for PY resistance in laboratory presented decreased longevity [63-65]. Pyrethroid resistance also induced similar effects on the longevity of A. gambiae females, in this case presumably due to affected energy metabolism and oxidative stress [66]. Defenses to non neurotoxic compounds can also affect longevity, as observed in one A. aegypti lineage selected in the laboratory for diflubenzuron (a chitin synthesis inhibitor) resistance [67]. As resistance mechanisms vary among species and populations, especially when metabolic, the life span of the resistant insects is not always affected, even when high resistance ratios are observed. This was the case of two Brazilian field populations of A. aegypti resistant to both OP and PY insecticides [68].

The time to complete the larval development is also of particular interest, since the longer it takes the higher is the exposure to adverse conditions of the breeding site and to natural predators and pathogens. Likewise longevity, resistance to several insecticides can affect this parameter. Increased developmental time was observed in Culex quinquefasciatus and A. aegypti selected in the laboratory for PY resistance [64, 65], and also to an A. aegypti field population with high resistance level to OP [65]. Natural populations of C. pipiens harboring the resistance alleles ace-1R (modified AChE), Ester1 and Ester4 (overproduction of EST) also presented a longer larval developmental time [69]. The kdr mutation was also the prime cause for a delay in the larval development of A. aegypti, especially when mutant and PY susceptible larvae were reared together and under more stringent conditions [57]. Again, impacts on this parameter were not restricted to neurotoxic insecticides, as demonstrated for an A. aegypti laboratory strain resistant to Bti toxins, which presented impairment on the larval development time [70].

Some behavioral aspects can also be affected by resistance, as the ability to detect a potential host. Under laboratory conditions, for example, fewer OP resistant A. aegypti females responded to the blood meal stimuli, compared to their susceptible counterparts [68]. Similar results were observed in lineages of the same vector selected for resistance to a chitin synthesis inhibitor. Additionally, these blood-fed females ingested 18-26% less blood than the susceptible lineage [67]. The blood meal acceptance and the amount engorged can directly influence the pathogen loads ingested, potentially influencing the vector competence. These parameters are also directly connected with fecundity, since blood feeding is related to the production of eggs. Indeed, the reduction in the amount of ingested blood in resistant A. aegypti mosquitoes was directly proportional to a lower number of eggs [67, 68]. Several studies evidenced the impact of insecticide resistance in blood-feeding aspects [64, 71, 72].

Besides longer developmental time, lower longevity, and problems with blood feeding, reproductive traits are potentially stronger parameters against dispersion and maintenance of resistance in the field. Some studies have addressed these aspects with laboratory-resistant lineages. Aedes aegypti populations resistant to OP and an IGR showed lower reproductive capacity, where resistant males were able to fecundate a lower number of females [67, 68]. In the same way, susceptible C. pipiens males had a mating advantage when competing with Ester-4, Ester-1, and Ace-1R resistant individuals [47].

Some advantageous resistance side effects also occur. A D. melanogaster with increased expression of GST enzymes lived longer. The authors suggested that this alteration also promoted a tissue protection against reactive oxygen species [73]. In the same context, the resistance allele Cyp6g1, also in D. melanogaster, conferred resistance to DDT and was associated with a higher adult fecundity and increased viability of eggs and larvae in absence of insecticide [74]. Females of the mosquito C. quinquefasciatus resistant to PY by MFO overexpression survived longer when maintained with sugar solution [75].

6. CONCISE SUMMARY

The concept of "evolution-proof insecticides" poses a problem for the implementation of novel compounds. A potential technique to mitigate the development of pesticide resistance involves administering substances targeting older mosquitoes, specifically after females have predominantly deposited their eggs. In this context, there exists minimal selection pressure on resistance genes, as nearly all progeny of both susceptible and resistant people are produced in each generation [76]. This is particularly pertinent to the management of vector-borne diseases, as numerous viruses possess an inherent incubation period within the insect host, during which the insects can feed on blood and oviposit multiple times prior to becoming infective. However, they are unable to survive long enough to have an infective blood feeding. Mathematical calculations indicate that this technique to combating mature insects might significantly influence the trajectory of pesticide resistance. [77].

Innovative techniques are presently undergoing field trials, including the deployment of genetically modified mosquitoes that reduce the natural population and a strain harboring endosymbiotic bacteria that decreases the vectorial capacity of mosquitoes. However, until these tools are accessible for large-scale application and considering various vectors, the use of pesticides must remain pivotal, particularly during epidemic outbreaks. In this context, the physiological, molecular, and evolutionary dimensions of pesticide resistance require examination subjected to more analysis and discourse to enhance the management of undesirable insect populations.

FOOTNOTES

1. A complete review about insecticides and their mode of action can be found at Sparks and Nauen (2015).
2. This denomination refers to the aminoacid in the position 119 of the AChE protein (AChE1), based on the Torpedo nomenclature (Toutant, 1989).
3. In the case of the voltage gated sodium channel (NaV), the M domestica aminoacid sequence is most commonly taken as reference.
4. NaVR2 is the kdr allele mutant in both 1016 (Val to Ile) and 1534 (Phe to Cys) of the voltage gated sodium channel (NaV), found in American populations of A. aegypti.

REFERENCES

1. WHO. Monitoring of Insecticide Resistance in Malaria Vectors. Cairo: World Health Organization, Eastern Mediterranean Region; 1997. 54pp
2. Braga IA, Valle D. Aedes aegypti: inseticidas, mecanismos de ação e resistência. Epidemiol Serv Saúde. 2007;16(4):277-91.

3. Sparks TC, Nauen R. IRAC: Mode of action classification and insecticide resistance management. Pestic Biochem Phys. 2015;121:122-8. PubMed PMID: WOS: 000356109700016.

4. WHOPES. WHOPES – recommended compounds and formulations for control of mosquito larvae: WHOPES; 2013 [cited 2015 05/08/2015]. Available from: http:// www.who.int/whopes/Mosquito_Larvicides_25_Oct_2013.pdf?ua=1.

5. WHOPES. WHO recommended insecticides for indoor residual spraying against malaria vectors Geneva: World Health Organization; 2014 [cited 2015 02/01/2015]. Available from: http://www.who.int/whopes/Insecticides_IRS_17_Nov_2014.pdf?ua=1.

6. Ware GW, Whittacre DM. The Pesticide Book. Willoughby, Ohio: Meister Media Worldwide; 2004. 386pp

7. Belinato TA, Martins AJ, Lima JB, Lima-Camara TN, Peixoto AA, Valle D. Effect of the chitin synthesis inhibitor triflumuron on the development, viability and reproduction of Aedes aegypti. Mem Inst Oswaldo Cruz. 2009;104(1):43-7. PubMed PMID: 19274375.

8. Graf JF. The role of insect growth regulators in arthropod control. Parasitol Today. 1993;9(12):471-4. PubMed PMID: 15463697.

9. Lacey LA. Bacillus thuringiensis serovariety israelensis and Bacillus sphaericus for mosquito control. J Am Mosq Control Assoc. 2007;23(2 Suppl):133-63. doi: 10.2987/8756-971X(2007)23[133:BTSIAB]2.0.CO;2. PubMed PMID: 17853604.

10. Hemingway J, Ranson H. Insecticide resistance in insect vectors of human disease. Annu Rev Entomol. 2000;45:371-91. Epub 2000/04/13. doi: 10.1146/annurev.ento. 45.1.371. PubMed PMID: 10761582.

11. Hemingway J, Hawkes NJ, McCarroll L, Ranson H. The molecular basis of insecticide resistance in mosquitoes. Insect Biochem Mol Biol. 2004;34(7):653-65. Epub 2004/07/10. doi: 10.1016/j.ibmb.2004.03.018. PubMed PMID: 15242706.

12. Wafford KA, Lummis SC, Sattelle DB. Block of an insect central nervous system GABA receptor by cyclodiene and cyclohexane insecticides. Proc R Soc Lond B Biol Sci. 1989;237(1286):53-61. PubMed PMID: 2474166.

13. Domingues LN, Guerrero FD, Becker ME, Alison MW, Foil LD. Discovery of the Rdl mutation in association with a cyclodiene resistant population of horn flies, Haematobia irritans (Diptera: Muscidae). Vet Parasitol. 2013;198(1-2):172-9. Epub 2013/09/24. doi: 10.1016/j.vetpar.2013.08.023. PubMed PMID: 24055107.

14. Weill M, Malcolm C, Chandre F, Mogensen K, Berthomieu A, Marquine M, et al. The unique mutation in ace-1 giving high insecticide resistance is

easily detectable in mosquito vectors. Insect Mol Biol. 2004;13(1):1-7. Epub 2004/01/20. doi: 452 [pii]. PubMed PMID: 14728661.

15. Martins AJ, Valle D. Nota Técnica N.O 2/2011/Ioc-Fiocruz/Diretoria - Assunto: RecomendaçÃo Técnica Sobre A InterrupçÃo Do Uso De Inseticidas Piretróides No Controle Do Aedes Aegypti no Brasil. Rio de Janeiro: Instituto Oswaldo Cruz; 2011. 2pp

16. Bass C, Nikou D, Donnelly MJ, Williamson MS, Ranson H, Ball A, et al. Detection of knockdown resistance (kdr) mutations in Anopheles gambiae: a comparison of two new high-throughput assays with existing methods. Malaria J. 2007;6. PubMed PMID: WOS:000249300900001.

17. Salgado VL, Saar R. Desensitizing and non-desensitizing subtypes of alpha-bungarotoxin-sensitive nicotinic acetylcholine receptors in cockroach neurons. J Insect Physiol. 2004;50(10):867-79. doi: 10.1016/j.jinsphys.2004.07.007. PubMed PMID: 15518655.

18. Puinean AM, Lansdell SJ, Collins T, Bielza P, Millar NS. A nicotinic acetylcholine receptor transmembrane point mutation (G275E) associated with resistance to spinosad in Frankliniella occidentalis. J Neurochem. 2013;124(5):590-601. PubMed PMID: WOS: 000315105000003.

19. Baxter SW, Chen M, Dawson A, Zhao JZ, Vogel H, Shelton AM, et al. Mis-spliced transcripts of nicotinic acetylcholine receptor alpha6 are associated with field evolved spinosad resistance in Plutella xylostella (L.). PLoS Genet. 2010;6(1):e1000802. doi: 10.1371/journal.pgen.1000802. PubMed PMID: 20062520; PubMed Central PMCID: PMC2792709.

20. ffrench-Constant RH, Pittendrigh B, Vaughan A, Anthony N. Why are there so few resistance-associated mutations in insecticide target genes? Philos Trans R Soc Lond B Biol Sci. 1998;353(1376):1685-93. Epub 1999/02/18. doi: 10.1098/rstb.1998.0319. PubMed PMID: 10021768; PubMed Central PMCID: PMC1692388.

21. Poupardin R, Riaz MA, Jones CM, Chandor-Proust A, Reynaud S, David JP. Do pollutants affect insecticide-driven gene selection in mosquitoes? Experimental evidence from transcriptomics. Aquat Toxicol. 2012;114-115:49-57. doi: 10.1016/j.aquatox. 2012.02.001. PubMed PMID: 22406618.

22. Dongus S, Nyika D, Kannady K, Mtasiwa D, Mshinda H, Gosoniu L, et al. Urban agriculture and Anopheles habitats in Dar es Salaam, Tanzania. Geospat Health. 2009;3(2):189-210. doi: 10.4081/gh.2009.220. PubMed PMID: 19440962.

23. Meyer UA. Overview of enzymes of drug metabolism. J Pharmacokinet Biopharm. 1996;24(5):449-59. PubMed PMID: 9131484.

24. Dermauw W, Van Leeuwen T. The ABC gene family in arthropods: comparative genomics and role in insecticide transport and resistance. Insect Biochem Mol Biol. 2014;45:89-110. doi: 10.1016/j.ibmb.2013.11.001. PubMed PMID: 24291285.

25. Frova C. Glutathione transferases in the genomics era: new insights and perspectives. Biomol Eng. 2006;23(4):149-69. doi: 10.1016/j.bioeng.2006.05.020. PubMed PMID: 16839810.

26. Misra JR, Horner MA, Lam G, Thummel CS. Transcriptional regulation of xenobiotic detoxification in Drosophila. Genes Dev. 2011;25(17):1796-806. doi: 10.1101/gad. 17280911. PubMed PMID: 21896655; PubMed Central PMCID: PMCPMC3175716.

27. Xu C, Li CY, Kong AN. Induction of phase I, II and III drug metabolism/transport by xenobiotics. Arch Pharm Res. 2005;28(3):249-68. PubMed PMID: 15832810.

28. Ranson H, Claudianos C, Ortelli F, Abgrall C, Hemingway J, Sharakhova MV, et al. Evolution of supergene families associated with insecticide resistance. Science. 2002;298(5591):179-81. Epub 2002/10/05. doi: 10.1126/science.1076781298/5591/179 [pii]. PubMed PMID: 12364796.

29. Montella IR, Schama R, Valle D. The classification of esterases: an important gene family involved in insecticide resistance--a review. Mem Inst Oswaldo Cruz. 2012;107(4):437-49. Epub 2012/06/06. PubMed PMID: 22666852.

30. Ffrench-Constant RH, Daborn PJ, Le Goff G. The genetics and genomics of insecticide resistance. Trends Genet. 2004;20(3):163-70. Epub 2004/03/24. doi: 10.1016/j.tig. 2004.01.003S0168952504000174 [pii]. PubMed PMID: 15036810.

31. Zhu G, Zhong D, Cao J, Zhou H, Li J, Liu Y, et al. Transcriptome profiling of pyrethroid resistant and susceptible mosquitoes in the malaria vector, Anopheles sinensis. Bmc Genomics. 2014;15(1):448. doi: 10.1186/1471-2164-15-448. PubMed PMID: 24909924.

32. Strode C, Wondji CS, David JP, Hawkes NJ, Lumjuan N, Nelson DR, et al. Genomic analysis of detoxification genes in the mosquito Aedes aegypti. Insect Biochem Mol Biol. 2008;38(1):113-23. doi: 10.1016/j.ibmb.2007.09.007. PubMed PMID: 18070670.

33. Riaz MA, Chandor-Proust A, Dauphin-Villemant C, Poupardin R, Jones CM, Strode C, et al. Molecular mechanisms associated with increased tolerance to the neonicotinoid insecticide imidacloprid in the dengue vector Aedes aegypti. Aquat Toxicol. 2013;126:326-37. Epub 2012/10/13. doi: 10.1016/j.aquatox.2012.09.010. PubMed PMID: 23058251.

34. Witzig C, Wondji CS, Strode C, Djouaka R, Ranson H. Identifying permethrin resistance loci in malaria vectors by genetic mapping. Parasitology. 2013:1-10. Epub 2013/03/02. doi: 10.1017/S0031182013000024. PubMed PMID: 23448678.

35. Black WC, Saavedra-Rodriguez K, Strode C, Suarez AF, Salas IF, Ranson H, et al. Quantitative trait loci mapping of genome regions controlling permethrin resistance in the mosquito Aedes aegypti. Genetics. 2008;180(2):1137-52. doi: 10.1534/genetics. 108.087924. PubMed PMID: ISI:000260284400036.

36. Irving H, Riveron JM, Ibrahim SS, Lobo NF, Wondji CS. Positional cloning of rp2 QTL associates the P450 genes CYP6Z1, CYP6Z3 and CYP6M7 with pyrethroid resistance in the malaria vector Anopheles funestus. Heredity. 2012;109(6):383-92. Epub 2012/09/06. doi: 10.1038/hdy.2012.53. PubMed PMID: 22948188; PubMed Central PMCID: PMC3499844.

37. Faucon F, Dusfour I, Gaude T, Navratil V, Boyer F, Chandre F, et al. Unravelling genomic changes associated with insecticide resistance in the dengue mosquito Aedes aegypti by deep targeted sequencing. Genome Res. 2015. doi: 10.1101/gr.189225.115. PubMed PMID: 26206155.

38. Crow JF. Genetics of insect resistance to chemicals. Annu Rev Entomol. 1957;2:227-46. doi: Doi 10.1146/Annurev.En.02.010157.001303. PubMed PMID: WOS:A1957WQ98000013.

39. Ffrench-Constant RH. Which came first: insecticides or resistance? Trends Genet. 2007;23(1):1-4. PubMed PMID: WOS:000243709500001.

40. Ffrench-Constant RH. The molecular genetics of insecticide resistance. Genetics. 2013;194(4):807-15. doi: 10.1534/genetics.112.141895. PubMed PMID: 23908373; PubMed Central PMCID: PMC3730913.

41. Roush RT, Mckenzie JA. Ecological genetics of insecticide and acaricide resistance. Annu Rev Entomol. 1987;32:361-80. PubMed PMID: WOS:A1987F630700017.

42. Raymond M, Berticat C, Weill M, Pasteur N, Chevillon C. Insecticide resistance in the mosquito Culex pipiens: what have we learned about adaptation? Genetica. 2001;112-113:287-96. Epub 2002/02/13. PubMed PMID: 11838771.

43. Labbe P, Lenormand T, Raymond M. On the worldwide spread of an insecticide resistance gene: a role for local selection. J Evol Biol. 2005;18(6):1471-84. Epub 2005/11/30. doi: 10.1111/j.1420-9101.2005.00938.x. PubMed PMID: 16313460.

44. Labbe P, Sidos N, Raymond M, Lenormand T. Resistance Gene Replacement in the mosquito Culex pipiens: fitness estimation from long-term cline series. Genetics. 2009;182(1):303-12. Epub 2009/03/19. doi: 10.1534/genetics.109.101444. PubMed PMID: 19293141; PubMed Central PMCID: PMC2674826.

45. McKenzie JA, Clarke GM. Diazinon resistance, fluctuating asymmetry and fitness in the Australian sheep blowfly, lucilia cuprina. Genetics. 1988;120(1):213-20. PubMed PMID: 17246476; PubMed Central PMCID: PMC1203491.

46. Davies AG, Game AY, Chen Z, Williams TJ, Goodall S, Yen JL, et al. Scalloped wings is the Lucilia cuprina Notch homologue and a candidate for the modifier of fitness and asymmetry of diazinon resistance. Genetics. 1996;143(3):1321-37. PubMed PMID: 8807304; PubMed Central PMCID: PMC1207401.

47. Berticat C, Boquien G, Raymond M, Chevillon C. Insecticide resistance genes induce a mating competition cost in Culex pipiens mosquitoes. Genet Res. 2002;79(1):41-7. doi: 10.1017/S001667230100547x. PubMed PMID: ISI:000174721100004.

48. Berticat C, Duron O, Heyse D, Raymond M. Insecticide resistance genes confer a predation cost on mosquitoes, Culex pipiens. Genet Res. 2004;83(3):189-96. doi: 10.1017/ S0016672304006792. PubMed PMID: ISI:000223331700004.

49. Alout H, Djogbenou L, Berticat C, Chandre F, Weill M. Comparison of Anopheles gambiae and Culex pipiens Acetycholinesterase 1 biochemical properties. Comp Biochem Physiol B Biochem Mol Biol. 2008;150(3):271-7. Epub 2008/05/06. doi: S1096-4959(08)00061-4. 10.1016/j.cbpb.2008.03.008. PubMed PMID: 18455457.

50. Labbe P, Berthomieu A, Berticat C, Alout H, Raymond M, Lenormand T, et al. Independent duplications of the acetylcholinesterase gene conferring insecticide resistance in the mosquito Culex pipiens. Mol Biol Evol. 2007;24(4):1056-67. Epub 2007/02/07. doi: 10.1093/molbev/msm025. PubMed PMID: 17283366.

51. Aronstein K, Ode P, Ffrenchconstant RH. Direct comparison of Pcr-based monitoring for cyclodiene resistance in Drosophila populations with insecticide bioassay (Vol 48, Pg 229, 1994). Pestic Biochem Phys. 1994;50(1):106-. PubMed PMID: WOS:A1994PN15900011.

52. Hansen KK, Kristensen M, Jensen KM. Correlation of a resistance-associated Rdl mutation in the German cockroach, Blattella germanica (L), with persistent dieldrin resistance in two Danish field populations. Pest Manag Sci. 2005;61(8):749-53. doi: 10.1002/ps.1059. PubMed PMID: 15834842.

53. Kwiatkowska RM, Platt N, Poupardin R, Irving H, Dabire RK, Mitchell S, et al. Dissecting the mechanisms responsible for the multiple insecticide resistance phenotype in Anopheles gambiae s.s., M form, from Vallee du Kou, Burkina Faso. Gene. 2013;519(1):98-106. doi: 10.1016/j.gene.2013.01.036. PubMed PMID: 23380570; PubMed Central PMCID: PMC3611593.

54. Mckenzie JA. Ecological and Evolutionary Aspects of Insecticide Resistance. Austin: R.G. Landes Company; 1996. 185 pp

55. Rowland M. Behavior and fitness of Gamma-Hch Dieldrin resistant and susceptible female Anopheles-Gambiae and an-Stephensi Mosquitos in the

absence of insecticide. Med Vet Entomol. 1991;5(2):193-206. PubMed PMID: WOS:A1991FT12700006.

56. Koffi AA, Ahoua Alou LP, Adja MA, Chandre F, Pennetier C. Insecticide resistance status of Anopheles gambiae s.s population from M'Be: a WHOPES-labelled experimental hut station, 10 years after the political crisis in Cote d'Ivoire. Malaria J. 2013;12:151. doi: 10.1186/1475-2875-12-151. PubMed PMID: 23641777; PubMed Central PMCID: PMC3658939.

57. Brito LP, Linss JG, Lima-Camara TN, Belinato TA, Peixoto AA, Lima JB, et al. Assessing the effects of Aedes aegypti kdr mutations on pyrethroid resistance and its fitness cost. PloS One. 2013;8(4):e60878. doi: 10.1371/journal.pone.0060878. PubMed PMID: 23593337; PubMed Central PMCID: PMC3620451.

58. Saavedra-Rodriguez K, Suarez AF, Salas IF, Strode C, Ranson H, Hemingway J, et al. Transcription of detoxification genes after permethrin selection in the mosquito Aedes aegypti. Insect Mol Biol. 2012;21(1):61-77. doi: 10.1111/j.1365-2583.2011.01113.x. PubMed PMID: 22032702; PubMed Central PMCID: PMC3540788.

59. Paris M, David JP, Despres L. Fitness costs of resistance to Bti toxins in the dengue vector Aedes aegypti. Ecotoxicology. 2011;20(6):1184-94. doi: 10.1007/ s10646-011-0663-8. PubMed PMID: 21461926.

60. Raymond B, Sayyed AH, Wright DJ. Genes and environment interact to determine the fitness costs of resistance to Bacillus thuringiensis. Proc Biol Sci. 2005;272(1571): 1519-24. doi: 10.1098/rspb.2005.3103. PubMed PMID: 16011928; PubMed Central PMCID: PMCPMC1559826.

61. Zhang LJ, Wu ZL, Wang KF, Liu Q, Zhuang HM, Wu G. Trade-off between thermal tolerance and insecticide resistance in Plutella xylostella. Ecol Evol. 2015;5(2):515-30. doi: 10.1002/ece3.1380. PubMed PMID: 25691976; PubMed Central PMCID: PMCPMC4314281.

62. Hardstone MC, Lazzaro BP, Scott JG. The effect of three environmental conditions on the fitness of cytochrome P450 monooxygenase-mediated permethrin resistance in Culex pipiens quinquefasciatus. BMC Evol Biol. 2009;9:42. doi: 10.1186/1471-2148-9-42. PubMed PMID: 19228410; PubMed Central PMCID: PMCPMC2661048.

63. Jaramillo ON, Fonseca-Gonzalez I, Chaverra-Rodriguez D. Geometric morphometrics of nine field isolates of Aedes aegypti with different resistance levels to lambda- cyhalothrin and relative fitness of one artificially selected for resistance. PloS One. 2014;9(5):e96379. doi: 10.1371/journal.pone.0096379. PubMed PMID: 24801598; PubMed Central PMCID: PMCPMC4011790.

64. Li X, Ma L, Sun L, Zhu C. Biotic characteristics in the deltamethrin-susceptible and resistant strains of Culex pipiens pallens (Diptera: Culicidae) in China. Appl Entomol Zool. 2002;37(2):305-8. doi: 10.1303/aez.2002.305.

65. Martins AJ, Ribeiro CD, Bellinato DF, Peixoto AA, Valle D, Lima JB. Effect of insecticide resistance on development, longevity and reproduction of field or laboratory selected Aedes aegypti populations. PloS One. 2012;7(3):e31889. doi: 10.1371/ journal.pone.0031889. PubMed PMID: 22431967; PubMed Central PMCID: PMC3303777.

66. Belinato TA, Valle D. The impact of selection with Diflubenzuron, a chitin synthesis inhibitor, on the fitness of two Brazilian Aedes aegypti field populations. PloS One. 2015;10(6):e0130719. doi: 10.1371/journal. pone.0130719. PubMed PMID: 26107715; PubMed Central PMCID: PMC4481264.

67. Otali D, Novak RJ, Wan W, Bu S, Moellering DR, De Luca M. Increased production of mitochondrial reactive oxygen species and reduced adult life span in an insecticide-resistant strain of Anopheles gambiae. Bull Entomol Res. 2014;104(3):323-33. doi: 10.1017/S0007485314000091. PubMed PMID: 24555527; PubMed Central PMCID: PMC4008687.

68. Belinato TA, Martins AJ, Valle D. Fitness evaluation of two Brazilian Aedes aegypti field populations with distinct levels of resistance to the organophosphate temephos. Mem Inst Oswaldo Cruz. 2012;107(7):916-22. PubMed PMID: 23147149.

69. Bourguet D, Guillemaud T, Chevillon C, Raymond M. Fitness costs of insecticide resistance in natural breeding sites of the mosquito Culex pipiens. Evolution. 2004;58(1): 128-35. Epub 2004/04/03. PubMed PMID: 15058725.

70. Paris M, David J-P, Despres L. Fitness costs of resistance to Bti toxins in the dengue vector Aedes aegypti. Ecotoxicology. 2011;20(6):1184-94. doi: 10.1007/ s10646-011-0663-8.

71. Kumar S, Pillai MK. Correlation between the reproductive potential and the pyrethroid resistance in an Indian strain of filarial vector, Culex quinquefasciatus Say (Diptera: Culicidae). B Entomol Res. 2011;101(01):25-31.

72. Kumar S, Thomas A, Samuel T, Sahgal A, Verma A, Verma A, Pillai MK. Diminished eproductive fitness associated with the deltamethrin resistance in an Indian strain of dengue vector mosquito, Aedes aegypti L. Trop Biomed. 2009; 26(2):155-164. PubMed PMID:19901902.

73. McElwee JJ, Schuster E, Blanc E, Piper MD, Thomas JH, Patel DS, et al. Evolutionary conservation of regulated longevity assurance mechanisms. Genome Biol. 2007; 8: R132.

74. McCart C, Buckling A, ffrench-Constant RH. DDT resistance in flies carries no cost. Curr Biol. 2005;15(15):R587-R9. PubMed PMID: WOS:000231230200009.

75. Hardstone MC, Huang X, Harrington LC, Scott JG. Differences in development, glycogen, and lipid content associated with cytochrome P450-mediated permethrin resistance in Culex pipiens quinquefasciatus (Diptera: Culicidae). J Med Entomol. 2010. 47(2):188-198. PubMed PMID: 20380299.76.Read AF, Lynch PA, Thomas MB. How to make evolution-proof nsecticides for malaria control. PLoS Biol. 2009;7(4):e1000058. doi: 10.1371/journal.pbio.1000058. PubMed PMID: 19355786; PubMed Central PMCID: PMC3279047.77.

76. Gourley SA, Liu R, Wu J. Slowing the evolution of insecticide resistance n mosquitoes: a mathematical model. Proc Royal Soc Lond A: Math Physic Engin Sci. 2011;467(2132):2127-48. doi: 10.1098/rspa.2010.0413.

77. Speranca MA, Capurro ML. Perspectives in the control of infectious diseases by transgenic mosquitoes in the post-genomic era–a review. Mem Inst Oswaldo Cruz. 2007;102(4):425-33. Epub 2007/07/07. PubMed PMID: 17612761.

78. Kidwell MG, Ribeiro JM. Can transposable elements be used to drive disease refractoriness genes into vector populations? Parasitol Today. 1992;8(10):325-9. Epub 1992/10/01. PubMed PMID: 15463527.

79. Maciel-de-Freitas R, Aguiar R, Bruno RV, Guimaraes MC, Lourenco-de-Oliveira R, Sorgine MH, et al. Why do we need alternative tools to control mosquito-borne diseases in Latin America? Mem Inst Oswaldo Cruz. 2012;107(6):828-9. Epub 2012/09/20. PubMed PMID: 22990977.

80. Moreira LA, Iturbe-Ormaetxe I, Jeffery JA, Lu G, Pyke AT, Hedges LM, et al. A Wolbachia symbiont in Aedes aegypti limits infection with dengue, Chikungunya, and Plasmodium. Cell. 2009;139(7):1268-78. Epub 2010/01/13. doi: 10.1016/j.cell. 2009.11.042. PubMed PMID: 20064373.

13 Chapter

AN EXAMINATION OF INSECTICIDE RESISTANCE CONDITIONS IN BOTSWANA

1. INTRODUCTION: AN OVERVIEW

1.1 What is insecticide resistance?

Resistance is characterized as 'the acquisition of a capability in an insect strain to withstand toxicant doses that would be lethal to the majority of individuals in a typical population of the same species' [1], and more recently as a 'genetic alteration in response to selection by toxicants that may hinder field control.' [2]. The resistance status indicates the reduced vulnerability of a pest population to a pesticide that was before effective, resulting from natural selection where genetic features for resistance are transmitted to following generations.

The emergence of pesticide resistance is contingent upon the genetic makeup of a species population. The insecticide is preadaptive, as it often does not cause heritable alterations but rather chooses advantageous mutations that enable the insect to endure the treatment [3]. The resistant strains consequently arise from the survival and reproduction of individuals with one or more mechanisms that facilitate survival post-exposure to an insecticide, each governed by one or more resistance (R) genes. Strains generally revert to susceptibility without insecticide application until they have become homozygous for the R genes [1, 4, 5]. Insecticide resistance is a natural phenomenon governed by genes that induce the biochemical, physiological, or behavioral alterations underlying resistance.

Resistance may diminish the long-term efficacy of a certain insecticide against a species population, necessitating the application of an alternate insecticide to which resistance has not developed; nevertheless, this frequently serves as a transient remedy. Cross-resistance may develop among drugs within a group that share a similar mechanism of action, particularly if their metabolism and target site binding are closely analogous [6].

Cross-resistance may arise among insecticide groups with distinct modes of action and can be governed by a single gene, indicating a monogenic basis due to one defense mechanism acting against many toxicants. It may also be polygenic, with several pathways that may not exert equal effects against various toxicants. The emergence of multiple resistances, which implicate many genes, could represent a significant concern if it manifests in the field [6].

2. HISTORY OF RESISTANCE TO INSECTICIDES

Resistance to insecticides by insect pests has been documented for over 75 years, but its greatest impact has occurred during the last 30 years following the discovery and extensive use of synthetic organic insecticides [7]. Insect resistance was first observed in 1908, reported by Melander [8] in the San Jose scale insects Aspidiotus perniciosus, found to have become insensitive to lime-sulphur. Thirty years later, there were further reports of insect resistance towards numerous other pesticides.

When dichlorodiphenyl-trichloro-ethane (DDT) was introduced in 1946, insect resistance to the compound appeared quickly and worldwide. The first sign of resistance towards DDT was shown in the housefly Musca domestica [9]. Thereafter, cases from different locations were reported: Aedes sollicitans in Florida, Culex pipiens in Italy, and Cimex lectularius in Hawaii [1]. New insecticides that were later introduced did not last long with regard to their usage as the number of species showing resistance to one or more toxicants doubled every six years between 1948 and 1983 [10].

A number of resistant species are also reported in other agriculturally important orders such as Lepidoptera (67 species, representing 15%), Coleoptera (66 species, representing 15%), Acarina (58 species representing 13%), Homoptera (46 species, representing 4%), and Heteroptera (20 species, representing 4%) [11]. However, studies have shown that resistance develops faster in insects with many generations per year rather than only one, at higher selection pressures than at lower ones. Sawicki [12] noted that resistance is regarded as a problem only when the cost of control becomes unjustified or when excessive use of the control agent presents health and environmental hazards.

3. INSECTICIDE USE IN BOTSWANA

The economy of Botswana is mainly dependent on agriculture and mining. The agricultural sector in Botswana covers both crops and livestock production. The industrial growth has brought about awareness in farming systems for both livestock and arable farming. However, this has also brought about an increase in the use of chemicals for pests on animals and crops. Insect pests are very important in crop production because they pose a serious problem to farmers. They reduce the yield and quality of crops resulting in lower prices for the crops and lower returns to the farmer.

Since the introduction and use of DDT in Botswana in the 1950s, other types of insecticides such as organophosphates, pyrethroids, and carbamates have been used in various aspects of agriculture. In crop production, these were used to target pests diamond back moth, aphids, locusts, and armyworms; fruit flies, diamond back moth, aphids, and leaf miners; American bollworm, diamond back moth, aphids cutworms, and bagrada bug, respectively [13].

From the results of experiments carried out during the 1970s in Botswana, carbaryl proved to be the most effective insecticide against Helicoverpa armigera on cotton, sorghum, and cowpea when tested against insecticides such as DDT, endosufan, monocrotophos, and tetrachlorvinphos [14]. However, the current pest management option for H. armigera in Botswana is the use of pyrethroids from recommendations based on the information from manufacturers and recommendations from other countries [14, 15].

Organophosphates are commonly used for the control of infestations of parasites for livestock and may also be applied as sprays and dips in form of acaricides. The same application of organophosphates has extended to spraying of the quelea birds by the Plant Protection Unit of the Ministry in Botswana [16].

Several control methods have been employed in the management of tsetse fly in Northern Botswana, and all of these methods involved the use of chemicals (Table 1). After the spraying of 2001 and 2002 in the Okavango Delta and 2006 in the Kwando-Linyanti systems, tsetse fly has not been found [17]. There were reports, however, that the deltamethrin spraying negatively affected other nontargeted organisms such as Cyrtobagous salvinae, with recovery in abundance after spraying [18].

Table 1. Insecticides used for the different control methods for tsetse fly.

Year of control	Method of control	Insecticide used
1960–1972	Residual ground spraying	DDT
1970–1990	Nonresidual aerial spraying	Endosulfan and pyrethroids
1990–2000	Traps and targets	Deltamethrin
2000 onwards	Aerial spraying	Deltamethrin
Source: Ingram [17].		

4. INSECTICIDE RESISTANCE STUDIES IN BOTSWANA

4.1 Mosquitoes

4.1.1 Esterases in Culex mosquitoes

The global spread of resistant genes acts as an example of evolution in action showing how selective forces, genetic variability, gene flow, migration, and life history can interact to produce changes in gene frequency. Two types of esterases are known and coded for loci est- A and est-B, corresponding to the production of esterases A and B, respectively [19]. These two elevated esterases have been shown to be overproduced as a result of amplification. This being said, several copies of one gene found on the same genome of the structural genes coding for them may happen in isolation or together [20, 21]. However, this is the major mechanism associated with organophosphate resistance in Culicine mosquitoes [22]. The association of these esterases has been found to be globally widespread in Culex pipiens complex mosquitoes, with highly active esterases A2 and B2 being shown to be strongly associated with organo-phosphate resistance in strains of Culex quinquefasciatus from California [23], West Africa [24], Kenya [25], Thailand and South Africa [26], and Vietnam [27]. Esterase B1 has been linked to North America and the Far East in Culex mosquitoes collected in Foshan [28]. Resistance by this massive overproduction of these esterases is conferred in a way that enzymes are able to detoxify the ester-based organophosphorus insecticides by hydrolysis to produce a nontoxic ionic metabolite. At the same time, the increased production of esterases can effectively sequester the insecticide and prevent it from reaching its target site [29].

A study was conducted in order to establish the concept of migration and the widespread of these esterases and whether they are present in mosquitoes sampled in Botswana. Their presence will be a clear indication of the possibility of resistance demonstrated by the mosquitoes due to the selection pressure from the use of insecticides.

The mosquito larvae were collected from two areas in the southern part of Botswana: Gaborone and Molepolole. The areas are at least 50 km apart and have different economic activities. Gaboronc is a city whilst Molepolole is a village. Single larvae at third and fourth instar and the adults were used for the experiment. The adults were identified as that of Culex specie from both collection sites. The presence of esterases was determined by subjecting each homogenised larvae and adult to a nondenaturing 7.5% PAGE gel to reveal the esterase bands.

Esterases identified as A2–B2 were revealed in the Molepolole strain, in both the adults and the larvae (Figure 1) Esterase activity was determined by carrying out esterase and protein assays on the same homogenates of the single larvae and adults. The method used was described by Callaghan [30]. The results showed that esterase activity also varied in the larval and the adult stages for these mosquitoes from the two areas. In the Gaborone strain, the esterase activity for the Culex larvae was up to 50 μmol of β -naphthol per microgram of protein, whilst that of the adults ranged from 101 to 150 μmol of β -naphthol per microgram of protein. The results revealed that esterase activity increased in almost 35% of the population tested upon reaching the adult stage. However, contrary to the results of the Gaborone strain, the esterase activity was found to be high at the larval stage.

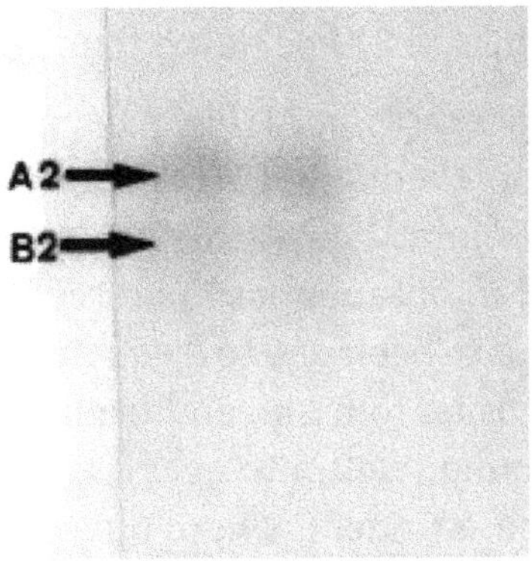

Figure 1. Gel showing the presence of A2B2 esterases from the adult Molepolole strain of Culex mosquitoes. The A and B esterases preferentially hydrolyse α- and β -naphthyl acetate, respectively. In the Gaborone strain, both the adults and the larvae revealed the presence of one esterase band B1, which preferentially hydrolysed β -naphthyl acetate (Figure 2a and b).

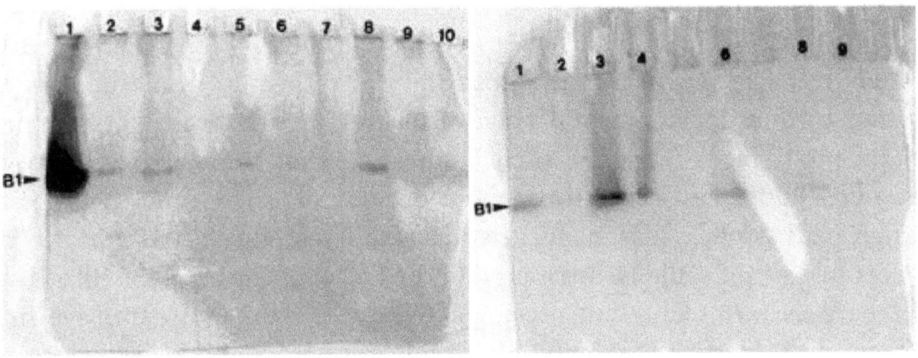

(a) (b)

Figure 2. (a). Gel showing the presence of B1 esterases from the adult Gaborone strain of Culex mosquitoes. (b). Gel showing the presence of B1 esterases from the larvae of the Gaborone strain of Culex mosquitoes.

The origin of the amplified A2–B2 esterases is yet to be identified. The appearance of resistance to organophosphates among populations of C. pipiens complex mosquitoes in Africa was reported in 1967 [31] and in Asia around 1968 [32]. On the basis of earlier reports of organoOphosphate resistance, Africa is thought to be the origin of amplification, but the first use of organophosphates in the continent is not well documented. According to Raymond et al. [33], the occurrence of overproduced esterase B1 is relegated to Asia and North America, but not in Africa. The distribution of B1 unlike the A2 and B2, which have been shown to be closely linked, has not yet been studied to elucidate the cause of its geographical spread and importance in organophosphate insecticide resistance [20, 34]. However, the presence of B1 esterases as shown in the Culex mosquitoes collected from Molepolole indicated that esterase B1 has slowly found its way into Southern Africa.

The distribution of A2–B2 esterases in Africa, Asia, and North America may have been attributed to migration events [33]. This spread was also inevitable in the city of Gaborone. Botswana shares borders with South Africa where the presence of A2–B2 has been reported in organophosphate resistant strain of Culex pipiens [26]. This spread could have also been due to frequent migration of people between the two countries through visits, tourism, and trade. It is also not surprising that the spread of esterase B1 found itself into every town and village in the country through the Asian and Chinese traders, where only esterase A2–B2 was known to be dominant, such as in Africa. Thus, the presence of esterase B1 in Gaborone is as a result of migration of the esterase genes from areas of known prevalence such as North America, Asia, and China. Gaborone is an urban area where domestic and industrial pesticide spraying is randomly carried out. The high esterase activity displayed by the adults in the Gaborone strain may have been acquired initially at the larval stage and increased in the adults due to increase in the resistance from the kill of these chemicals.

The study has also shown that the possibility of resistance in mosquitoes is not restricted to one developmental stage. However, the esterase patterns in the developmental stages were the same, indicating that the same esterase genes are responsible for resistance throughout development from the larvae to adult stages, in both strains. Reasons for the increase in esterase activity in the larvae of Molepolole strain could be attributed to the fact that the area is a village within which there are farms lands and rearing of livestock. There is an extensive amount of agricultural practices whereby the application of insecticides on crops or acaricides on livestock is bound from time to time. These may have found their way into the nearby streams and gutters, which make good breeding sites for the mosquitoes.

4.1.2 Susceptibility tests on malaria vector, Anopheles arabiensis

Malaria is distributed in the northern part of the country, and this is a disease that is of public health priority to the government of Botswana, as it accounts for over 95% of malaria cases in Botswana [36]. A. arabiensis is the main malaria vector, and studies conducted in 2006 revealed that the species is distributed in all malaria areas of Botswana.

In order to reduce malaria transmission, the government of Botswana has engaged in what is called integrated vector management (IVM), which involves the utilisation of different interventions, including environmental management, safe, careful, and thoughtful use of insecticides. One such intervention is the indoor residual spraying (IRS) of insecticides, which goes back to the 1940s when spraying of human dwellings was initiated [37]. In the 1950s, the use of diethyl-dichloro-trichloroethane (DDT) started in Botswana for the malaria vector control using IRS [35]. In 1997, Botswana then introduced insecticide-treated nets (ITNs) to complement IRS as part of the IVN initiative [38]. Between 1971 and 1973, fenitrothion, which is an organophosphate, briefly replaced DDT. However, due to the poor efficacy of fenitrothion, DDT was reinstated as the main insecticide to serve together with IRS as Botswana's principal vector control intervention against malaria.

The WHO global strategy for the Malaria control is to break the malaria parasite transmission by using indoor residual spraying or pyrethroid impregnated materials such as bed nets. It is during such programmes that the annual vector susceptibility studies are carried out in Botswana.

Similar studies were conducted [39] to confirm the presence of pyrethroid resistance among Anopheles gambiae from West Africa (Benin and Burkina Faso), Central Africa (Cameroon), and A. arabiensis from Southern Africa (Botswana). WHO test kits for resistance tests were used with the adult mosquitoes being subjected to exposure to permethrin, deltamethrin, and DDT. From the results, permethrin resistance were detected in Benin, Burkina Faso, and deltamethrin resistance was detected also in Cote d' Ivoire. Botswana, on the other hand, showed susceptibility of A. arabiensis towards permethrin and DDT. The results obtained by Coetzee [40] on the susceptibility status of A. Arabiensis in Botswana using the same three insecticides were found to be in agreement with the susceptibility tests conducted annually for vector susceptibility by the Ministry of Health in Botswana. Table 2 presents the summary of the results.

Table 2. Percentage susceptibility levels of Anopheles arabiensis towards insecticides.

Insecticide	Source: Chandre et al. [39]	Source: Botswana National Strategic Plan 2006–2011 [36]
DDT	99.6	99.09
Permethrin	86.3	90.71
Deltamethrin	-	92.47

Both studies have been able to show that the malaria vector A. arabiensis does not have an indication of resistance towards the insecticides used in Botswana as it is fully susceptible to DDT and pyrethroids. DDT still remains the most sensitive insecticide, when tested against pyrethroids. Baseline studies were carried out on insecticide resistance in five Southern African countries including Botswana. The results showed that there was also complete susceptibility of A. arabiensis to DDT and pyrethroids [40].

5. STUDIES ON THE USE OF ALTERNATIVES TO CHEMICALS IN BOTSWANA

Application of insecticides indiscriminately on agricultural crops can reduce or kill the natural enemies of insect pests. Continuous use of insecticides as we have already seen can also induce the resistance development in the targeted pests as well as killing beneficial nontargeted organisms. However, the detriment can also extend to human health through dietary exposure of contaminated crops. This great concern has brought about the need for alternatives to chemical insecticides that can be safe to human and the environment and at the same time affordable to farmers. Most of these natural insecticides are derived from plants and botanical insecticides, and some are of microbial type.

5.1 Microbial and spinosyns

Reports in Botswana have indicated that most insect pests found on agricultural crops have been subjected to chemical control. Diamondback moth Plutella xylostella, which is a pest of cabbage, is one such example of insect pest whose control relies heavily on the application of pyrethroids. It has also been demonstrated that DBM quickly develops resistance to many new insecticides [41].

Studies were conducted using Bacillus thuringiensis (Bt) [41] and Spinosad [42] as alternatives to insecticides to demonstrate their efficacy on the diamondback moth (DBM). Bt is a soil dwelling bacterium and is largely used in agriculture worldwide. It is a natural insecticide that produces crystals protein (cry proteins), which are toxic to many species of insects but nontoxic to humans.

Spinosad is derived by fermentation from the soil actinomycete and is effective by both contact and ingestion to numerous insect species [43]. Bioassays using both natural insecticides were carried out on the eggs and 2nd instar larvae of DBM. The results using Bt indicated that Bt was effective against both the eggs and the larvae, whereas spinosad was shown to be more effective against the eggs than against the larvae. These results were able to demonstrate that both natural insecticides used in the experiments can achieve effective control of the developmental stages: eggs and larvae of DBM. However, more bioassays still remain to be done on other insect pests that cause damage to various agricultural crops commonly grown in Botswana.

5.1.1 Resistant crop varieties

One other option to using insecticides on crops is to plant crop varieties that are found to be insect resistant. However, resistant varieties are usually only resistant to one or a limited number of insect pests. Genetic engineering has been able to allow the transfer of desired genes from one species to another, resulting in a quicker development of pest-resistant varieties or transgenic crops. On the evaluation of nine cabbage varieties for resistance to the cabbage aphid, Munthali [44] concluded that the most resistant cabbage variety would be the one that has a combination of low aphid numbers and low percentage of damaged leaves per plant. Notwithstanding, the use of these partially resistant varieties would also be recommended for use in combination with a low dose of insecticide.

6. CONCISE SUMMARY

The resistance levels in the Gaborone and Molepolole strains for both esterases B1 and A2–B2 have to be clarified through bioassays conducted against the susceptible strains. This method will ascertain the association between esterase levels and pesticide resistance in these strains. This will also indicate the type of resistance mechanism that may be given in these strains. The origin and migration path of these esterases into Botswana may be traced at the DNA level.

The ongoing application of DDT and pyrethroids on insecticide-treated nets (ITNs) and indoor residual spraying (IRS) has demonstrated that A. arabiensis is vulnerable to both insecticides. Reports indicate the presence of DDT resistance in South Africa (ANVR, 2005). The assessment of the present susceptibility status of the malaria vector Anopheles arabiensis to alternative insecticides should be promoted as Botswana advances its malaria eradication program.

Despite the prevalent application of pesticides in Botswana's agriculture sector, it is promising that research is concentrating on alternative insecticides that would not endanger the environment or human health. This approach enables the agriculture and health sectors to effectively manage the development of pesticide resistance in crop pests and disease vectors, respectively.

REFERENCES

1. Brown, A.W.A. Pest resistance to pesticides. In: White Stevens, R., editors. Pesicides in the environment. 1st ed. New York: Marcel Dekker; 1971. p. 457–552.
2. Sawicki, R.M. Definition, detection and documentation of insecticide resistance. In: Ford, M.G.; Holloman, D.W., Khambay, P.B.S and Sawicki, R.M., editors. Combating resistance to xenobiotics; biological and chemical approaches. Chischester, England: Ellis Horwood; 1987. p. 105–117.
3. Moberg, W.K. Understanding and combating agrochemical resistance. In: Managing resistance to agrochemicals from fundamental research to practical strategies; 1990; Acs symposium series; 1990. p. 2–15.
4. Wood, R.J. Insecticide Resistance: genes and mechanisms. In: Bishop, J.A. and Cook, L.M., editors. Genetic consequences of man-made change. New York: New Academic; 1981. p. 53–94.

5. Raymond, M., Poulin, E., Boiroux, V., Dupont, E and Pasteur, N. Stability of insecticide resistance due to amplification of esterase genes in Culex pipiens. Heredity. 1993;70:301–307.
6. Hassall, K.L. Insect resistance to insecticides. In: Hassall, K.L., editors. The biochemistry and uses of pesticides. 2nd ed. England: Macmillan Press; 1990. p. 237–259.
7. Georghiou, G.P. and Taylor, C.E. Genetic and biological influences in the evolution of insectide resistance.Journal of Economical Entomology. 1977;70:319–323.
8. Melander, A.L. Can insects become resistant to sprays?. Journal of economic Entomology. 1914; 7(75):167–173.
9. Forgash, A.J. History, evolution and consequences of insecticide resistance. Pesticide Biochemistry and physiology. 1984;22:178–186.
10. Georghiou, G.P. and Mellon, R.B. Pesticide resistance in time and space. In: Georghiou, G.P. and Saito, T., editors. Pest resistance to pesticides. New York: Plenum Press; 1983. p. 1–46.
11. Georghiou, G.P. The magnitude of the resistance problem. In: Pesticide resistance: strategies and tactics for management. Washington D.C.: Nature Academy of Sciences; 1986. p. 14–43.
12. Sawicki R.M. Resistance to pesticides. In: Resistance of insects to insecticides; 1989; Kuala Lumpur, Malaysia. Nagoya: Nagoya University Press; 29 Oct–1 Nov 1996. p. 50–52.
13. Obopile, M., Munthali, D.C. and Matilo, B. Farmer's knowledge, perceptions and management of vegetable pests and diseases in Botswana. Crop Protection. 2008. 27:1220–1224.
14. Ingram, W.R. The use of carbaryl as an alternative to DDT for the control of Heliothis armigera (Hb) in the field crops in Botswana. In Proceedings of the Cotton Insect Control Conference, 24–27 March 1971. Blantyre, Malawi.
15. Obopile, M. and Mosinkie, K.T. Intergrated pest management for African bollworm (Helicoverpa armigera (Hubner) in Botswana: review of past research and future perspectives. 2007. Vol 1: issue 2: 1–9.
16. Serumola, O and Mbongwe, B. Chemical Substances in Botswana. In. Conference proceedings. 2000. Lusaka, Zambia.
17. Kurungudla, C.N., Kgori, P.M. and Moleele, N. Management of Tsetse Fly Using Insecticides in Northern Botswana. 2012. 449–476.
18. Kurungudla, C.N., Bonyongo, M.C and Serumola, O. Impact of deltamethrin aerial sprays on adult Cyrtobagous salvinae in Botswana. 2007. Journal of Aquatic Plant Management. 45: 124–129.

19. Hemingway, J. and Ranson, H. Insecticide resistance in insect vectors of human disease. Annual Review of Entomology. 2000;45:371–391.

20. Rooker, S., Guillemaund, T., Berge, J., Pasteur, N. and Raymond, M. Coamplification of esterases A and B as a single unit in Culex pipiens mosquitoes.Heredity. 1996;77:555–561.

21. Callaghan, A., Boiroux, V., Raymond, M. and Pasteur, N. Prevention of changes in the electrophoretic mobility of overproduced esterases from organophosphate resistant mosquitoes of the Culex pipiens complex. Medical and Veterinary Entomology. 1994;8:391–394.

22. Pasteur, N., Iseki, A. and Georghiou, G. Genetic and biochemical studies of the highly active esterases A and B associated with OP resistance in mosquitoes of the Culex pipiens complex. Biochemical Genetics. 1981;19(9/10):909–919.

23. Raymond, M., Pasteur, N., Georghiou, G.P. Mellon, R.B., Wirth, M.C., and Hawley, M. Detoxification of esterases new to California in Organophosphate resistant Culex quinquefasciatus (Diptera:Culicidae). Journal of Medical Entomology. 1987;24:24–27.

24. Magnin, M., Marboutin, E. and Pasteur, N. Insecticide resistance in Culex quinquefasciatus (Diptera: Culicidae) in West Africa. Journal of Medical Entomology. 1988;25:99–104.

25. Curtis, C.F. and Pasteur, N. Organophosphate resistance in vector populations of the complex of Culex pipiens L. (Diptera:Culicidae). Bulletin of entomological Research. 1981;71:153–161.

26. Guillemaund, T., Rooker, S., Pasteur, N. and Raymond, M. Testing the unique amplification event and the worldwide migration hypothesis of insecticide resistance genes with sequence data. Heridity. 1996;77:535–543.

27. Pasteur, N., Marquine, M., Rousset, F., Failloux, S.A., Chevillon, C. and Raymond, M. The role of passive migration in dispersal of resistance genes in Culex pipiens quinque fasciatus within French Polynesia. Genetical Research. 1995;66:139–146.

28. Qiao, C. and Raymond, M. The same esterase B haplotype is amplified in insecticide resistant mosquitoes of the Culex pipiens complex from the Americas and China. Heredity. 1995; 74:339–345.

29. Callaghan, A.Temperature-related activity loss and mobility changes of esterases associated with insecticide resistance in Culex pipiens mosquitoes. Medical and Veterinary Entomology. 1993; 7:287–290.

30. Callaghan, A. Genetics and biochemical stdies of elevated and non-elevated esterases of Culex pipiens [thesis]. University of London:1989. 125 p.

31. Hamon, J. and Mouchet, J.la resistance aux insecticides chez Culex pipiens fatigans Weidemann. WHO. 1967;37:277–286.

32. Yasutomi, K. Studies on organophosphate resistance and esterase activity in the mosquito of the Culex pipiens group. Journal of Sanitation Zoology. 1970;21:41–45.

33. Raymond, M., Callaghan, A, Fort, P. and Pasteur, N.Worlwide migration of amplified insecticide resistance genes in mosquitoes. Nature. 1991;350 (6314):151–153.

34. Callaghan, A., Malcolm, C.A. and Hemingway, J. Biochemical studies of A and B carboxylesterases from organophosphorus-resistant strains of an Italian Culex pipiens mosquito. Pesticide Biochemistry and Physiology. 1991;41:98–206.

35. Botswana Ministry of Health. Botswana malaria indicator survey 2012 report. Botswana National Malaria Programme. 2012; Botswana Ministry of Health. Botswana Malaria Strategic Plan 2006–2011.

36. WHO: Implementation of Indoor residual Spraying of Insecticides for Malaria Control in the WHO African Region Report. 2007. 11–13.

37. Botswana Ministry of Health. Guidelines for Malaria Vector Control. 2007.

38. Chandre, F., Darrier, F., Manga, L., Akogbeto, M. Faye, O., Mouchet, J. and Guillet, P. Status of pyrethroid resistance in Anopheles gambiae sensu lato. Bulletin of World Health Organisation. 1999;77 (3):230–234.

39. Coetzee, M. Malaria and dengue vector biology and control in Southern and Eastern Africa. In: Knols, B.G.J. and Louis, C, editors. Bridging lab and field research for genetic control of disease vectors. Wageningen UR Frontis Series. No. 11 pp101–109.

40. Legwaila, M.M., Munthali, D.C., Kwerepe, B.C. and Obopile, M. Efficacy of Bacillus thuringiensis (var. kurstaki) against diamondback moth (Plutella xylostella L.) eggs and larvae on cabbage under semi-controlled greenhouse conditions. International Journal of Insect Science. 2015. 7. 39–45

41. Legwaila, M.M., Munthali, D.C., Obopile, M. and Kwerepe, B.C. Effectiveness of spinosad against diamondback moth (Plutella xylostella L.) eggs and larvae on cabbage under Botswana conditions. International Journal of Insect Science. 2014: 6, 15–21.

42. Hertleina, M.B., Thompsona, G.D., Subramanyamb, B. and Athanassiouc, C.G. Spinosad: a new natural product for stored grain protection. Journal of Stored Products Research. 2011. Vol.47 issue 3, 131–146

43. Munthali, D.C. Evaluation of cabbage varieties for resistance to the cabbage aphid. African Entomology. 2009. 17 (1): 1–7

14 Chapter

IMPACT OF IMIDACLOPRID ON THE BACTERIAL SOIL ISOLATE BACILLUS WEIHENSTEPHANENSIS

1. INTRODUCTION: AN OVERVIEW

Insecticides have been utilized for the previous 40 years to safeguard crops against insects. Most pesticides contribute to the contamination of air, soil, and water due to their use through extensive spraying. Pesticide use has become essential to contemporary farming practices. An estimated 4 million tons of pesticides are administered to global crops each year for insect control [1]. Residual pesticides can serve as pollution sources, posing a significant threat to soil and groundwater through rainfall infiltration. Certain pesticides influence biochemical processes prevalent among various animals, plants, and microbes, hence posing a heightened risk to non-target creatures. Imidacloprid is a systemic chloronicotinyl insecticide employed for soil, seed, and foliar applications to manage sucking insects, such as rice hoppers, aphids, thrips, whiteflies, termites, turf insects, soil insects, and some beetles. The active compound disrupts the passage of signals throughout the insect's nervous system [3]. Imidacloprid is classed as a Category II acute toxicant and is therefore designated as a General Use Pesticide. Imidacloprid is detrimental to insects, particularly honeybees, as well as to fish, avian species, and algae.

In humans it is linked to reproductive and mutagenic effects and is considered neurotoxic. Reproductive toxicity testing also showed that imidacloprid is an agonist to the acetylcholine receptors that regulates the endocrine system in the brain [4]. The over accumulation of this pesticide in environment requires higher awareness about this pesticide. Imidacloprid is reported to have different impacts on soil bacterial community and also cluster analyzing clearly showed that imidacloprid has significant negative impact on soil bacterial diversity in highly polluted farms and soil microbial balance has been gradually upset by the application of more pesticide.

Oxidative stress is a misbalance between reactive oxygen species (ROS) generation and detoxification, resulting in increased levels of enzyme activity. ROS are of increasing interest in environmental toxicity as they may provide insights to toxicity mechanisms and may identify novel biomarkers. ROS can modify and inactivate proteins in a variety of ways. It is commonly recognized that Escherichia coli is the most suitable model system for the investigation of the cell response to oxidative stress. When organisms or cells are exposed to low levels of certain harmful physical or chemical agents, the organisms acquire an induced tolerance against the adverse effects. The effect of hydrogen peroxide on the activity of Sox RS and Oxy R regulon enzymes in different strains of Escherichia coli has been studied. Exposure to acetamiprid in Escherichia coli, Pseudomonas and Bacillus subtilis resulted in synthesis of stress enzymes [5].

The term "genomics" was first used by Winkler to describe the haploid set of chromosomes and the genes associated with them. Genomics includes many scientific disciplines [6]. Toxicogenomics is the subdiscipline combining the fields of genomics and toxicology [7]. It has also been described as the study of genes and their products important in adaptive responses to chemical-derived exposures. The toxicogenomic approach provides opportunities to improve understanding of the molecular mechanisms underlying toxic responses to environmental contaminants [8]. Therefore, the present investigation was carried out to study the effect of 10–7 to 10–3 molar concentrations of imidacloprid for a period of 24, 48, 72 and 96 h on three antioxidant enzymes superoxide dismutase, catalase and peroxidase in Bacillus weihenstephanensis. Further, the genes for the three enzymes involved in the antioxidant defense process in soil isolate Bacillus weihenstephanensis were sequenced and identified.

2. MATERIALS AND METHODS

2.1 Laboratory experiment

The experiment was carried out during the summer of 2011 at the laboratory of Department of Biotechnology and Microbiology, Karanatak University, Dharwad, Karnataka. The soil samples were collected from cotton fields around Hubli city. These fields did not have a history of imidacloprid applications for the past 5 years. Soil was collected at a depth of 15 cm and samples were passed through a sieve of 2 mm to remove stones and plant debris. One gram of soil was mixed with 9 ml of sterilized water and mixed by shaking for even distribution of soil in water. And 1 ml of solution from this test tube was then added to another test tube with 9 ml sterilized water. This gives a dilution of 10–2 and in the same pattern dilutions up to 10–7 were prepared. And 100 µl of solution from 10–6 dilution was spread on nutrient plates containing different concentration (125, 250, 500 and 1,000 ppm) of imidacloprid. These plates were incubated at 37°C for 48 h. After incubation, bacterial colonies were counted using colony counter and results were expressed as the number of bacteria in 1 g of soil [9].

2.2 Field experiment

Imidacloprid was applied to experimental field at recommended rates and at 1.5× rates on two plots on same field in replicates, the plot without application served as control. Soil samples were taken on 7, 14, 21 and 28th day of application. About 1 g of sample was suspended in 9 ml of sterilized water. Serial dilutions were done as mentioned earlier. Then, 100 µl of solution from 10–6 dilution was spread-plated on nutrient agar plates. These plates were incubated at 37°C for 48 h. After incubation, colonies of bacteria were counted using colony counter and results were expressed as the number of bacteria in per gram of soil [9].

2.3 Preparation of stock solution of imidacloprid

The stock solution of one molar imidacloprid was prepared and further diluted to give 10–3, 10–4, 10–5, 10–6 and 10–7 molar. Soil isolate was isolated and identified from soil as described in our previous publication. The bacterium was maintained at 4°C on nutrient agar and sub- cultured every fortnight. The medium used for toxicity testing was an optimized medium (dextrose – 0.65 g /l: yeast extract – 1.05 g /l; K HPO –0.30 g/l; NaCl – 0.25 g /l) [9].

2.4 Preparation of inoculum

Pre-inoculum was prepared by inoculating a loop full of bacteria from the overnight incubated nutrient agar slant cultures on a 100 ml sterilized optimized growth medium and incubated for 24 h at 37°C under static conditions [9].

2.5 Identification of bacterial isolate

Imidacloprid tolerant colonies isolated and identified morphological, cultural and biochemical characters and 16S rDNA identification as described in our previous publication. The pure culture was grown on nutrient agar medium [10].

2.6 Experimental procedures

Five millilitres of the pre-inoculum was inoculated to 250 ml Erlenmeyer's flask containing 100 ml of sterilized optimized growth medium amended with different molar concentrations of imidacloprid. The flasks were incubated at 37°C for 96 h under shaking conditions at 120 rpm on a rotary shaker. At regular intervals, sample was taken out from each flask aseptically for analysis [10].

2.7 Extraction of enzymes

The cells were centrifuged at 8,000 rpm for 3 min and the pellet was dissolved in 0.2 ml of lysis buffer (50 mM tris-cl and 10 mM lysozyme). The tubes were incubated at 37°C for 10 min and centrifuged at 10,000 rpm for 10 min. Supernatant was used as the source of enzyme.

2.8 Estimation of stress enzyme activity

The activity of SOD, catalase and peroxidase were assayed using the supernatant from centrifugation (15,000 rpm) for 12 min at 4°C homogenate by standard methods [11–12].

2.9 Isolation of genomic DNA

The genomic DNA was isolated from soil isolate SP-03 by CTAB method. The DNA stock samples were quantified using Nanodrop spectrophotometer at 260 and 280 nm using the convention that one absorbance unit at 260 nm wavelength equals 50 µg DNA per ml. Quality and purity of DNA were checked by agarose gel electrophoresis. The DNA was used further for PCR. And 250 µl of the isolated genomic DNA was taken and treated with 1 µl of RNase enzyme and incubated at 37°C in a water bath for 30 min and further incubated at 60°C for 10 min in the water bath and used as a template with PCR mix [13].

Table 1. Composition of PCR mix

Contents	Volume (µl)
10× Taq Assay buffer	1.0
$MgCl_2$	0.5
Template	2.0
Primers (Forward + Reverse) (10 pM)	1 + 1
dNTP mix	0.4
Taq DNA polymerase	0.3
HPLC grade water	3.8

Table 2. Primers of stress enzymes

SL. No.	Antioxidant enzyme	Primer
1	Superoxide dismutase (SOD)	Forward- 5-atagcttggcagagcgacat Reverse- 5-tatacgcctcattgcagcag
2	Catalase (CAT)	Forward -5-ggaaaccactggcaggtaaa Reverse- 5-ctgccgatctcacttcatca
3	Peroxidase (POX)	Forward- 5-tcacaaaccgttcatttcca Reverse- 5-ccagagctgcttcgtaatcc

Table 3. PCR Conditions

Steps	Temperature (°C)	Time	Cycles
Initial denaturation	95	2 min	1
Final denaturation	94	30 sec	
Annealing*		30 sec	30
Extension	72	90 sec	
Final extension	72	10 min	1
*mentioned in text			

The annealing temperatures for catalase peroxidase and for superoxide dismutase were 46°C.

2.10 Statistical analysis

Statistic significance between the control and the experimental data were subjected to analysis of variance (ANOVA) followed by post-hoc Dunnett's test ($P \leq 0.05$) [9].

3. RESULTS

3.1 Effect of imidacloprid on soil bacterial populations in laboratory studies

The result of the laboratory study is given in Table 1. Application of imidacloprid at 125, 250, 500 and 1,000 ppm resulted in 9.80, 08.40, 6.73 and 5.60 × 10–6 colonies ($P < 0.05$) when compared to 11.05 × 10–6 in control plates (Graph 1).

3.2 Effect of imidacloprid on soil bacterial populations in field studies

In the field studies, imidacloprid was applied in both recommended and ×1.5 rates and the results are given in Graph 2. The results showed that imidacloprid-treated fields at recommended and ×1.5 rates showed significant ($P < 0.05$) decline in bacterial counts at different post-application intervals when compared to control, but the colony count increased with the time.

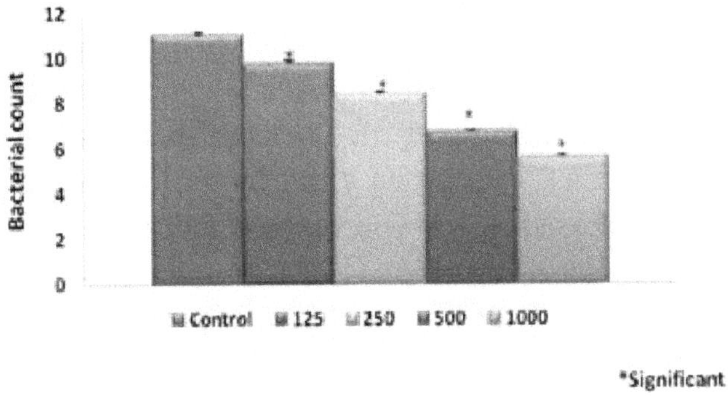

Graph 1. Effect of imidacloprid on bacterial populations in the soil under laboratory conditions

3.3 Identification of soil isolate

The bacterial strain SP-03 isolated from soil was a rod-shaped, Gram-positive bacterium, facultatively anaerobic, grows at 5–40°C, at pH 6–7; produce subterminal ellipsoidal endospores; white-colored colonies; positive for catalase activity, Voges-Proskauer, starch hydrolysis and oxidase; and negative for methyl red, gelatin liquefaction, production of indole and citrate. The 16S rDNA gene of SP-03 was isolated and sequenced. This 16S rDNA gene sequence was then compared with previously published 16S rDNA gene sequences and based on matches the strain was classified as a member of the genus Bacillus. The sequence of strain SP-03 displayed the highest identity (100%) with the 16S rDNA gene of Bacillus weihenste phanensis KBAB4 (GenBank Accession Number: HG 486214.1) (Figure 1, 2). The Bacillus weihenstephanensis showed highest growth at 22°C and at pH of 7.0.

3.4 Effect of imidacloprid on antioxidant enzymes in Bacillus weihenstephanensis

On exposure of Bacillus weihenstephanensis to various molar concentrations (10–3 to 10–7) of imidacloprid for 24, 48, 72 and 96 h, there was a significant ($P \leq 0.05$) increase in the activity of antioxidant enzymes studied. There was a significant increase ($P \leq 0.05$) in the activity of superoxide dismutase (Graph 3), catalase (Graph 4) and peroxidase (Graph 5) in all the treated groups. The antioxidant enzyme activity increased with an increase in the concentration of imidacloprid.

3.5 Gene isolation and sequencing of the stress enzymes of Bacillus weihenstephanensis on exposure to imidacloprid

In the present gene sequencing study, different markers of 400, 600, 1,000 and 1,200 bp (Figure) 6) were run along with our test sample. The superoxide dismutase gene was corresponding to 624 bp. This gene was isolated, eluted and sent for sequencing. The sequence was received and this was subjected to NCBI BLAST which revealed that the nucleotide sequence was of 624 letters bearing accession number YP_001648011.1 that showed 100% query coverage with Bacillus weihenstephanensis Fe/MnSOD (sod A) (Figure 3). The gene for catalase corresponded band at 800 bp. The isolated gene was eluted and sent for sequencing. The received sequence, subjected to NCBI BLAST revealed that the nucleotide sequence was of 800 letters bearing accession number YP_001647388.1 and showed 100% query coverage with Bacillus weihenste-phanensis Hydroxyperoxidase HP(II) (Kat E) (Figure 4). The gene for peroxidase that corresponded to 480 bp was isolated, eluted and sent for sequence analysis (Table 2, 3). The sequence report was received, which was subjected to NCBI BLAST which revealed the nucleotide sequence of 480 letters bearing accession number YP_001644822.1 and showed 100% query coverage with glutathione peroxidase of Bacillus weihenstephanensis (Figure 5).

4. DISCUSSION

4.1 Effect of imidacloprid on soil bacterial populations in laboratory and field studies

Results obtained in laboratory studies showed significant ($P < 0.05$) decrease in bacterial count when compared to that of control. A gradual decrease in bacterial count is observed with increase in concentration of imidacloprid, with minimal count reported at 1,000 ppm. The results obtained were similar to results reported earlier in a study involving five other pesticides [14].

The results indicate toxic effect of imidacloprid on bacterial populations.

Results obtained from bacterial enumeration of imidacloprid-treated soils at recommended rate showed significant ($P < 0.05$) decrease in bacterial numbers, proving negative effect of imidacloprid on bacteria. This negative effect reduced after 14 days of treatment. The negative effect of imidacloprid was vanished by 28th day of application, indicated by bacterial count, which was almost similar to pre-treatment count. Similar results were reported in a study involving imidacloprid and five other pesticides; in the study the toxic effect was vanished by 21st day of imidacloprid application [14].

The different studies have shown that the impact of pesticides application microorganisms present in soil is variable. The impact depends on interaction between microorganisms and active substances and formulation. It also depends on surfacing of specific group of micro organisms [15]. The microorganisms can develop the ability to use an applied pesticide as a source of energy and growth [16].

The initial decrease in bacterial count is expected as pesticides are known to affect the microbial populations by controlling the survival and reproduction of individual species. Initial reduction in microbial count is also reported in studies involving different pesticides such as endosulphan, cypermirithin thiodan, etc. [12, 17–18], and herbicides like glyphosate, atrazine, simazin and alachlor [19–22] when applied at recommended rates. It has been observed in many studies that pesticides stimulated the mineralization rate of organic carbon in comparison with control samples [23–24]. Microorganisms susceptible to toxic effects of pesticides are removed from the population of soil microflora. The pesticides kill the bacterial cells by penetration and disturbing the cell metabolism.

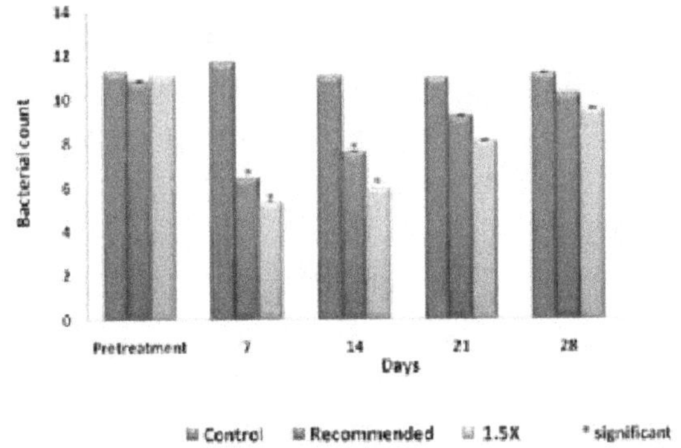

Graph 2. Effect of imidacloprid on bacterial populations in soil under field conditions

The reduction in the number of sensitive microorganisms and increase in resistant micro-organisms lead to reduced soil microbial biodiversity. The increase in bacterial numbers after 14th day may be due to the ability of bacteria to degrade toxic compounds like pesticides [25]. The growth of pesticide-resistant microorganisms may compensate the loss of pesticide- sensitive microorganisms in the population [24]. The addition of fungicide leads to increase in bacterial populations due to no competition with fungi or antagonistic inhibition by fungi [26]. Bacteria are known to become resistant to toxic compound with production of specific degrading enzymes [27].

The application of 1.5× of imidacloprid showed significant ($P < 0.05$) decrease in bacterial number. The results were similar to recommended rates, but bacterial numbers increased slowly. These results were comparable to the results reported in similar studies with pesticides like metoalchlor, atrazine, dimethoate and endosulfan [20, 22, 28–29].

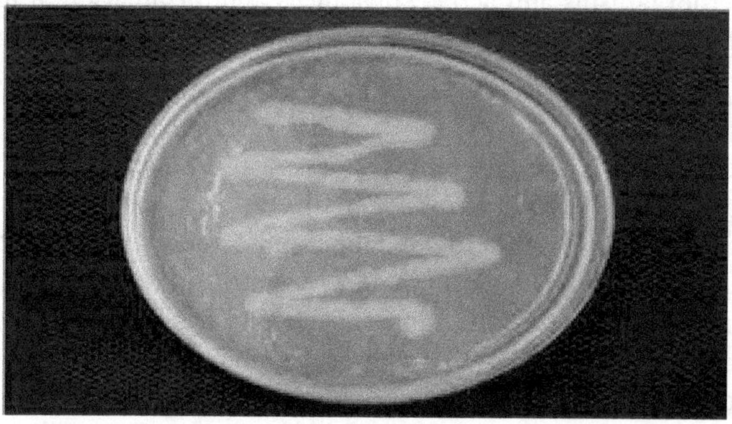

Figure 1. Bacillus weihenstephanensis

4.2 Identification of soil isolate

The bacterial strain SP-03 isolated from soil was a rod-shaped, Gram-positive bacterium, facultatively anaerobic, grows at 5–40°C, at pH 6–7; produce subterminal ellipsoidal endospores ; white-colored colonies; positive for catalase activity, Voges-Proskauer, starch hydrolysis and oxidase; and negative for methyl red, gelatin liquefaction, production of indole and citrate. The 16S rDNA gene of SP-03 was isolated and sequenced. This 16S rDNA gene sequence was then compared with previously published 16S rDNA gene sequences and based on matches the strain was classified as a member of the genus Bacillus. The sequence of strain SP-03 displayed the highest identity (100%) with the 16S rDNA gene of Bacillus weihenstephanensis KBAB4 (GenBank Accession Number: HG 486214.1). The Bacillus weihenstephanensis showed highest growth at 22°C and at pH of 7.0.

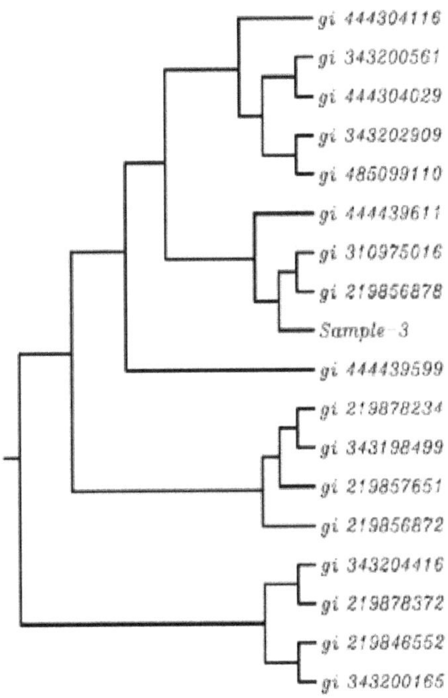

Figure 2. Phylogenetic tree of Bacillus weihenstephanensis

4.3 Antioxidant enzymes

Partial reduction of oxygen to water during microbial respiration gives rise to reactive oxygen intermediates, e.g. superoxide radicals, hydrogen peroxide and hydroxyl radicals. Micro-organisms have developed efficient enzymatic and nonenzymatic mechanisms to eliminate these toxic and mutagenic reactive oxygen species. Superoxide is eliminated by dismutation to H2O2 catalyzed by superoxide dismutase and accumulation of H2O2 is prevented by the action of catalases and peroxidases [30].

Numerous pesticides such as paraquat, DDT, PCB, Arochlor, etc. have been used as model factors inducing oxidative stress both in vivo and in vitro [31]. The tissue damage occurs due to conversion of pesticides to free radicals or superoxide radical during their metabolism. Organisms exposed to different concentrations of xenobiotics have the risk of carcinogenic effect, neurological actions and brain damage [32]. The organisms have developed some mechanisms to control the amount of hydroxyl and superoxide radicals generated to overcome the toxic effects of xenobiotics. Antioxidants quickly scavenge the hydroxyl and superoxide radicals generated. The antioxidants can be enzymatic or nonenzymatic which safely interact with free radicals and terminate the chain reactions before vital molecules are damaged. The antioxidant enzymes include catalase, superoxide dismutase (SOD), glutathione reductase, glutathion-S-transferase and glutathione peroxidase [33].

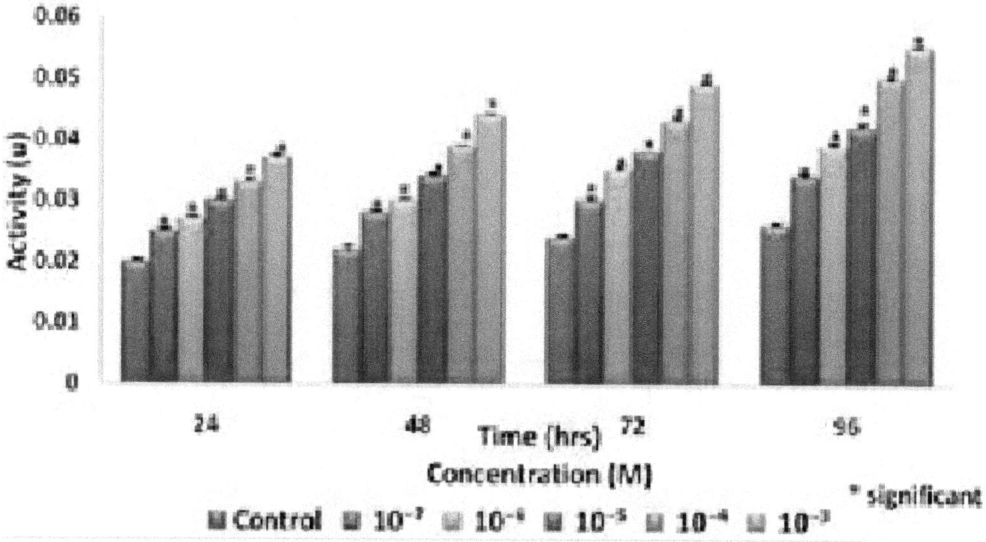

Graph 3. Effect of imidacloprid on SOD activity in Bacillus weihenstephanensis

Significant increase in activity of SOD compared with the control may be due to the toxic effects of imidacloprid. ROS depends on the oxidative metabolism of xenobiotics or endogenous compounds. The antioxidant defense systems work by lowering the concentrations of xenobiotics rather than complete elimination. When the ROS generated exceeds the antioxidants' capability of that cell, it results in oxidative stress [34].

The stress-mediated cytotoxicity results due to oxidative processes and loss of key antioxidant enzymes. Escherichia coli, Salmonella typhimurium and mammalian cells induce antioxidant proteins in response to oxidative stress [35]. It is suggested that an increase in SOD and CAT might be in response to increased oxidative stress or might be due to compensatory response to oxidative stress induced by this xenobiotic. Superoxide dismutase, catalase and peroxidase are the enzymes that participate in the protection against reactive oxygen species.

Catalase is one of the most efficient antioxidants known so far. It is present in peroxisomes of nearly all aerobic cells and protects the cells from the toxic hydrogen peroxide effects by catalyzing its decomposition into molecular oxygen and water without the production of free radicals. In addition, catalase is known to act on toxic compounds by per oxidative reactions.

It is demonstrated that acetamiprid-induced oxidative stress on Escherichia coli, Pseudomonas sp and Bacillus subtilis resulted in elevated superoxide dismutase and catalase activities to antagonize oxidative stress [5].

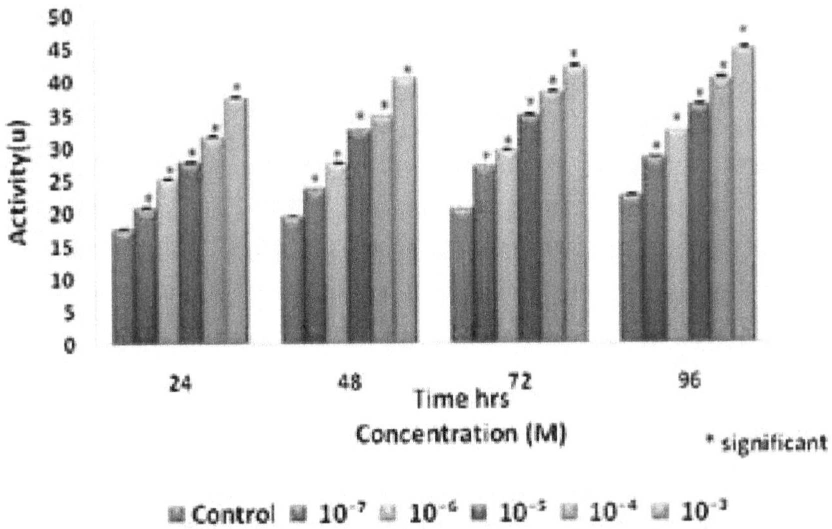

Graph 4. Effect of imidacloprid on catalase activity in Bacillus weihenstephanensis

The present study revealed that the catalase activity was significantly increased in all the groups with increase in the dose and durational exposure of imidacloprid to Bacillus weihen-stephanensis. Similarly, it has been reported that induction of major antioxidant enzymes, such as superoxide dismutase and catalase, were observed after their exposure to a single oxygen generating system in Escherichia coli. It is suggested that response to low concentrations of hydrogen peroxide induces catalase in Escherichia coli during logarithmic growth [36].

Peroxidase is found among animals, plants and microorganisms, where they perform essential roles in the metabolism. To prevent the lethal effects of such metal-ion-catalyzed oxidation (MCO), bacterial cells have evolved protective mechanisms to neutralize the formation of toxic oxygen radicals. For instance, small molecule antioxidants, such as catalases and peroxidases, have been reported to play protective roles in the enteric bacteria in Pseudomonas sp. and in Bacteroides sp. [37].

The enzyme peroxidase is an important antioxidant enzyme, which plays a pivotal role in plant growth and development. The presence of phenol substances leads to enhanced activity of peroxidase (POD). The POD helps in providing resistance to stress and self-defense by increasing the rate of respiration under stress conditions [38].

The present study revealed that the peroxidase activity in the treated groups increased significantly in higher dose (10–5, 10–4 and 10–3 M) of exposure and there was no significant increase observed in the lower dose (10–7 and 10–6 M) of imidacloprid in Bacillus weihenstephanensis. Similar results were reported in other organisms which suggest that a gradual increase of catalase or peroxidase production in aging cultures is not surprising since catalase and/or CP is one of the radical-scavenging enzymes in cells in response to oxidative stress [39]. On the other hand, several organisms produce two or more catalase peroxidase, whereby one enzyme was expressed at the end of exponential growth and during the stationary phase. This behavior was observed in Escherichia coli, Pseudomonas putida, Streptomyces coelicolor and Arcobacter nitrofigilis [40]. It is also reported that superoxide dismutase and peroxidase form the first line of defense against reactive oxygen species [41].

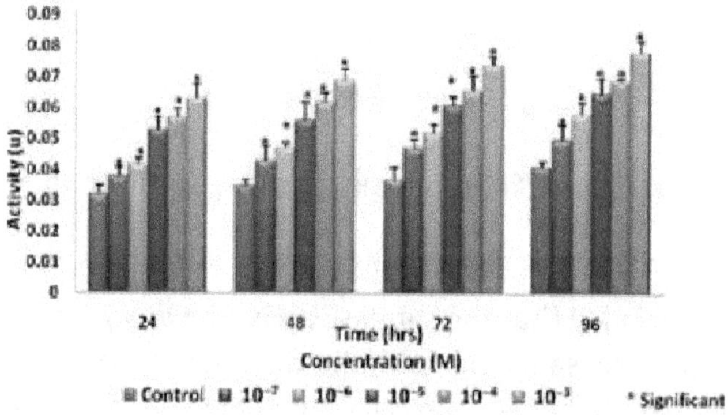

Graph 5. Effect of imidacloprid on peroxidase activity in Bacillus weihenstephanensis

The significant increase in the antioxidant enzymes activity observed in the present study may be due to synthesis of these enzymes as a response to chemical stress induced by imidacloprid or due to inhibition of the membrane-bound enzymes by affecting the enzyme complex, oxidative stress-mediated cytotoxicity enzymes, induction of antioxidant proteins in response to oxidative stress [42].

4.4 Gene sequencing of stress enzymes of Bacillus weihenstephanensis on exposure to imidacloprid

The research in life sciences is affected significantly by the mapping of the genes and genomes of organisms. The related mapping technology is changing the current understanding of biological systems [43]. The application of life science areas of toxicology, genetics, molecular biology and environmental health to describe the response of organisms to environmental stimuli is called toxicogenomics. The toxicogenomics is developed in the past 15 years and will help in advancing the scientific basis of risk assessments for the environmental contaminants [44].

In the present study, the exposure of Bacillus weihenstephanensis to imidacloprid resulted in the expression of manganese containing superoxide dismutase (sod A) gene. MnSOD and FeSOD have an extremely broad phylogenetic distribution, being expressed in both prokaryotic (eubacterial and archaeal) and eukaryotic cells and are quite homologous [45]. Expression of sod A gene has also been reported for other bacterial species. A strain of Sulfolobus sulfataricus produced Fe-Mn SOD with half-life of 2 h at 100°C [46]. In a study, superoxide dismutase producing Bacillus sp. was isolated from Bulgarian thermal spring [47]. In another study, Thiobacillus denitrificans strain "RT" Fe-superoxide dismutase has been purified with a molecular weight of 43,000, and is composed of two identical subunits. Aerobically and anaerobically grown Thiobacillus denitrificans cells contain the same Fe-enzyme with similar levels of activity. Manometric sulfite oxidation measurements suggest for the enzyme a protective function of sulfite against the auto-oxidation initiated by superoxide free radicals [48]. Escherichia coli when grown under anaerobic conditions contained only Fe-SOD, but exposure to oxygen induced the synthesis of Mn-SOD and New-SOD [49].

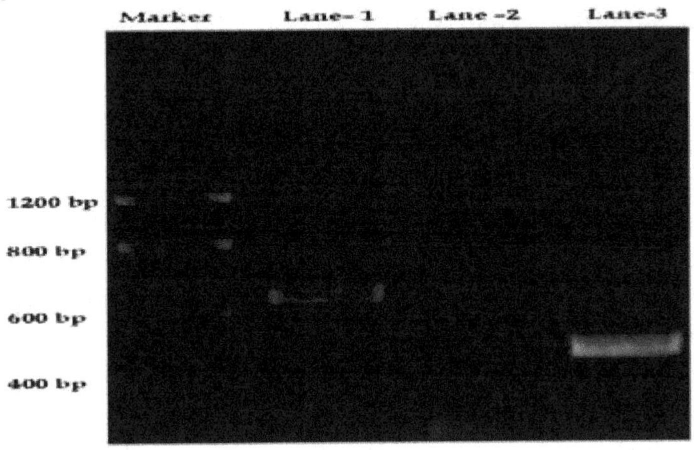

Figure 3. Gel image of stress enzymes amplicon

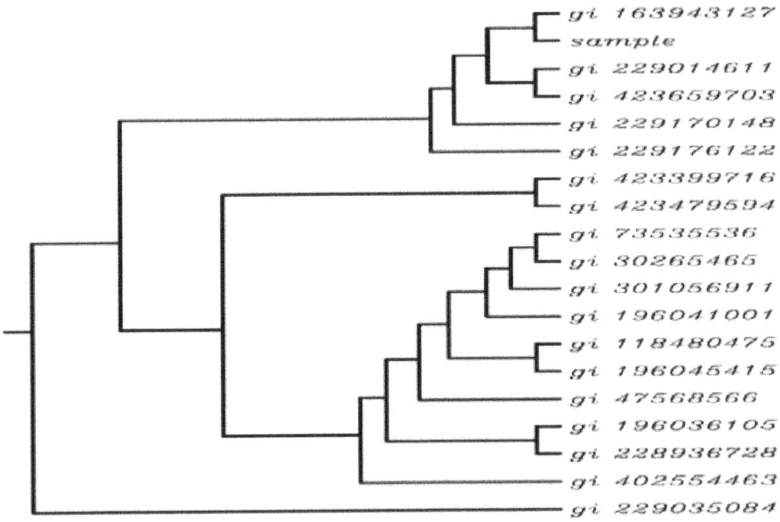

Figure 4. Phylogenetic tree of SOD

SOD of B. subtilis is manganese associated as indicated by a high similarity of the putative amino acid sequence of B. subtilis SodA to those of Mn-Sod from B. caldotenax and B. stearothermophilus, and presence of four conserved metal-binding sites. This SOD was found in vegetative cells and in spores [50]. A new, thermostable superoxide dismutase (SOD) from Bacillus licheniformis M20, is isolated from Bulgarian mineral springs. It is reported that B. subtilis contains a cytosolic Mn-superoxide dismutase [51]. Xenobiotic degrading bacteria experience oxidative stress, both as directly from the pollutants themselves and from intermediates generated during biodegradation processes [52]. Depending on the type of oxidative stress, not only different amounts of proteins can be modified but also different species may appear. It was shown that the set of oxidized proteins depended on the method of induction of oxidative stress.

The present study reveals that the genome of catalase encoded in our Bacillus weihenstephanensis culture on exposure to imidacloprid was Kat E (HPII). HPII and catalase-2 monofunctional catalases of E. coli and B. subtilis, expressed in the stationary phase, have D-isomer prosthetic groups with six haem. Catalase from E. coli HPI R., Halobacterizlm halobitlm and facultative alkalophilic Bacillus species have bifunctional catalase-peroxidases [53–56]. Several organisms produce two or more catalase/peroxidases, whereby one enzyme was expressed at the end of exponential growth and during the stationary phase. This behavior was observed in Escherichia coli, Pseudomonas putida, Streptomyces coelicolor and Arcobacter nitrofigilis [40]. It has been reported that various bacteria such as Citrobacter freundii, Edwardsiella tarda, Enterobaacter aaerogenes, Klebsiella pneumonia and Salmonella typhimurium exhibited patterns of catalase activity similar to that of HPI and HPII bands of Escherichia coli.

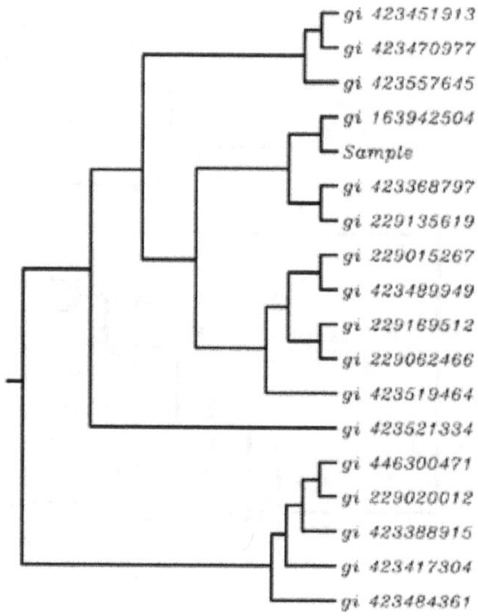

Figure 5. Phylogenetic tree of catalase

Bacterial monofunctional catalases of E. coli HPII [53] and B. subtilis catalase-2, both of which are expressed in the stationary phase, contain six haem D-isomer prosthetic groups in a hexameric structure of larger subunits. From the present study of exposure of Bacillus weihenstephanensis to imidacloprid, it can be concluded that the catalase enzyme can be encoded by the Kat E (HPII) gene.

The present sequence analysis of the peroxidase gene suggests that Bacillus weihenstephanensis subjected to imidacloprid expressed the glutathione peroxidase (Tpx) gene. The findings of study also support a thiol-dependent antioxidant activity for thiol peroxidase in Streptococcus parasanguis, which protects the organism from stress [57]. It is reported that the bacterial thiol peroxidases include a pair of cysteine residues and comprise part of the functional group for the peroxidase activity [58]. Escherichia coli thiol peroxidase is part of an oxidative stress defense system that uses reducing equivalents from thioredoxin (Trx1) and thioredoxin reductase to reduce alkyl hydroperoxides [59]. The specific mechanism(s) by which thiol peroxidase protects Streptococus parasanguis from the toxicant may be similar to those described for other thiol-specific antioxidants of Escherichia coli. The Mycobacterium sp. strain PYR-1 degrades polycyclic aromatic hydrocarbons, environmental pollutants. It was shown that inducible catalase-peroxidase of katG gene of this culture is involved in molecular mechanisms of degradation of these pollutants [60].

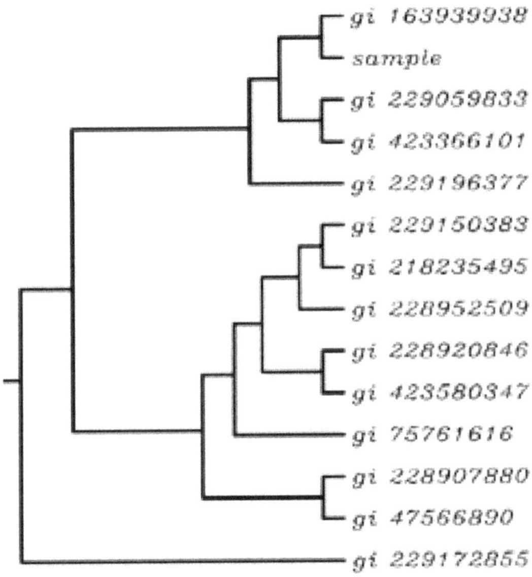

Figure 6. Phylogenetic tree of glutathione peroxidase

5. CONCISE SUMMARY

The current study aimed to evaluate the impact of imidacloprid on the antioxidant enzymes superoxide dismutase, catalase, and peroxidase in the soil isolate Bacillus weihenstephanensis, which was isolated following field studies on the effects of imidacloprid at both recommended and 1.5× rates, revealing an increase in the activity of all three antioxidant enzymes. The enzyme activity augmented with rising pesticide concentration, demonstrating that the inhibitory impact is dose-dependent. Additionally, sequencing disclosed that Fe/MnSOD (sod A), The genes for hydroxyperoxidase HP(II) (Kat E) and glutathione peroxidase were expressed in response to stress induced by imidacloprid treatment in Bacillus weihenstephanensis. The current study demonstrates that imidacloprid generates stress, leading to the expression of antioxidant enzymes in the soil isolate Bacillus weihenstephanensis to safeguard cellular components from oxidative damage. The study indicates that the soil isolate Bacillus weihenstephanensis has acquired resistance to imidacloprid toxicity through the manufacture of antioxidant enzymes. Additional research may be conducted to utilize it for pollution monitoring and risk assessment related to imidacloprid contamination in soil, thereby investigating the potential application of soil-isolated imidacloprid-resistant Bacillus weihenstephanensis in the examination of intricate biological processes and in the remediation of fields contaminated with imidacloprid.

REFERENCES

1. Pimentel D. Effects of pesticides on the environment. 10th International Congress on Plant Protection. Crydon 1983;2:685–91.
2. Nauen R, Hungenberg H, Tollo B, Tietjen K, Elbert A. Antifeedant effect, biological efficacy and high affinity binding of imidacloprid to acetylcholine receptors in Myzus persicae and Myzusnicotianae. Pest Manage Sci 1998;53:133–40.
3. Okazawa A, Akamatsu M, Ohoka A, Nishiwaki H, Cho W, Nakagawa N, Ueno, T. Prediction of the binding mode of imidacloprid and related compounds to house-fly head acetylcholine receptorsusing three-dimensional Qsar analysis. Pestic Sci 1998;54:134–44.
4. Kapoor U, Srivastava, MK, Srivastava LP. Toxicological impact of technical imidacloprid on ovarian morphology, hormones and antioxidant enzymes in female rats. Food Chem Toxicol 2011;49(12):3086–9.
5. Yao XH, Min H, Zhen-Mel LV. Response Of Superoxide Dismutase, Catalase, And Atpase Activity In Bacteria Exposed To Acetamiprid Biomed Environ Sci 2006;19:309–14.
6. Snape JR, Maund SJ, Pickford DB, Hutchinson TH. Ecotoxicogenomics: the challenge of integrating genomics into aquatic and terrestrial ecotoxicology. Aquat Toxicol 2004;Apr 14;67(2):43–54.

7. Nuwaysir EF, Bittner M, Trent J, Barrett JC, Afshari CA. Microarrays and toxicology: the advent of toxicogenomics. Mol Carcinogenesis 1999;24:153–9.
8. Bradley B, Theodorakis C. The post-genomic era and ecotoxicology. Ecotoxicology 2002;11:7–9.
9. Shetti AA, Kaliwal BB. Imidacloprid induced intoxication in soil isolate Brevundimonas Sp. MJ 15. Int J Life Sci Pharma Res 2012;2(3):L-105–17.
10. Shetti AA, Kulkarni A G, Kaliwal RB, Shivasharana CT, Kaliwal, BB. Influence of imidacloprid on biochemical parameters in soil isolate Escherichia Coli. Int J Pharm Bio Sci 2012;3(4)(B):1155–63.
11. Beauchamp C, Fridovich I. Superoxide dismutase: improved assays and an assay applicable to acrylamide gels. Anal Biochem 1997;144:276–87.
12. Sadasivam S, Manickam A. Bio-Chemical Methods, II Edition, New Age International (P) Limited, Publishers, New Delhi; 1996; pp.107–108.
13. Sambrook J, Maccallum P, Russel D. Molecular Cloning: A Laboratory Manual, 3 nd ed. Cold Springs Harbour Press, NY, ISBN 0- 87969-577-3. 2001; p. 2344.
14. Ahmed S, Ahmad M. Effect of insecticides on the total number of soil bacteria under laboratory and field conditions. Pak Entomol 2006;28:63–8.
15. Nowak A, Zbiec I, Gawińska H, Hrebien T. The influence of some pyridine herbicides on microorganisms count and amount of microbial biomass in soil. Zesz Nauk Akad Rol Szczecinie 1999;78:243–52.
16. Johansen K, Jacobsen CS, Torsvik V. Pesticide effects on bacterial diversity in agricultural soils – a review. Biol Fertil Soils 2001;33:443–53.
17. Ekundayo EO. Effect of pesticides used in the Niger delta basin of southern Nigerian soils microbial population environmental monitoring and assessment. Environ Monitor Assess 2006;89:35–41.
18. Adebayo TA, Ojo OA, Olaniran OA. Effect of two insecticides Karate and Thiodan on population dynamics of four different soil microorganism. RJBS 2007;2(5):557–60.
19. Weaver MA, Krutz JL, Zablotowicz RM, Reddy KN. Effects of glyphosate on soil microbial communities and its mineralization in a Mississippi soil. Pest Manag Sci 2007;63:388–93.
20. Ayansina ADV, Oso BA. Effect of two commonly used herbicides on soil microflora at two different concentrations. AJB 2006;5(2):129–32.
21. Fantroussi S, Verschuere L, Verstraete W, Top EM. Effects of phenylurea herbicides on soil microbial community's estimated by analysis of 16s rRNA gene fingerprinting and community level physiological profiles. AEM 1999;28;982-88.

22. Ismail BS, Shamsuddin N. Effects of Alachlor and Metolachor on microbial populations in the soil. MJM 2005;(1):36–41.

23. Bhuyan S, Sreedharan B, Adhya TK, Sethunathan N. Enhanced biodegradation of γ- hexachlorocyclohexane (γ- HCH) in HCH (commercial) acclimatized looded soil: factors affecting its development and persistence. Pest Sci 1993;38:49–55.

24. Das AC, Mukherjee D. Soil application of insecticides influences microorganisms and plant nutrients. Appl Soil Ecol 2000;14:55–62.

25. Cremlyn RJ. Pesticide preparation and mode of action. John Wiley and Sons, New York. 2006; pp: 50–130.

26. Chen SK, Subler S, Edwards CA. Effects of the fungicides benomyl, captan and chlorothalonil on soil microbial activity and nitrogen dynamics. Soil Biol Biochem 2001;33:1971–80.

27. Kulkarni AG, Kaliwal BB. Methomyl induced effects on free and immobilized Escherichia coli. Int J Biotechnol Res (IJBR) 2009;2(2):97–101.

28. Mandal A, Lahiri P. Insecticide induced haematological changes in pigeons. Proc Ind Acad Anim Sci 1989;98(2):133–7.

29. Dıgrak M, Ozcelik S. Effect of some pesticides on soil microorganisms. Bull Environ Contam Toxicol 1998;60(1):916–22.

30. Fridovich I. The biology of oxygen radicals. Science 1978;201:875–80.

31. Palmeira CM. Herbicide – induced mitochondrial and cellular liver toxicity: a review of paraquat, Dinoseb and 2 4-D effects. J Tox Substs Mech 1999;18:187–204.

32. Bus JS, Gibson JE. Paraquat model for antioxidant – initiated toxicity. Environ Health Persp 1984;55:37–46.

33. Al-Omar MA, Beedham C, Alsarra IA. Pathological roles of reactive oxygen species and their defense mechanisms. Saudi Pharm J 2004;12:1–17.

34. Sevanian A, Peterson AR. Cholesterol epoxide is a direct- acting mutagen. Proc Natl Acad Sci (USA) 1984;81:4198–202.

35. Kim SY, Kim EJ, Park JW. Control of singlet oxygen-induced oxidative damage in Escherichia coli. J Biochem Molecul Biol 2002;35(4):353–7.

36. Zaid T, Sukumaran T, Srikumar, Benov L. Growth of Escherichia coli in iron-enriched medium increases HPI catalase activity. J Biochem Molecul Biol 2003;36(6):608–10.

37. Cha MK, Kim HK, Kim IH. Mutation and mutagenesis of thiol peroxidase of Escherichia coli and a new type of thiol peroxidase family. J Bacteriol 1996;178:5610–4.

38. Weimberg R. Enzyme level in pea seedlings grown on highly salinized media. Plant Physiol 1970;46:466–70.

39. Yun EJ, Lee YN. Production of two different catalase-peroxidases by Deinococcus radiophilus FEMS Microbiol Lett 2000;184:155–59.

40. Park YB, Hah YC. J Microbiol 1997;35:239.

41. McCord JM, Fridovich I. Superoxide dismutase: an enzymatic function for erythrocuprein (Hemocuprein). J Biol Chem 1969;244:6049–55.

42. Christman MF, Morgan RW, Jacobson FS, Ames BN. Positive control of a regulon for defenses against oxidative stress and some heat-shock proteins in Salmonella typhimurium. Cell 1985;41:753–62.

43. U.S. EPA. Ecological effects preliminary review. NTN 33893-2 systemic insecticide. 2004. Janus JA. Integrated criteria document Zic; ecotoxicity, RIVM Report 2011;710401019.

44. Whittaker JW. The irony of manganese superoxide dismutase. Biochem Soc Transac 2003;31(6):1318-21.

45. Russo AD, Ruilo R, Nitti G, Masuilo M, Bochini, V. Biochim Biophys Acta 1997;1343:23–30.

46. Bojadzhieva I, Elka E. Characteristics of bacterial strain, superoxide dismutase producer, isolated Rom Bulgarian thermal spring. J Culture Collect 2005;4:61–7.

47. Baldensperger JB. An Iron-containing superoxide dismutase from the C. hemolithotrophic Thiobacillus denitrifieans "RT" strain. Arch Microbiol 1978;119:237–44.

48. Hassan HM, Fridovich I. Enzymatic defenses against the toxicity of oxygen and of streptqnigrin in Escherichia coli. J. Bacteriol 1977;129:1574–83.

49. Inaoka T, Yoshinobu M, Tetsuaki T. Molecular cloning and nucleotide sequence of the superoxide dismutase gene and characterization of its product from Bacillus subtilis A novel, thermostable manganese-containing superoxide dismutase from Bacillus licheniformis. J Bacteriol 1998;180:3697–703.

50. Koji T, Kido T, Shimasue Y, Soda K. Isolation of manganese-containing superoxide dismutase from Bacillus subtilis. Agric Biol Chem 1983;47(12):2865–70.

51. Park W, Jeon CO, Cadillo H, DeRito C, Madsen EL. Survival of naphthalene-degrading Pseudomonas putida NCIB 9816-4 in naphthalene-amended soils: toxicity of naphthalene and its metabolites. Appl Microbiol Biotechnol 2004;64:429–35.

52. Loewen PC, Switala L. Multiple catalases in Bacillus subtilis. J Bacteriol 1987;169:3601- 7.

53. Claiborne A, Fridovich L. Purification of the o-dianisidine peroxidase from Escherichia coli B. J Biol Chem 1979;254:4245–52.

54. Hochman A, Shemesh A. Purification and characterization of a catalase-peroxidase from the photosynthetic bacterium Rbodopseudomonas capsulata. J Biol Chem 1987;262:6871–6.

55. Yumoto I, Fukumori Y, Yamanaka T. Purification and characterization of catalase from a facultative alkalophilic Bucillus. J Biochem 1990;108:583–7.

56. Spatafora G, Hoeven NV, Katherine W, Taylor P. Evidence that ORF3 at the Streptococcus parasanguis fimA locus encodes a thiol-specific antioxidant. Microbiology 2002;148:755–62.

57. Cha MK, Kim HK, Kim IH. Mutation and mutagenesis of thiol peroxidase of Escherichia coli and a new type of thiol peroxidase family.. J Bacteriol 1996;178:5610–14. [PubMed: 8824604]

58. Laura M, Baker S, Leslie B. Poole catalytic mechanism of thiol peroxidase from Escherichia coli: sulphonic acid formation and overoxidation of essential cys61. J Biol Chem 2003;278(11):1–25.

59. Wang RF, Wennerstrom D, Cao WW, Khan AA, Cerniglia CE. Cloning, expression, and characterization of the katG gene, encoding catalase-peroxidase, from the polycyclic aromatic hydrocarbon-degrading bacterium Mycobacterium sp. Strain PYR-1. Appl Environ Microbiol 2000;66:4300–04.

15 Chapter ANTIMICROBIAL RESISTANCE IN BACTERIA

1. INTRODUCTION: AN OVERVIEW

Resistance arises from bacteria acquiring new genes in response to pesticides and antibiotics, enabling them to persist in their environment. A variety of xenobiotics and antibiotics are being introduced as a consequence of bacterial resistance development. The effective utilization of any therapeutic substance is acknowledged to be compromised by the potential emergence of tolerance or resistance to that molecule from its initial application. These substances are indeed utilized in the treatment of parasite and microbial infections, as well as chronic ailments such as cancer and diabetes, and they are applicable to conditions affecting any living organisms, including plants, insects, humans, animals, and fish. Numerous alterations in physiological and biochemical systems may account for the occurrence of resistance. The intricate complexities associated with the development of antimicrobial agents and insecticides, which contribute to the emergence and proliferation of resistance, cannot be overstated. Additionally, the frequent absence of fundamental information and understanding regarding these matters is a primary factor in the minimal progress achieved in the effective prevention and control of resistance development in organisms. [1]

2. PESTICIDES RESISTANCE IN BACTERIA

Agriculture is the most essential of the Indian economy. It ensures food security for the biggest population with shrinking cultivable land resource necessitates use of high yielding variety of seeds, balanced use of fertilizers, and judicious use of quality pesticides. The indiscriminate and unplanned use of agrochemicals influences microbial processes that are an essential component of carbon, nitrogen, and sulfur cycles.

Pesticides are added to the environment for the purpose of killing or injuring the pests. Pesticides are the chemical substances that kill pests like fungi, insects, worms, nematodes, etc., which cause damage to field crops. It is almost impossible to control the spreading of pesticides effect. Even though it is applied in very much small area, it is dissolves in water, absorbed by soil, and spreads through the air. The contaminated water ultimately reaches much bigger and new area. These pesticides also frequently trickle into ground water, which we humans consume, and as a result, these pesticides poison us over a period of time. However, in addition to these, residual pesticides that remain on the plants are sometimes consumed by animals and also humans, leading to serious many disorders like cancer and even sometimes death. The excessive use of pesticides leads to an accumulation of a huge amount of pesticide residues in the food chain and drinking water environment that further leads to a substantial health hazard for the current and future generations due to uptake and accumulation of these toxic compounds [2]. Majority of agrochemicals devoid of mutagenic activity and induce their effects by genotoxic or non genotoxic modes of action. However, in some cases, the modes of action are known and they give a clear indication of the likely human hazards, but in many cases, the data are lacking or incomplete resulting in a more conventional approach toward human hazard and risk assessment [3].

Pesticides can be classified according to their toxicity, chemical group, environmental persistence, target organism, or other features. Classes of organic pesticides (consisting of organic molecules) include organochlorine, organophosphate, organometallic, pyrethroids, and carbamates among others [4]. The use of explosives, refrigerants, pesticides, solvents, and many dyes in urban, industrial, and agricultural applications results into the release of xenobiotic compounds into the environment, and the problem of such toxic waste disposal has become enormous due to the proliferation of these xenobiotic compounds. Many xenobiotic compounds, particularly those used as insecticides, are toxic. Insecticides have been reported to affect the microbial populations by controlling the survival and reproduction of species.

Biodegradation and bioremediation are identical processes to an extent that both of these are based on the conversion or metabolism of many chemicals and pesticides by microorganisms. Here, the difference is that biodegradation is a natural process while bioremediation is a technology. A successful bioremediation technique requires an efficient bacterial strain that can degrade largest pollutant to minimum level [5]. Soil microorganisms play a key role in biodegradation. Soil microorganisms are of great concern for using in biotechnology because they are able to metabolize and degrade many of pollutants and pesticides. Simultaneously, on the other hand, microbial degradation can lead to formation of more toxic and persistent metabolites. Although the soil microbial population are characterized by their adaptability to the changed environmental condition, fast flexibility and the application of pesticides in long term as well as in short term can cause significant irreversible changes in their population.

Bioremediation constitutes an attractive alternative to physicochemical methods of remediation, as it is less expensive and can selectively achieve complete destruction of organic pollutants. In bioremediation, microbes that can degrade the pesticides in-situ are used. For a successful bioremediation technique, an efficient bacterial strain that can degrade largest pollutant to minimum level is required. In predicting the persistence of synthetic chemicals in soil, sediment, and natural water, it is necessary to determine the role of endogenous micro-organisms in the overall degradation process.

The ability of organisms to bioremediate pesticides is mainly based on their biodegradation activity. Although bioremediation has been firstly achieved using microorganisms (bacteria or fungi), other organisms like plants or algae can be used. To eradicate undesirable effects of pollutants from the environment the strategy of bioremediations are used. However, this method is not always possible and sometimes it would be appropriate to eliminate pollutants, although some organisms could restrict or immobilize them. For an instance, the used organisms can accumulate contaminants and to some extent decrease their presence and their environmental effect, but they do not completely eliminate them from the environment. Those organisms able to bioremediate would be called bioremediators. Traditionally, bioremediation has been achieved by using microorganisms [6].

Pesticide resistance describes the decreased susceptibility of a pest population to a pesticide that was previously effective at controlling the pest. Pest species develop pesticide resistance via natural selection; as a result, the highest resistant varieties survive and continuously transfer their genetic traits to their next generation [7]. Many times, when pesticide degradation happens, it usually involves more than one microorganism, i.e., each microorganism contributes to the process of biodegradation reactions on pesticides. However, there is no evidence, and an example of mineralization by a single strain has been described. It shows that the presence of different microorganisms is essential for an adequate and significant biodegradation. Reported micro-biodegraders belong to Basidiomycetes or to bacterial classes: gamma- proteobacteria (Pseudomonas, Aerobacter, Acinetobacter, Moraxella, and Plesiomonas), beta- proteobacteria (Burkholderia and Neisseria), alpha-proteobacteria (Sphingomonas), actinobacteria (Micrococcus), and flavobacteria (Flavobacterium) [6].

Many studies have focused on the employment of bacteria, consortia or on the search for biotransformation enzymes. Bacteria have been used extensively for bioremediation purposes due to their fast growth, easy handling, and low cost, making them suitable for bioremediation. However, some disadvantages are there, such as the pathogenicity of bacteria, disposal of bacterial biomass, and bioactivation, among others. Bacteria can be found in the environment everywhere such as soil, water, or even in particles dispersed in air. Unfortunately, only a small segment of bacteria (<10% from soil) can be cultured in laboratory conditions. Because of this, a number of studies about pesticide biodegradation mechanisms are less than those about biodegrader's isolation and then slight information on biochemical mechanisms or enzymes is accessible.

Pollutants might undergo biodegradation reactions like cleavage, oxidation, dechlorination, reduction by different enzymes. Since biodegradation capability is based on enzymes, which are promiscuous and have evolved to detoxifying enzymes, the shorter the duplication time of organism and the more sufficient the organism, the easier to obtain biodegraders. Therefore, bacteria with replication time around or less than minutes are much admirable to respond to natural and artificial pollutant-induced evolutionary pressure; this response is involved in the selection of biotransformation enzymes able to degrade them. These promiscuous enzymes are present in the organisms; even before the exertion of the evolutionary pressure, the induced genetic recombination or mutation in the organisms could lead to enzymes with better biodegradation ability [6].

The carbamates are mainly used in agriculture, as insecticides, fungicides, herbicides, nematocides, or sprout inhibitors; also, it is used as a potential in public health vector control. In addition, these carbamates are used as biocides for industrial or other applications and in household products. As a result, these chemicals are part of the large group of synthetic pesticides that have been developed, produced, and used on a large scale in the society and environment. They are derivatives of carbamic acid and like organophosphates; the mechanism of action of these chemicals is that inhibiting the vital enzyme acetyl cholinesterase is reversible as compared to organophosphates, which is irreversible. Exposure to cholinesterase inhibiting agents is considered as a major health problem for the farm workers throughout the world.

Three classes of carbamate pesticides are generally known. The carbamate herbicides have the general structure R1NHC(O)OR2, in which R1 and R2 are aromatic and/or aliphatic moieties. The carbamate ester derivatives, used as insecticides (and nematocides), are usually stable and have a low vapor pressure and low water solubility. Fungicide carbamate contains a benzimidazole group. Carbamates are metabolized by microorganisms, plants, and animals or broken down in water and soil. Soil microorganisms have the capacity of metabolizing (hydrolyzing) carbamates and can easily acclimate themselves to metabolize the different types of carbamates. However, at the high dose levels of carbamates, their metabolites cause changes and can significantly affect the microflora, which may be of importance in soil productivity. Some of the common names of the carbamate insecticide are aldoxycarb, allyxycarb, aminocarb, bufencarb, butacarb, carbanolate, carbaryl, carbofuran, aldicarb, methomyl, oxamyl, thiofanox, thiodicarb, etc.

2.1 Methomyl

Methomyl is widely used throughout the world since it is effective as "contact insecticide" as well as "systemic insecticide." The IUPAC name of methomyl is

S-methyl N-(methylcarbamoyloxy) thioacetimidate. Methomyl belongs to a class of compounds known as oxime carbamates, and it is widely used for the control of insects and nematode pests by inhibiting the enzyme acetylcholinesterase, which hydrolyzes the neurotransmitter acetylcholine [8]. Methomyl has been classified as a pesticide of category-I toxicity [9]. Methomyl is a metabolite of thiodicarb, and acetimidate is a suspected oncogen, which is metabolite in animal tissues [10]. Methomyl is endocrine disruptor and also potent genotoxic, capable of inducing structural and numerical chromosomal aberration in mammalian cells [11]. The World Health Organization (WHO), Environment Protection Agency (EPA), and European Chemical Classification (ECC) classify methomyl as a very toxic and a most hazardous pesticide. Methomyl is highly soluble in water (57.9 g/liter, at 25°C [8]), and since the sorption affinity of soils for this pollutant is rather low, it can easily cause contamination of both ground and surface water resources [12].

The study of methomyl-induced alteration in mice hepatic-oxidative status and methomyl- induced gonadal dysfunction, biochemical contents, and enzyme activities in mice suggests that chronic exposure to methomyl insecticide has deleterious effect on mouse liver and also showed effect on reproductive system in mice. Therefore, the application of such insecticide for designed program should be limited or special care should be taken to minimize its hazards [13, 14].

Many pesticide degradation genes present in soil bacteria have been shown to reside on plasmids, a common location for other degradation genes [15, 16]. Some plasmids are known as catabolic plasmids because they bear genes encoding for enzymes capable of degradation and such plasmids have been of great attraction. The organisms containing the catabolic plasmids have the ability to degrade certain compounds. Many catabolic plasmids have been found in species of Actinobacter, Flavobacterium, Pseudomonas, Alcaligenes, Klebsiella, Moraxel□ la, and Arthrobacter [17].

The innovation of microorganisms capable of tolerating or growing in high concentrations of pesticides provides a potentially interesting possibility for treating hazardous wastes [18]. Some investigations resulting in the identification of microbial isolates, which are apparently responsible for the accelerated degradation of individual pesticides, is necessary [19]. Many xenobiotic degradation genes present in soil bacteria have been shown to reside on plasmid, a common location for other degradation genes [20]. The study of plasmid curing suggests that the degrading ability for methomyl is encoded in the plasmid for the genus Pseudomonas aeruginosa [21]. Soil microbial populations, particularly members of Pseudomonas, Bacillus, and Escherichia coli have evolved the considerable nutritional versatility and are capable of the degradation of a range of complex, naturally occurring aromatic and aliphatic compounds. An interesting feature of these degradative plasmids is that they have been isolated almost exclusively from the species of the genus Pseudomonas, Bacillus, and E. coli [22].

It is suggested that the detoxification metabolism occurs when a microorganism uses the pesticide as a carbon and energy source and the process is assisted by resistant microorganisms [23]. The plasmid-coded biodegradation of methomyl may be due to the broad host range plasmids and selection pressures on spontaneous mutants due to the presence of xenobiotics, vertical gene transfer, or horizontal gene transfer, including transposons and broad host range plasmids and selection pressures on spontaneous mutants due to the presence of xenobiotics or due to strains that harbor a single plasmid with a role in pesticide biodegradation [24].

The proteomic profiling of E. coli in response to carbamate pesticide–methomyl research work was conducted by Kulkarni and Kaliwal [25]. The study suggests that the proteomic profiling is a sensitive tool for environmental stress diagnosis and that the stress proteins could be used as biomarkers for environmental pollution identification. Thus, biological decontamination methods are preferable to conventional approaches because, in general, microorganisms degrade numerous environmental pollutants without producing toxic intermediates [26, 27].

2.2 Imidacloprid

Imidacloprid is the first synthetic neonicotinoid insecticide used against sucking pests, such as rice hoppers, aphids, thrips, and whiteflies. Imidacloprid has been used widely for foliar and seed treatment, soil drench, and stem application [28]. Today imidacloprid is used in over 120 countries to treat more than 140 different crops [29]. It is most commonly used on cotton, rice, cereal, maize, sunflowers, potatoes, and vegetables. The active chemical in imidacloprid works by interfering with the transmission of stimuli in the insect's nervous system. Imidacloprid is a neurotoxic insecticide, which belongs to the class of the neonicotinoid pesticides. Imidacloprid is registered to control insect pests on agricultural and nursery crops, structural pests, and parasites on companion animals. Imidacloprid is an agonist of the nicotinic acetylcholine receptor (nAChR) at the neuronal and neuromuscular junctions in insects and vertebrates. It is structurally and functionally related to nicotine. The toxicity of imidacloprid is largely due to interference of the neurotransmission in the nicotinic cholinergic nervous system. Extended activation of the nAChR by imidacloprid causes desensitization and blocking of the receptor and leads to incoordination, tremors, decreased activity, reduced body temperature, and death.

The ability of some microorganisms to grow in the presence of pesticides may result in the compensation of an adverse effect by the increased activity of remaining part of soil community. Bacteria's are known to become resistant to toxic compound with production of specific degrading enzymes [30]. Three imidacloprid tolerant strains were isolated and identified based on morphology, biochemical characters and 16s rDNA identification as E. coli (Fig. 1), Brevundimonas sp. MJ 15 (Fig. 2), and Bacillus weihenstephanensis (Fig. 3), and they were evaluated for their toxicity toward imidacloprid using standard methods to determine their biochemical contents, growth, and enzyme parameters on exposure to the toxicant and concluded that imidacloprid-induced toxicity and stress in bacterial soil isolates. The knowledge about intoxication effects and conservation of toxicity mechanisms in organisms will enable to choose appropriate model organisms for relevant monitoring of specific environmental toxicants [31].

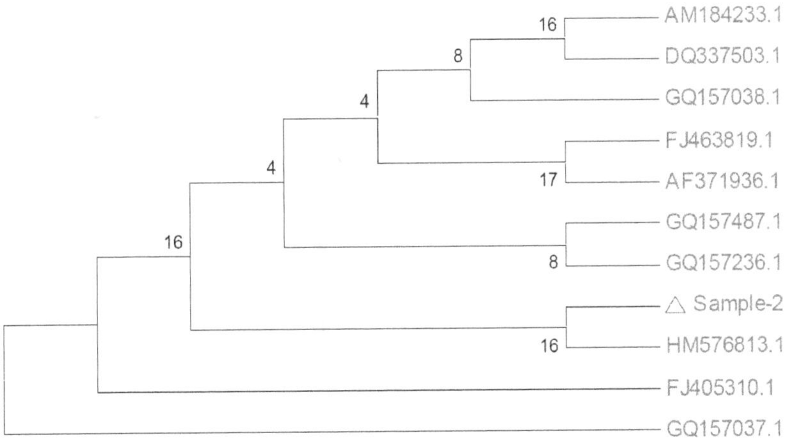

Figure 1. Phylogenetic tree of E. coli.

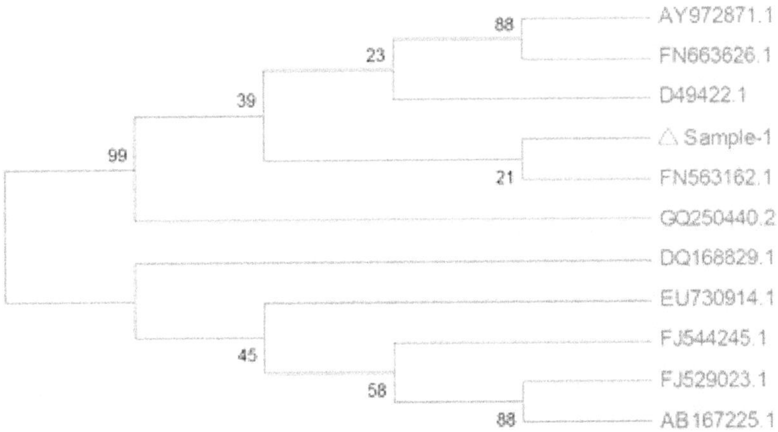

Figure 2. Phylogenetic tree of Brevundimonas sp. MJ 15.

Genes responsible for pesticide degradation in microorganism have been found to be located on plasmids, transposons, and chromosomes. Some microbial strains possess genetic determinants that confer resistance. In bacteria, these determinants are often found on plasmids, which have facilitated their study at the molecular level [32]. The involvements of plasmids in degradation of organic xenobiotics were first reported 20 years ago, and the list is increasing continuously till today [33]. Such type of plasmids have mostly found in xenobiotic degrading bacteria, after selective isolation of the strain by many methods. Also, direct exogenous plasmid isolation has yielded several novel catabolic plasmids more recently. The involvement of plasmids in degradation of pesticides in reported in several studies. Plasmid encode the genes were found responsible for degradation of anthracene, fenitrothion, carbofuran, dimethoate, phenanthrene, and dichlorophenoxyacetic acid by Pseudomonas sp., Burkholderia sp., Sphingomonas sp., Bacillus licheniformis, Flavobacterium sp., and Alcaligenes paradoxus, respectively [34, 35].

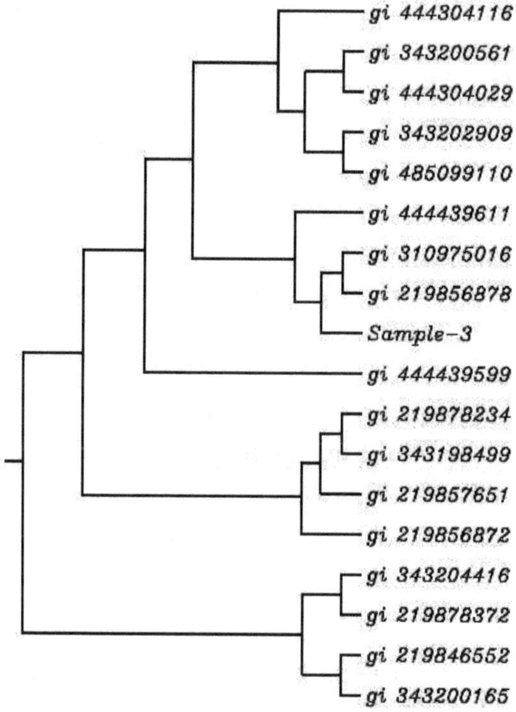

Figure 3. Phylogenetic tree of Bacillus weihenstephanensis.

3. ANTIBIOTIC RESISTANCE IN BACTERIA

Antibiotic use promotes the development of antibiotic-resistant bacteria. Every time a person takes antibiotics, sensitive bacteria are killed, but resistant bacteria may be left to grow and multiply. Bacteria can be described as being susceptible to, tolerant of, or resistant to specific antibiotics. When an antibiotic attacks a group of bacteria, those cells that are susceptible will die.

The primary cause of increase in drug-resistant bacteria is mainly because of repeated and improper uses of antibiotics. The keen use of antibiotics is the key to controlling the spread of resistance bacteria. For example, antibiotics should be used for the treatment of bacterial infections, but many times, they are given for the viral infections such as common cold, most sore throats, and the flu, for which they have not prepared. Extensive and widespread use of antibiotics promotes the spread of antibiotic resistance. Antibiotic resistance occurs when bacteria changes in some way that decreases or eliminates the effectiveness of drugs or other agents designed to cure or prevent infections. As a result, the bacteria survive and continue to multiply causing more harm to the host. This can be performed by bacteria by several mechanisms. Some bacteria develop the ability to neutralize the antibiotic before it can do harm, others can rapidly impel the antibiotic out, and still others can change the antibiotic attack site so it cannot affect the normal function of the bacteria.

An increase in the frequency of antibiotic resistance in bacteria since the 1950s has been observed for all major classes of antibiotics used to treat a wide variety of diseases. There may be many reasons such as resistance is the result of bacteria evolving new genes in response to the presence of antibiotics, or the antibiotic-resistant bacteria selected for in the environment by possessing antibiotic resistance genes and also several factors involved in antibiotic resistance will show that resistance is a designed feature of preexisting genes enabling bacteria to compete with the antibiotic producers in their environment.

3.1 Antibiotic resistance in humans

The development of resistance in bacteria is a major concern for another reason of human health. Historically, infectious diseases have killed billions of people and have the reason for most devastating chapters in the history of humankind. Scientists have been so successful in the earlier century in preventing and curing infectious diseases that only a few years ago, it was thought that modern science had at last enabled us to "close the book on infectious diseases." Bacteria that have become resistant to several antibiotics, said to be multidrug resistant, are often called super bugs by the media. An important mechanism by which bacteria become resistant is by obtaining one or more specific resistance genes from other bacteria. This type of resistance can be acquired by the transfer of a plasmid, already existing in the bacterium gene pool that carries a gene for an enzyme which either destroys or inactivates the antimicrobial substance.

Staphylococcus aureus is a major cause of hospital-acquired infections, causing high morbidity and mortality throughout the world. Vancomycin has been the drug of choice for 30 years for the treatment of methicillin-resistant S. aureus (MRSA). Over the last decade, methicillin- resistant S. aureus (MRSA) strains have become endemic in hospitals worldwide. In addition, it is now incipient community pathogen in many geographical regions. The emergence of high levels of penicillin resistance followed by development and spread of strains resistant to the semisynthetic penicillins (methicillin, nafcillin, and oxacillin), macrolides, tetracyclins, and amino glycosides has made therapy of staphylococcal disease a global challenge. Now a day's resistance to semisynthetic penicillin's had spread throughout the world, compromising the use of these drugs for empiric therapy for staphylococcal infections in a number of regions. This had led to increased reliance on vancomycin for the treatment of documented MRSA infections. As a consequence, selective pressure was established that eventually lead to the emergence of strains of S. aureus and other species of staphylococci with decreased susceptibility to vancomycin and other glycopeptides (Fig. 4).

Vancomycin is the only effective agent against some pathogens, and even in recent years, it has been lost some of its effectiveness. This problem has many causes, including the improper and misuse of antibiotics and the result of transfer of resistance genes from one bacterium to another. This might be possible because many bacteria have an integral natural resistance to a number of antibiotics, and the genes that provide this resistance can be passed on to other bacteria by a variety of means. Multidrug-resistant S. aureus (MDRS) is a much more serious threat, particularly to hospitalized patients in globally, and it represents a challenge for public health, as community-acquired infections seem to be on the increase in adults and children. S. aureus colonization has been shown to be a major risk factor for community-acquired and nosocomial infections. In epidemiological study, Unakal and Kaliwal [36] reported the multidrug resistance and vancomycin resistance S. aureus isolated from Health Care Centres in India for the first time and detected Van A gene in 4 strains, Van B gene in one strain from the 14 VRSA strains procured from the clinical samples pus, infected blood, and cerebrum spinal fluid.

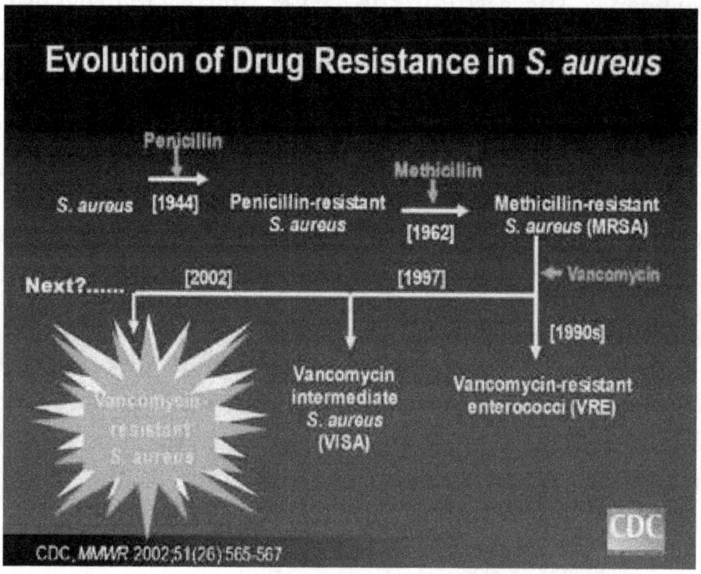

Figure 4. Evolution of drug resistance in Staphylococcus aureus.

The first report of the S. aureus with reduced susceptibility to vancomycin was from Japan [37]. This report was quickly followed by similar ones from other countries, including the United States [38] and Belgium [39]. The extensive longitudinal study of the current situation of vancomycin resistance has reported the first incidence of VRSA emergence from northern part of India [40]. The first clinical infection with vancomycin-resistant S. aureus was reported from Michigan, with second case in Pennsylvania. Further, the second confirmed VRSA from Pennsylvania was reported, which represents the VRSA isolate from patient in the United States. The emergence of decreased vancomycin susceptibility in MRSA strains presents a significant clinical problem with few therapeutic options. The rapid evolution of antibiotic resistance is of considerable concern. Considering high prevalence of MRSA and increased use of vancomycin, the development of vancomycin resistance (VRSA) in clinical strains seems likely to occur. In 1996, the documented infection caused by S. aureus with reduced susceptibility to vancomycin (vancomycin-intermediate S. aureus [VISA]) was reported in Japan [37]. Thereafter, about 20 cases of VISA infections have been reported in several countries, including Korea [41].

Vancomycin is a glycopeptide antibiotic, and glycopeptide resistance has emerged in S. aureus with the interspecies transfer of resistant gene from nonpathogenic Enterococcus faecalis in vitro [42]. Moreover, cellular modification due to prolonged use of vancomycin results in an increased extracellular material, which cause thickening of cell wall [43, 44].

3.2 Antibiotic resistance in bovines

Bovine mastitis is a common disease entity of dairy cows, accompanied by physical, chemical, pathological, and bacteriological changes in milk and glandular tissue [45]. It is a harmful disease affecting the dairy industry worldwide and is a matter of great concern for leading milk-producing countries like India because of the losses incurred due to high morbidity, discarded milk, treatment costs, and reduced milk production, thus drawing in more attention toward its treatment and control [46]. Mastitis is produced by a variety of pathogenic micro-organisms. The majority of cases in bovines are infectious, and it has been estimated that up to 200 microbial species are potential causative agents [47].

The antibiotic efficacy of tetracycline, cefixime, ofloxacin, amoxicillin, and ampicillin was investigated in S. aureus, E. coli, and B. subtilis-induced mastitis in mice by Chinchali [48], and the results were evidenced by the bacterial count and inflammatory enzyme activities of mammary gland tissue against the induced bacterial pathogens in comparison with control group. In the results, there was an orderly decrease in the bacterial counts of S. aureus, E. coli, and B. subtilis, which showed susceptibility for the antibiotics used in the study and indicated their effectiveness of bactericidal activity and their efficacy level against the induced inflammatory reaction. This study demonstrates the effectiveness of the antibiotic in the treatment of the disease. However, continued use and overuse of the antibiotics without an in vitro study will lead to the antibiotic resistance strains.

The usage of antibiotics in the bovine mastitis correlates with the emergence and maintenance of antibiotic-resistant traits within pathogenic strains [49]. These traits are coded by particular genes that may be carried on the bacterial genome or plasmids [50]; hence, these are easily transferred among isolates. The prevalence of antibiotic resistance usually varies between isolates from the different sampled area, environment, and even between isolates from different herds on the same farm or environment [51].

The evolution of antibiotic resistance in S. aureus strains is a serious cause of concern in dairy animals [52]. Antibiotic-resistant S. aureus isolates pose a severe challenge to both in veterinary and health professions and in dairy cattle producers because they have a serious negative impact on the therapy management. S. aureus has been and become the main issue of studies on antibiotic resistance because of its importance for all forms of mastitis in dairy cows [53]. Multiple antibiotic-resistant S. aureus strains have been isolated and screened from milk obtained from cattle, beef, and human samples from the various part of the world.

Antimicrobials have been used frequently as a conventional measure to prevention and control diseases in dairy farming. Especially in mastitis control programs, more and more antibiotics were applied even without any clinical symptoms in dairy cattle herds. However, long-term in-feed use of antibiotics on dairy farms has led to the alarming increase of antibiotic-resistant bacteria, which has become a public health issue worldwide, e.g., methicillin-resistant S. aureus (MRSA) from raw milk and environmental samples constitutes a great threat to food safety. In order to better understand the potential of dairy cattle as a reservoir for antibiotic-resistant bacteria, it is important to investigate the prevalence of antimicrobial resistance among bacteria isolated from raw milk [54].

References [55–59] reported many bovine mastitic bacteria and reported the isolated bacteria were resistant to many antibiotics. Sadashiv [60] isolated a total of 878 strains of bacteria from bovine mastitis (Table 1), which showed multidrug resistance to the antibiotics such as amikacin, amoxyclav, ampicillin, methicillin, oxacillin, penicillin G, cefaclor, cefixime, cefpodoxime, ceftriaxone, ciprofloxacin, norfloxacin, ofloxacin, gentamicin, azithromycin, erythromycin, streptomycin, tetracycline, and chloramphenicol. Further, the strains were identified by 16s rDNA method. Due to wide sampling area, the identified strains were subjected to random amplified polymorphic DNA (RAPD) polymorphism analysis and found huge diversity among the strains, demonstrating the migration of the antibiotic resistance strains (Fig. 5). The study concluded that the examination of the antibiotic resistance profiles of the isolates must be done earlier to the use of antibiotics in both to choose appropriate antibiotic for treatment and prevention of the disease.

Table 1. Prevalence of different bacterial isolates from the bovine mastitis milk.

Sl. no	Bacteria	Isolates	Percentage (%)
1	Staphylococcus aureus	210	23.91
2	Coagulase-negative *Staphylococcus*	165	18.79
3	*Bacillus* spp.	221	25.17
5	Escherichia coli	97	11.04
4	*Pseudomonas* spp.	72	8.20
7	*Aerococcus* spp.	34	3.87
6	Cronobacter spp.	23	2.61
8	Others	56	6.37
Total		878	100

It is possible that mastitogenic bacteria can lose the sensitivity to antibiotics over the time or even acquire sometimes this feature [61]. Important reasons for the failure of treatment of mastitis are the indiscriminate use of antibiotics without in vitro sensitivity of causal organisms. It is necessary to monitor mastitis pathogens to assess any changes in their antibiotic resistance patterns. Careful use of antibiotics can avoid the increase and dissemination in antimicrobial resistance arising from the use of antimicrobial drugs in animals.

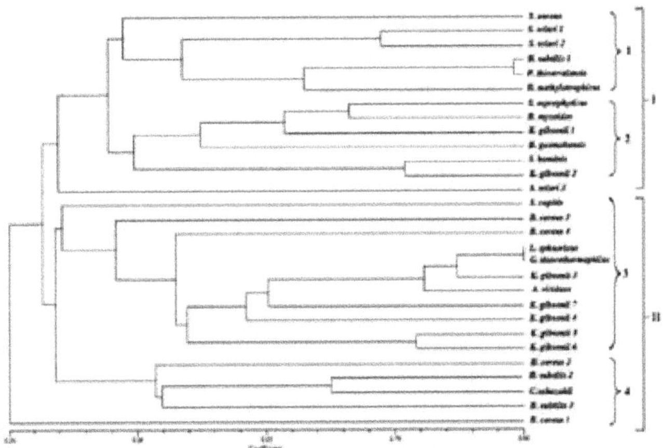

Figure 5. RAPD analysis of the bacterial strains. Dendrogram showing the relationship of the bacterial strains generated by neighbor-joining method using cluster analysis.

Jaccard's similarity (I and II clusters, 1, 2, 3, and 4 subclusters)

Some pharmaceutical companies have expanded their R&D efforts due to in response to the increased number of bacteria that have developed resistance to the one or more number of antibiotics. New drugs are being developed which interfere with the resistant cells and the method of new defence against certain antibiotics. For example, bacteria resistant to penicillin produce the enzyme penicillinase, which breaks up the penicillin before it can perform its work. Presently, a penicillinase inhibitor is available that is taken in tandem with the penicillin, therefore by preventing the penicillinase from destroying the penicillin and by allowing the antibiotic to perform its work [62].

The increased use, and sometimes misuse, of antibiotic drugs has resulted in bacterial resistance to a large and growing number of these drugs. However, much more research into newer and newer antibiotics continues, and measures can and must be taken to reverse the practices that usually promote the development of antibiotic resistance in bacteria.

4. CONCISE SUMMARY

Currently, the emergence of resistance in all microorganisms constitutes a significant global worry and challenge. Microorganisms can gain resistance through many mechanisms. The judicious and strategic application of insecticides, herbicides, and antibiotics will significantly aid in mitigating the emergence of resistance. Nonetheless, screening novel microorganisms and employing advanced recombinant technologies on these microbes will assist in diminishing tolerance to xenobiotic chemicals.

REFERENCES

1. Davies J, Davies D. Origins and evolution of antibiotic resistance. Microbiol. Mol. Biol. Rev., 2010: 417–433. doi:10.1128/MMBR.00016-10.
2. Naphade SR, Durve AA, Bhot M, Varghese J, Chandra N. Isolation, characterization and identification of pesticide tolerating bacteria from garden soil. Euro. J. Exp. Biol., 2012; 2(5): 1943–1951.
3. Green T, Toghill A, Lee R, Waechter F, Weber E, Peffer R, Noakes J. Toxicol Sci., 2005; 86: 36–47.
4. Gilden RC, Huffling K, Sattler B. Pesticides and health risks. J. Obstet. Gynecol. Neonatal Nurs., 2010; 39(1): 103–110.
5. Singh DK. Biodegradation and bioremediation of pesticide in soil: concept, method and recent developments. Indian J. Microbiol. 2008; 48: 35–40.
6. Velázquez-Fernández JB, Martínez-Rizo AB, Ramírez-Sandoval M, Domínguez-Ojeda D. Biodegradation and bioremediation of organic pesticides, pesticides—

recent trends in pesticide residue assay. Dr. R.P. Soundararajan (Ed.), InTech, (2012). ISBN: 978-953-51-0681-4, DOI: 10.5772/48631.

7. PBS. Pesticide Resistance. 2007. Retrieved September 15, 2007. Tomlin CDS. The Pesticide Manual. BCPC Publications.13th ed. 2003: 1344; ISBN 978-1-90139613-3.

8. Baron RL, Hayes WJ, Laws ER (Eds.). San Diego, Calif, Carbamate insecticides. In Handbook of Pesticide Toxicology. Vol 3. Academic Press, New York. 1991; 1125- 1190.

9. EPA. Drinking Water Regulations and Health Advisories, USEPA, 1996; 822-B-96-002, Washington, DC.

10. Andersen HR, Vinggaard AM, Rasmussen TH, Gjermandsen IM, Bonefeld-Jorgensen EC. Toxicol. Appl. Pharmacol., 2002; 179: 1–12.

11. Strathmann TJ, Stone AT. Reduction of the carbamate pesticides oxamyl and methomyl by dissolved FeII and CuI. Environ. Sci. Technol., 2001; 35(12): 2461–2469.

12. Manawadi S, Kaliwal BB. Methomyl induced alteration in mice hepatic-oxidative status. Int. J. Biotechnol. Appl., 2010 2(2): 11–19.

13. Manawadi SI, Kaliwal BB. Methomyl induced gonadal dysfunction, biochemical contents and enzyme activities in male albino mice. Int. J. Biotechnol. Appl., 2010; 2(2): 20–32

14. Chung MJ, Ka JO. Isolation and characterization of 2,4-dichlorophenoxyacetic acid- degrading bacteria from paddy soils. J. Microbiol., 1998; 36(4): 256–261.

15. Laemmli CM, Leveau JHJ, Zehnder AJB, van der meer JR. Characterization of a second tfd gene cluster for chlorophenol and chlorocatechol metabolism on plasmid pJP4 in Ralstonia eutropha JMP134 (pJP4). J. Bacteriol., 2000; 182(15); 4165–4172.

16. Sayler GS, Hooper SW, Layton AC, King JMH. Catabolic plasmids of environmental and ecological significance. Microb. Ecol., 1990; 19(1); 1–20.

17. Manisha DM, Shyamapada M, Nishith KP. Plasmid-mediated dimethoate degradation by Bacillus licheniformis isolated from a fresh water fish Labeo rohita. J. Biomed. Biotechnol., 2005; 3: 280–286.

18. Racke KD, Coats JR, Comparative degradation of organophosphorus insecticides in soil: specificity of enhanced microbial degradation. J Agric. Food Chem. 1988; 36: 193–199.

19. Altalhi AD, Plasmid-mediated detoxification of mycotoxin zearalenone in Pseudomonas Sp. ZEA-1. Am. J. Biotechnol. Biochem., 2007; 3: 150–158.

20. Kulkarni AG, Kaliwal BB. Bioremediation of methomyl by soil isolate—Pseudomonas aeruginosa. J. Bioremed. Biodeg. 2015; 6; 281. doi:10.4172/2155-6199.1000281

21. Don RH, Pemberton JM. Properties of six pesticide degradation plasmids isolated from Alcaligenes paradoxus and Alcaligenes eutrophus. J Bacteriol., 1981; 145: 681–686.

22. Matsumura F. Degradation of pesticides in the environment by microorganisms and sunlight. In Matsumura, F and Krishna Murti CR (eds), Biodegradation of pesticides. New York, Academic Press, 1988; 3: 67–87.

23. Bhagobaty RK, Malik A. Utilization of chlorpyrifos as a sole source of carbon by bacteria isolated from wastewater irrigated agricultural soils in an industrial area of Western Utter Pradesh. India. Res. J. Microbiol., 2008; 3: 293–307.

24. Kulkarni AG, Kaliwal BB. Proteomic Profiling of Escherichia coli in Response to Carbamate Pesticide—Methomyl. Insecticides—Basic and Other Applications. Intech Publication. ISBN 979-953-307, 2012; 200–216.

25. Pieper DH, Reineke W. Engineering bacteria for bioremediation. Curr. Opin. Biotechnol,, 2000; 11(3); 262–270.

26. Furukawa K. 'Super bugs' for Bioremediation. Trends Biotechnol., 2003; 21(5): 187– 190.

27. Nauen R, Hungenberg H, Tollo B, Tietjen K, Elbert A. Antifeedant effect, biological efficacy and high affinity binding of imidacloprid to acetylcholine receptors in Myzus Persicae and Myzusnicotianae. Pest Manag. Sci., 1998; 53: 133–140.

28. Krohn J, Hellpointner E. Environmental fate of imidacloprid. Pflanzenschutz-Nachr Bayer. 2002; 55: 3–26

29. Kulkarni AG, Kaliwal BB. Methomyl induced effects on free and immobilized Escherichia coli. Int. J. Biotechnol. Res. 2009; 2(2), 97–101.

30. Shetti AA, Kaliwal BB. Imidacloprid induced intoxication in soil isolate Brevundimonas sp. Mj 15. Int. J. Life Sci. Pharma Res., 2012; 2(3): 105–117.

31. Bahig AE, Aly EA, Khaleed AA, Amel KA. Isolation, characterization and application of bacterial population from agricultural soil at Sohag province, Egypt. Malays. J. Microbiol., 2008; 4(2), 42–50.

32. Top EM, Moenne-Loccoz Y, Pembroke T, Thomas CM. Phenotypic traits conferred by plasmids. In: The horizontal gene pool bacterial plasmids and gene spread (Thomas, C.M., Ed.). Harwood Academic Publishers. 2000; 249–285.

33. DebMandal M, Shyamapada M, Nishith K. Potential metabolites of dimethoate produced by bacterial degradation. World J. Microbiol. Biotechnol., 2008; 24; 69–72.

34. Kumar G, Singla R, Kumar R. Plasmid associated anthracene degradation by Pseudomonas sp. isolated from filling Station site. Appl. Environ. Microbiol., 2010; 8(4); 89.

35. Unakal CG, Kaliwal BB. Phenotypic characterization and risk factors of nosocomial Staphylococcus aureus from health care centers. Adv. Microbiol., 2012; 2: 122–128.

36. Hiramatsu K, Aritaka N, Hanaki H, Kawasaki S, Hosoda Y, Hori S, Fukuchiand Y, Kobayashi I. Dissemination in Japanese hospitals of strains of Staphylococcus aureus heterogeneously resistant to vancomycin. Lancet, 1997; 350: 1670–1673.

37. Smith TL, Pearson ML, Wilcox KR, Cruz C, Lancaster MV, Robinson-Dunin B, Ten over EC, Zervos Ml, Band ID, White E, Larvis WR. Emergence of vancomycin resistance in Staphylococcus aureus. N. Engl. J. Med., 1999; 340: 493–501.

38. Denis O, Nonhoff C, Byl B, Knoop C, Bobin-Dubreux S, Struelens MJ. Emergence of vancomycin-intermediate Staphylococcus aureus in a Belgian hospital: microbiological and clinical features. J. Antimicrob. Chemother., 2002; 50, 383–391.

39. Hare KT, Malay RS. Emergence of vancomycin resistant Staphylococcus aureus (VRSA) from a tertiary care hospital from northern part of India. BMC Infect. Dis., 2006; 6: 156.

40. Kim MN, Hwang SH, Pyo YJ, Munand HM, Pai CH. Clonal spread of Staphylococcus aureus heterogeneously resistant to vancomycin in a university hospital in Korea. J. Clin. Microbiol., 2002; 40; 1376–1380.

41. Noble WC, Virani Z, Cree RC. Co-transfer of vancomycin and other resistance genes from Enterococcus faecalis NCTC 12201 to Staphylococcus aureus. FEMS Microbiol. Lett., 1992; 72: 195–198.

42. Sieradzki K, Roberts RB, Haberand SW, Tomasz A. The development of vancomycin resistance in a patient with methicillin-resistant Staphylococcus aureus infection. N. Engl. J. Med., 1999; 340: 517–523.

43. Cui L, Ma X, Sato K, Okuma K, Tenover FC, Mamizuka EM, Gemmell CG, Kim MN, Ploy MC, Solh NE, Ferrazand V, Hiramatsu K. Cell wall thickening is a common feature of vancomycin resistance in Staphylococcus aureus. J Clin. Microbiol., January 2003; 41(1): 5–14.

44. Samad MA. Animal Husbandry and Veterinary Science, Volume II, LEP Pub No. 11, Bangladesh, 2008.

45. Mohanty NN, Das P, Pany SS, Sarangi LN, Ranabijuli S, Panda HK. Veterinary World, 2013; 6: 739–743.

46. Blowey R, Edmondson P. Mastitis Control in Dairy Herds: An Illustrated and Practical Guide. Ipswich, UK: Farming Press Books, 1995.

47. Chinchali JF. Studies on experimental bacterial mastitis in swiss albino mice [thesis]. Karnatak University, Dharwad, India, 2014

48. Shitandi A, Sternesjö A. Prevalence of multidrug resistant Staphylococcus aureus in milk from large and small-scale producers in Kenya. J. Dairy Sci., 2004; 87: 4145– 4149.

49. Rychlik I, Gregorova D, Hradecka H. Distribution and function of plasmids in Salmonella enterica. Vet. Microbiol., 2006; 112(1): 1–10.

50. Waage S, Bjorland J, Caugant DA. Spread of Staphylococcus aureus resistant to penicillin and tetracycline within and between dairy herds. Epidemiol. Infect., 2002; 129: 193–202.

51. Wang Y, Wu CM, Lu LM, Ren GWN, Cao XY, Shen JZ. Macrolide. Incosamide resistant phenotypes and genotypes of Staphylococcus aureus isolated from bovine clinical mastitis. Vet. Microbiol., 2008; 130:118–125

52. Malinowski E, Klossowska A, Kaczmarowski M, Lassa H, Kuzma K. Antimicrobial susceptibility of staphylococci isolated from affected with mastitis cows. Bull. Vet. Inst. Pulawy, 2002; 46: 289–294.

53. Holmes MA, Zadoks RN. Methicillin resistant S. aureus in human and bovine mastitis. J. Mammary Gland. Biol. Neoplasia, 2011; 16(4):373–382.

54. Kurjogi MM, Kaliwal BB. Prevalence and antimicrobial susceptibility of bacteria isolated from bovine mastitis. Adv. Appl. Sci. Res., 2011; 2(6): 229–235.

55. Sadashiv SO, Kaliwal BB, Kurjogi MM, Sanakal RD. Prevalence and antimicrobial susceptibility of coagulase-negative staphylococci isolated from bovine mastitis. Veterinary World, 2011; 4(4): 158–161.

56. Sadashiv SO, Kaliwal BB. Antibiotic resistance of Staphylococcus aureus and coagulase-negative staphylococci (CNS) isolated from bovine mastitis in the region of north Karnataka, India. World J. Pharm. Res., 2014; 3(01): 571–586.

57. Sadashiv SO, Kaliwal BB. Isolation, characterization and antibiotic resistance of Bacillus spps. from bovine mastitis in the region of north Karnataka India. Int. J. Curr. Microbiol. Appl. Sci., 2014; 3(4): 360–373.

58. Sadashiv SO, Kaliwal BB. Screening and antimicrobial resistance of Escherichia coli isolated from bovine mastitis in the region of north Karnataka, India. Indo

Am. J. Pharm. Res., 2015; 5(04): 1309–1316.

59. Sadashiv SO. Studies on isolation, molecular characterization of bacteria and their effects on milk in Bovine Mastitis [thesis]. Karnatak University, Dharwad. India; 2014

60. Malinowski E, Lassa H, Smulski S, Kłossowska A, and Michał Kaczmarowski, Antimicrobial susceptibility of bacteria isolated from cows with mastitis in 2006–2007. Bull. Vet. Inst. Pulawy., 2008; 52: 565–572

61. Norris S. Antibiotic Resistance. Library of Parliament, 2008.

16
Chapter

THE FUNCTION OF GLUTATHIONE TRANSFERASES IN THE EMERGENCE OF INSECTICIDE RESISTANCE

1. INTRODUCTION: AN OVERVIEW

This chapter will examine literature on glutathione S-transferases from a comprehensive perspective, focusing on their properties, functions, purifying procedures, and associated problems. The emphasis will be on the prevalence of GSTs in insects and the comprehension of their function in pesticide resistance. This review aims to examine the correlation between specific isoforms of GSTs and responses to commonly utilized agricultural pesticides.

2. GLUTATHIONE-DEPENDENT ENZYMES

Glutathione (GSH, γ-glutamylcysteinylglycine) is a low molecular weight sulfhydryl compound. It is a tripeptide with the sequence glutamic acid, cysteine, and glycine. GSH is a crystalline solid with a melting point of 192–195°C and molecular weight of 307.33. It dissolves readily in water. There are two peptide bonds, two carboxylic acid groups (pKCOOH = 3.53 and 2.12), one amino group (pKNH3+ = 8.66), and one thiol group (pHSH = 9.66). GSH is found intracellularly in all mammalian tissues and is the major nonprotein thiol compound present in the cell, with concentrations ranging from 0.1 to 10 mM. It is involved in a variety of metabolic processes, for instance, detoxication of xenobiotics, reduction of hydoperoxides, synthesis of leukotrienes and prostaglandins, maintenance of protein and membrane structures, and regulation of numerous enzyme activities. This functional diversity is due to the properties of the thiol group that participates in redox transitions, thiol exchange reactions, thioether formation, and radical scavenging.

A variety of different enzymes utilize glutathione in a variety of biotransformations [1]. Glutathione reductase (GR) catalyses the reduction of GSSG (oxidized glutathione) using NADPH as a reductant. GR is important in maintaining the high cellular reduction potential.

Selenium-dependent glutathione peroxidase (GPOX) is another type of GSH-requiring enzyme that catalyses the reduction of peroxides using GSH as the reducing agent. There are the glutathione S-transferases (GSTs) that are also GSH dependent enzymes with many catalytic activities including the conjugation of GSH to xenobiotics [2,3].

3. GLUTATHIONE S-TRANSFERASES (GSTS, E.C. 2.5. 1.18) WITH DIVERSE FUNCTIONS

GSTs are found in almost every species, including plants [4] microorganisms [5,6], and animals [7]. GSTs are divided into classes based on their amino acid sequence, immunological, kinetic, and structural properties. In mammals, at least nine classes of GSTs have been identified, namely, Alpha, Mu, Pi, Theta, Omega, Sigma, Zeta, Kappa, and a microsomal class. Human GSTs have been reviewed in reference [8, 9], and [10]. The majority of GSTs are found mainly in the cytosol. Each class consists of one or more protein isoforms. The classes are defined such that the amino acid identity between two isoforms of the same class is more than 50% but more than 30% if they are in different classes [11]. Human cytosolic GSTs are not only in cytoplasm but may also be localized in the mitochondria or the nucleus [12]. The microsomal family of membrane-bound GSTs is also reported and is different from cytosolic GSTs in molecular weight, subunit structure, and immunological reactivity [13-15]. The microsomal GSTs are trimeric, membrane-bound proteins. Mitochondrial GST 13-13 previously purified from rat liver [16] has been later characterized as GSTK1-1 of a Kappa class GST [17]. The Kappa class GSTs are located in mammalian mitochondria and peroxisomes [18,19] and are structurally distinct from the microsomal and cytosolic GSTs [20].

GSTs have a broad and overlapping specificity. Among the reactions catalyzed by GSTs are the substitution of halogens in halogenohydrocarbon, the addition to double bonds, the cleavage of epoxides, and the reduction of organic peroxides. 1-Chloro-2,4-dinitrobenzene (CDNB) is the most common substrate used to assay GSTs in the laboratory as most, but not all, GSTs show catalytic activity with it. Other substrates that have been commonly used to characterize the enzymes are 1,2-dichloro-4-nitrobenzene (DCNB), trans-4-phenyl-3-butene-2-one (PBO), ethacrynic acid (EA), 1,2-epoxy-3-nitrophenoxypropane (EPNP), p-nitrobenzyl chloride (NBC), and sulfobromophthalein (BSP) (Figure 1).

GSTs catalyze the nucleophilic attack by the thiol group of reduced glutathione (GSH) on a wide range of electrophilic substrates. GSTs play important roles in the development of resistance to a variety of exogenous xenobiotics, such as chemotherapeutic drugs [21], chemical carcinogens [22], herbicides [4], and insecticides [7].

3.1 Conjugation of exogenous toxins (Biotransformation)

GSTs play important roles in the protection of macromolecules from attack by reactive electrophiles. The enzymes generally exist in dimeric forms with a subunit molecular weight of approximately 26 kDa. GSTs occur both as homo- and heterodimers. The cytosolic isoenzymes have two active sites per dimer that behave independently of one another [23]. Each active site consists of at least two ligand-binding regions, namely, the GSH binding site (hydrophilic G-site), which is specific for GSH, and the electrophile-binding site (hydrophobic H-site), which is less specific and thus enables GSTs to react with a wide variety of xenobiotics [24]. A review in reference [25] listed xenobiotics that could be conjugated by GSTs. These include halogenonitrobenzenes, organophosphorus compounds, steroids, αβ-unsaturated carbonyl compounds, aryl halides epoxides, quinones, isothiocyanates, and arylnitro compounds.

The conjugations catalyzed by the GSTs occur between the nucleophilic GSH and the compounds possessing a sufficiently electrophilic center [25]. The GSTs function by decreasing the pKa of GSH, thereby allowing its deprotonation and the formation of a more reactive thiolate anion. In most GST classes (Pi, Mu, Alpha, and Sigma), a tyrosine residue in the N-terminal region interacts with GSH to stabilize the thiolate anion. In Theta and Omega classes, this role is carried out by serine and cysteine residues, respectively [9].

This GSH conjugation has been shown to occur in mammals, birds, reptiles, amphibians, fish, insects, and other invertebrates [26], and it is the first step of mercapturic acid formation that is one of the metabolic pathways for detoxication of xenobiotics in vivo. The glutathione conjugates, which are water soluble and generally nontoxic, may be converted to the corresponding cysteine conjugate following sequential removal of glutamate and glycine. The cysteine conjugate is either N-acetylated to be excreted as a mercapturic acid or cleaved to a mercaptan by β -lyase. The thiol can be then further metabolized, for example, to be excreted as a glucuronide [26, 27].

Prostaglandin D-synthase, the enzyme involved in production of the D and J series of prostanoids, was characterized as belonging to the Sigma class of GSTs [28]. GSTs also participate in the isomerization of biologically active molecules. A prostaglandin-HE-isomerase of Ascaridia galli [29] and prostaglandin-HD-isomerase of rat spleen [30] were identified as Sigma class GSTs. GSTs can also catalyze cis-trans isomerizations, for example, the isomerization of maleylacetoacetic acid to fumarylacetoacetic acid. The maleylacetoacetate isomerase activity of hGSTZ1-1 (human GST Zeta 1-1) has been investigated [31] using a spectrophotometric assay with (±)-2-bromo-3-(4-nitrophenyl)propanoic acid (BNPP) as substrate. Some GSTs possess keto-steroid isomerase activity and catalyze the conversion of Δ5-3-androstene-3,17- dione to Δ4-3-androstene-3,17-dione [21,32].

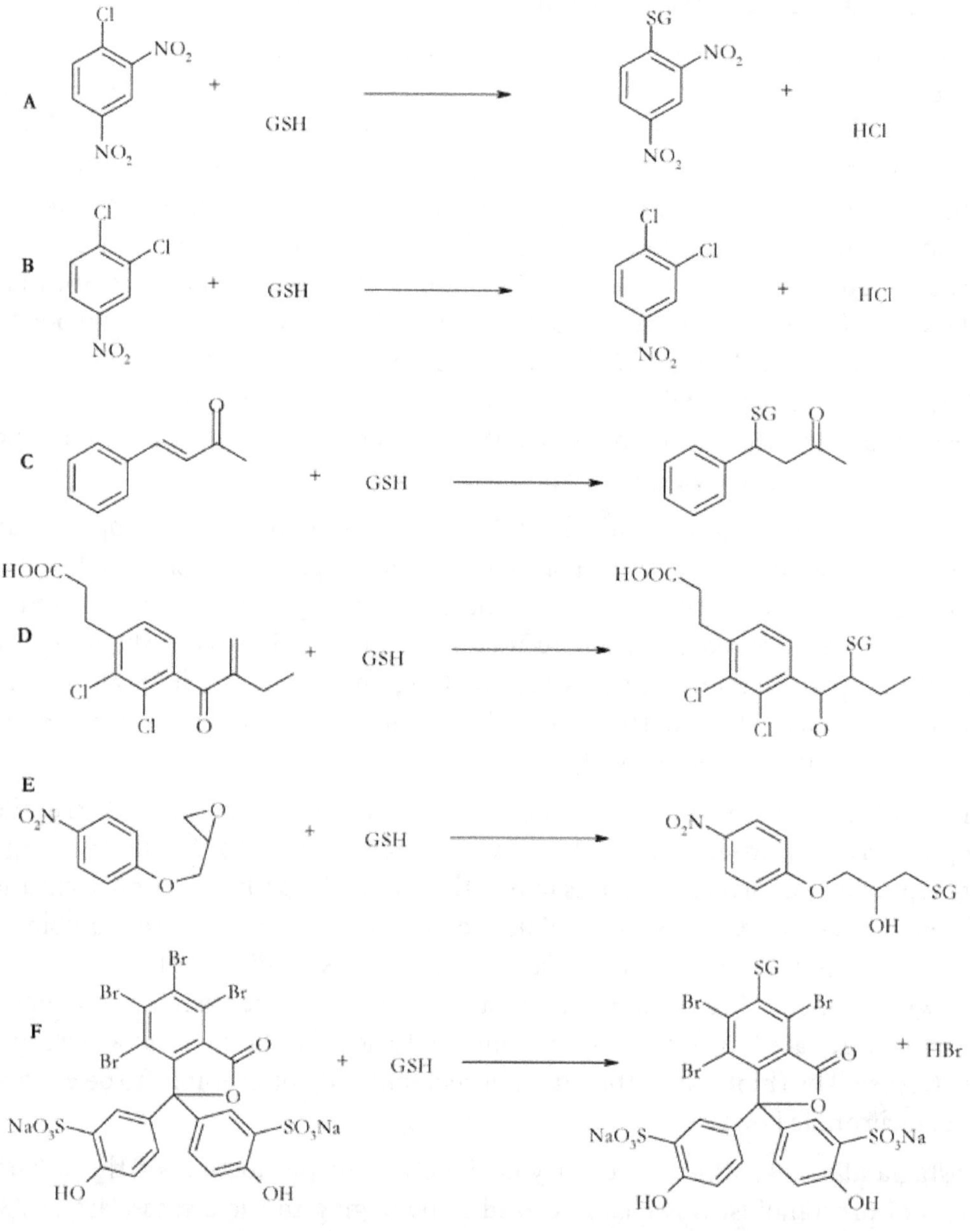

Figure 1. Enzymatic conjugation of common substrates by GST in the presence of GSH. (A) CDNB, (B) DCNB, (C) PBO, (D) EA, (E)EPNP, and (F) BSP.

Human GST A3-3 was shown to efficiently catalyze the double-bond isomerization of Δ5- androstene-3,17-dione and Δ5-pregnene-3,20-dione [12]. Human GSTs were reported to act as retinoic isomerases that catalyze the steric conversion of 13-cis-retinoic acid (13-cRA) to all- trans-retinoic acid (t-RA) [33].

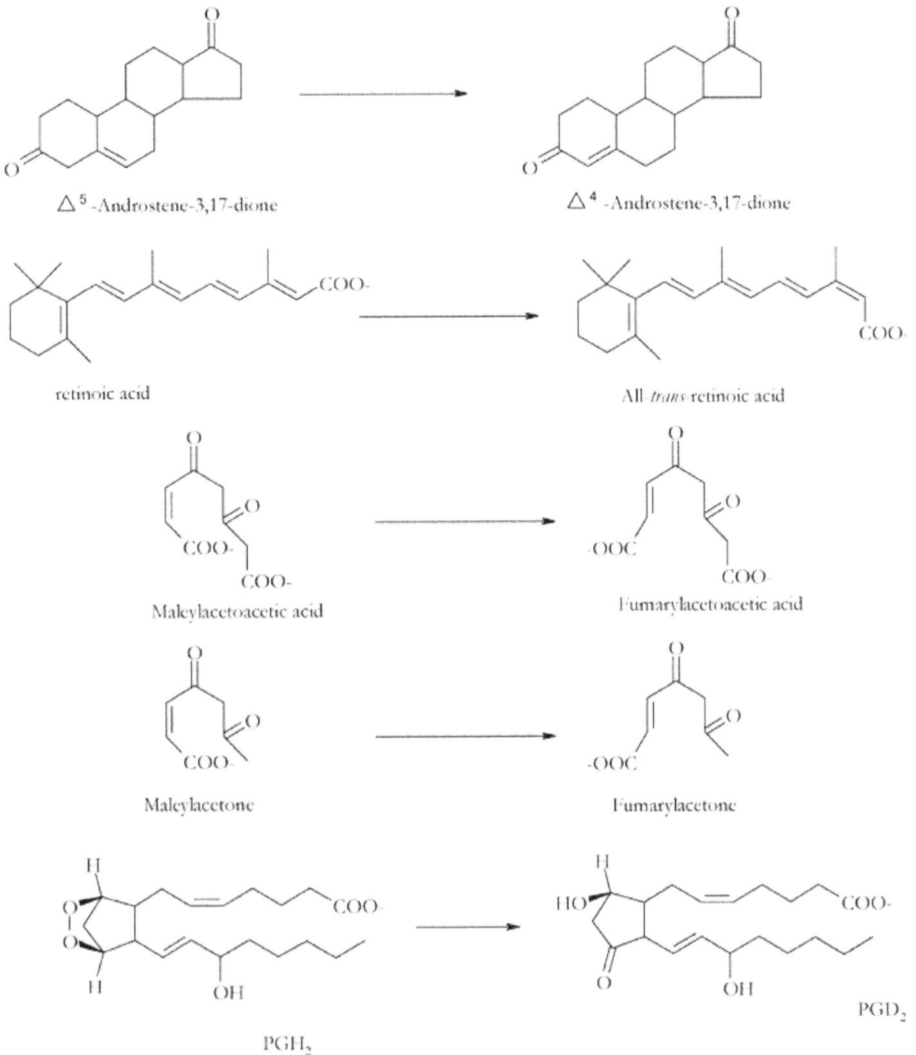

Figure 2. Some examples of GSTs with isomerase and synthase functions. (adapted from [9])

3.2 Participation in countering oxidative stress

Reactive oxygen species (ROS) such as superoxide anion, hydrogen peroxide, and the hydroxyl radical are constantly produced during normal aerobic metabolism. The free radicals may attack the polyunsaturated fatty acyl moieties and lead to the peroxidation of lipid bio- membranes. The cleavage of polyunsaturated fatty acids is known to be associated with formation of organic hydroperoxides and reactive aldehydes. These include alkanals, alkenals, and malondialdehyde. These may interfere with several biological processes such as DNA and protein synthesis by inhibition of specific enzyme reactions [34,35].

Peroxidized lipids, produced during oxidative stress, are substrates for GSTs. 4-Hydroxyalkenals are derived from membrane-bound phospholipid hydroperoxides. They were shown to be efficiently conjugated to GSH by GST A4-4 [36,37]. A membrane-bound mouse GSTA4-4 [38]and a rat GSTA8-8 [39,40] were also shown efficiently conjugating 4-hydroxyalkenals. It was reported that transfected mGSTA4 protects HL 60 [41] and K562 human erythroleukemia cells [42] against 4-hydroxynonenal-induced apoptosis.

Free radicals can also cause DNA peroxidation. The toxicity of thymine propenal, which is generated by oxidative damage to DNA, was shown to be substantially reduced when HeLa cells received GSTP1-1 and GSH [43]. The rat GST6-6 had previously shown active toward thymine hydroperoxide [44].

Within the cells, peroxides occur either as hydrogen peroxide (H2O2) or organic hydroperoxides, such as fatty acid and phospholipid hydroperoxides. GSTs protect tissues from endogenous organic hydroperoxides produced during oxidative stress [35,45]. Some GSTs have shown selenium-independent H2O2 and organic hydroperoxidase activity, which are involved in free radical reactions during oxidative stress [46]. A microsomal GST A1-1 of sheep liver exhibited peroxidase activity toward fatty acid hydroperoxides such as linoleic and arachidonic acid hydroperoxides [47]. Human GSTs, such as hGSTA1-1 and hGSTA2-2 [48,49], also exhibited glutathione peroxidase activity toward phospholipid hydroperoxide [50]. Other workers observed elevated GSTs in Nilaparvata lugens when treated with pyrethroids, which induces oxidative stress and lipid peroxidation in insects [51].

3.3 Involve in cells regulatory functions

Recent studies of GSTs have demonstrated that a Pi class GST is involved in regulation of c- Jun N-terminal kinase (JNK) signaling in mammals. GSTP interacts with c-Jun N-terminal kinase 1 (JNK1) suppressing the basal kinase activity [52,53]. A model of GST inhibition of JNK signaling was proposed [52]. Under a nonstressed condition, GSTp can exist as free dimeric enzyme or complexed with Jun–JNK thus inhibiting JNK. Upon stress, GSTP forms larger aggregates, which are unable to associate with the Jun–JNK complex, thus enabling the JNK phosphorylation of c-Jun. Phosphorylated Jun can act as a stable and active transcription factor. The accumulation of ROS in response to oxidative stress results in the activation of multiple stress kinase cascades and an elevated level of GSTp expression [54].

Apoptosis signal-regulating kinase 1 (ASK 1) can activate the JNK and the p38 signaling pathways. It plays important role in stress-induced apoptosis. Mouse GSTM1-1 was shown to physically interact with ASK1 and repress ASK1-mediated signaling [55,56].

It has also been reported that human GST class Omega, GSTO1-1, modulates calcium channel (ryanodine receptors, RyRs) protecting mammalian cells from apoptosis induced by calcium (Ca2+) mobilization [57]. It was suggested that RyRs has two binding sites for GSTO1-1. The mammalian protein Bax (21 kDa) is an inducer of apoptosis. A study [58] has reported a plant GST (Theta class) as one of the Bax-inhibiting plant proteins, which prevent apoptosis in plants. GSTP1-1 was proposed [59] interacting with physiological nitric oxide (NO) carriers such as S-nitrosoglutathione (GSNO) and dinitrosyl-diglutathionyl iron complex (DNDGIC). In the absence of GSH, GSNO interacts with and modifies Cys47 and Cys101 residues of GSTP1-1 by an S-nitrosylation reaction. Thus, in the cellular depletion of GSH, GSTP1-1 acts directly as an NO carrier without losing its detoxication activity. The expression of 'tissue' transglutaminase (tTG) is induced in cells programmed to die such as in cells undergoing apoptosis. It was reported that [60] the overexpression of tTG in human neuroblastoma cells increases apoptosis. The study showed that tTG interacts with β-tubulin, histone H2B, and GSTP1-1 to form a protein complex. It was proposed that the interaction with tTG resulted in oligomerization of GSTP1-1. The formation of multimers of GSTP1-1 leads to inactivation of the enzyme toward ROS.

4. GSTS IN INSECTS

In insects, GSTs are classified into two groups, class I and class II GSTs [61]. According to a recent proposed classification [62], an insect-specific Delta class GST is classified as a class I GST. This includes those from Drosophila melanogaster, DmGSTD1 to DmGST10 [62,63]; Musca domestica, MdGSTD1 to MdGSTD5 [49,61] and Anopheles gambie, AgGSTD1 to AgGSTD6 [64,65]. Class II consists primarily of Sigma class GSTs as identified in D. melanogaster, DmGSTS1 [66] and A. gambie, AgGSTS1 [67]. A third proposed class of insect's GSTs (class III) [68] that comprises GSTs classified as the Epsilon class in Drosophila and the AgGST3-1 and AgGST3-2 of Anopheles gambiae.

Studies on insect GSTs were reviewed in detail [7,23,69]. GSTs have been detected in Lepidoptera, Diptera, Coleoptera, Dictyoptera, and Hymenoptera [23]. At present, only few insect GST structures are known. These include that from the Australian sheep blowfly, L. cuprina (Theta classs) [66], mosquito, Anopheles dirus (isoenzymes 1–3 and 1–4 by[70,71] and fruit fly, D. melanogaster (GSTS1) [72,73].

4.1 Characterization of GSTs and its challenges

A problem faced during the extraction of insect GSTs is the possible presence of endogenous inhibitors [74]. Quinones and catecholamines released during homogenization can inhibit the GSTs' activity [75]. The last-named authors suggested the inclusion of GSH in the homogenizing buffer to protect the GSTs from inhibition. Polyphenol pigments have also been shown to inactivate the GSTs. The inclusion of 5–10 mM cysteine in the homogenate prevents the formation of polyphenol. The endogenous inhibitors can also be removed by initial loading of the crude homogenate through an ion exchange or gel filtration column [7]. There are many different strategies employed to purify the GSTs from insects. One of those is by using affinity chromatography with several different affinity matrices. A ligand, sulfobromophthalein- glutathione conjugate (BSP), has been immobilized to an agarose matrix by using either cyanogen bromide [76] or epichlorohydrin activation [77]. The enzymes bind selectively to the resulting matrices when a crude homogenate is applied directly to the columns. The bound enzymes can be eluted by using 0–5 mM of BSP [77] or 1–5 mM GSH solution [76]. The matrix has been used to purify a GST to homogeneity from Galleria mellonella [76,78], Costelytra zealandica [79], Musca domestica [77,80,81], Drosophila melanogaster [82] and Wiseana cervinata [83]. Another form of affinity matrix that has been used widely is the immobilized GSH-agarose matrix [84]. The bound enzymes are best eluted from this support with GSH solution. This technique has been used to purify GSTs from a number of insects. These include Drosophila [85]. In another instance, a study [86] had used another form of affinity column, namely immobilized S-hexylglutathione [87] to purify GSTs from Drosophila.

Investigation of multiple forms of GSTs with different isoelectric points could be performed by using isoelectrofocusing [78,88,89] or chromatofocusing [90]. Purification by affinity chromatography followed by isoelectrofocusing revealed the existence of multiple forms of GSTs [91], in house fly strains Rutgers, Cornell R, and Hirokawa. The presence of multiple isoenzymes of GSTs have also been reported in other species, such as Aedes aegypti [92], G. mellonella [78], Plutella xylostella [92], C. zealandica [79] and Tenebrio molitor [94].

4.2 GSTs and insecticide resistance

The majority of studies on insect GSTs have focused on their role in conferring insecticide resistance. A review [95] has indicated the importance of genetic and biochemical mechanisms in Drosophila in countering toxins and thus developing resistance. There are two types of biochemical mechanism of insecticide resistance outlined [96,97].The first is by selection of an altered target molecule so that no interaction between the molecules and the toxins occurs, resulting in target site resistance. The target sites of insecticides in insects include the GABA receptor, sodium channel, acetylcholine esterase, acetylcholine receptor, and juvenile hormone receptor. Second, through the process of metabolism, these toxins may be converted to less toxic and more easily excretable derivatives before reaching the target sites. This is a metabolic resistance. Several enzymes play important roles in these types of resistance, such as hydrolases, mixed function oxidases, and glutathione S-transferases and a variety of other conjugating (Phase II) enzymes.

High levels of GSTs have consistently been observed in resistant insect strains and play a major role in insecticide resistance [77,98,99]. Increased activity of GSTs in housefly was found to be associated with resistance to azinphosmethyl [100], parathion [98], phyrethroids [101], tetrachlorovinphos [102] and malathion [103]. D. melanogaster develops resistance toward several insecticides. These include malathion [104,105] and 1,1,1-trichloro-2,2-bis-(p-chlorophenyl)ethane (DDT) dehydrochlorinase [106]. Resistance to diazinon, chlorpyrifos, prope-tamphos, and dichlofenthion and their correlation to increased GST activity has also been reported in larval L. cuprina [107].

GSTs have been shown to play an important role in insecticide resistance. The catalysis of conjugation of insecticides, such as organophosphorus compounds, chlorinated hydrocarbons, and carbamate insecticides is shown in Figure 3. It was classified three types of reactions catalyzed by GSTs in metabolism of organophosphorus insecticides [108]. The detoxification of organophosphates (OP) occurs by the conjugation of GSH to OP via an O-dealkylation or O-dearylation conjugation, which later forms O-alkyl, O-aryl, and phosphonate conjugates which are all less toxic derivatives.

For the organochlorine insecticides the process involves dehydrochlorination and the GSH conjugation to the parent molecules [97]. Phyrethroids, however, trigger oxidative damage in cell. Therefore, GSTs role has been detoxifying the lipid peroxidation products resulted by the insecticide [51]. It is well known that some classes of GSTs have shown peroxidases activities. For example, a Delta class GST of N. lugens [51] and Epsilon class GST of A. aegypti mosquitoes [109,110] and Sigma class GST of D. melanogaster [111,112]. There was however report to suggest that protection against phyrethroids can be achieved through a passive sequestration process in which GSTs are capable to bind to various phyrethroids [113,114].

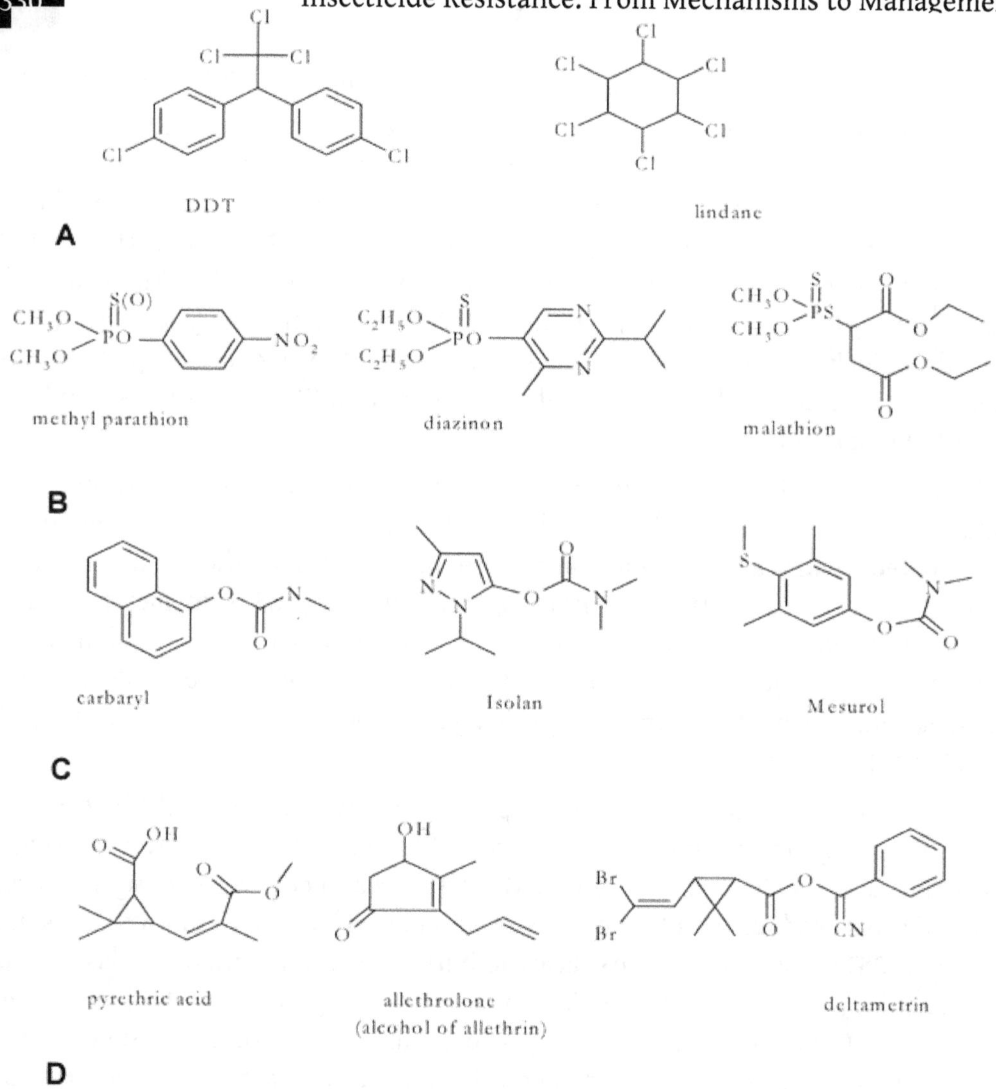

Figure 3. Structures of some (A) halogenated hydrocarbons, (B) organophosphorus, (C) carbamates, and (D) pyrethroids.

5. CONCISE SUMMARY

GSTs are enzymes with multifunctional roles. Research on insects has consistently highlighted the importance of GSTs in providing resistance to pesticides. Multiple categories of GSTs have been demonstrated to counteract pesticides by direct GSH conjugation and their capacity to react with lipid peroxidation products. This is attributable to the capacity of certain pesticides to induce oxidative damage. The direct isolation of the relevant GST from insects has posed a problem due to the restricted capacity of existing affinity matrices to capture all kinds of GSTs. The characterisation of recombinant GSTs may have facilitated a deeper comprehension of their mode of action and subsequent control in response to pesticide exposure. The accessibility of fully sequenced genomes in model insects such as D. melanogaster [115] and, more recently, M. domestica [116] may facilitate further research on glutathione transferase-related pesticide resistance.

REFERENCES

1. Fukami J-I. Metabolism of several insecticides by glutathione S-transferase. Pharmacol. Ther. 1979;10: 473–514.
2. Mannervik B. Glutathione and evolution of enzymes for detoxication of products of oxygen metabolism. Chemica Scripta. 1986;26: 281–284
3. van Bladere PJ. Glutathione conjugation as a bioactivation reaction. Chem. Biol. Interact. 2000;129: 61–76
4. Edwards R, Dixon P, Walbot V. Plant glutathione S-transferases: Enzymes with multiple functions in sickness and health. Trends Plant Sci. 2000;5: 193–198
5. Sheehan D, Casey JP. Microbial glutathione S-transferases. Comp. Biochem. Physiol. 1993;104: 1–6
6. Vuilleumier S. Bacterial glutathione S-transferases: What are they good for? J. Bacteriol. 1997;179: 1431–1441
7. Clark AG. The comparative enzymology of the glutathione S-transferases from non- vertebrate organisms. Comp. Biochem. Physiol. 1989;92: 419–446
8. Ketterer B. A bird's view of the glutathione transferase field. Chem. Biol. Interact. 2001;138: 27–42
9. Sheeha, D, Meade G, Foley VM, Dowd CA. Structure, function and evolution of glutathione transferases: implications for classification of non-mammalian members of an ancient enzyme superfamily. Biochem J. 2001;360: 1–16
10. Hayes JD, Flanagan JU, Jowsey IR. Glutathione transferases. Annu. Rev. Pharmacol.Toxicol. 2005;45: 51–88
11. Mannervik B, Awasthi YC, Board PG, Hayes JD, Di Ilio C, Ketterer B, Litowsky I, Morgenstern R, Muramatsu M, Pearson WR., Picket CB, Sat K, Widersten M, Wolf C.R. Nomenclature for human glutathione transferases. Biochem. J. 1992;282: 305–306
12. Johansson AS, and Mannervik, B. Human glutathione transferase A3-3, a highly efficient catalyst of double-bond isomerisation in the biosynthetic pathway of steroid hormones. J. Biol. Chem. 2001;276: 33061–33065
13. Morgenstern R, Meijer J, DePierre JW, Ernster, L. Characterization of rat-liver microsomal glutathione S-transferase activity. Eur. J. Biochem. 1980;104: 167–174

14. Morgenstern R, Guthenberg C, Depierre JW. Microsomal glutathione S-transferase. Purification, initial characterisation and demonstration that it is not identical to cytosolic glutathione S-transferase A, B, C. Eur. J. Biochem. 1982;128: 243–248

15. Morgenstern R, DePierre JW. Microsomal glutathione transferase : Purification in unactivated form and further characterization of the activation process, substrate specificity and amino acid composition. Eur. J. Biochem. 1983;134: 5911–597

16. Harri, JM, Meye DJ, Coles B, Ketterer, B. A novel glutathione transferase (13-13) isolated from the matrix of rat liver mitochondria having structural similarity to class theta enzymes. Biochem. J. 1991;278: 137–141

17. Pemble SE, Wardle AT, Taylor JB. Glutathione S-transferase class Kappa : characterisation by the cloning of rat mitochondrial GST and identification of a human homologue. Biochem J. 1996;319: 749–754

18. Morel F, Rauch C, Petit E, Piton, Theret N, Coles B, Guillouzo A. Gene and protein characteization of the human glutathione S-transferase Kappa and evidence for peroxisomal localization. J. Biol. Chem. 2004;279: 16246–16253.

19. Landler JE, Parsons JF, Rife C, Gillilard GL, Armstrong RN. Parallel evolutionary pathways for glutathione transferases: structure and mechanisms of the mitochondrial class Kappa rGSTK1-1. Biochemistry. 2004;43: 352–361

20. Robinson A, Huttley GA, Booth HS, Board, PG. Modelling and bioinformatics studies of the human Kappa-class glutathione transferase predict a novel third gluta thione transferase family with similarity to prokaryotic 2-hydroxychromene-2- carboxylate isomerase. Biochem J. 2004;379: 541–552.

21. Haye, JD, Pulford DJ. The glutathione S-transferase supergene family: Regulation of GST and the contribution of the isoenzymes to cancer chemoprotection and drug resistance. Crit. Rev. Biochem. Mol. Biol. 1995;30: 445–600

22. Coles B, Ketterer B. The role of glutathione and glutathione S-transferases in chemical carcinogenesis. Biochem. Mol. Biol. 1990;25: 47–70

23. Danielson UH, Mannervik B. Kinetic independence of the subunits of cytosolic glutathione S-transferase from the rat. Biochem. J. 1985;231: 263–267

24. Dir H, Reinemer P, Huber R. X-ray crystal structure of cytosolic glutathione S-transferases: implication for protein architecture, substrate recognition and catalytic function. Eur. J. Biochem. 1994;220: 645–661

25. Chasseaud LF. The role of glutathione and glutathione S-transferases in the metabolism of chemical carcinogens and other electrophilic agents. Adv. Cancer Res. 1979;29: 175–274

26. Boyland E, Chasseaud LF. The role of glutathione and glutathione S-transferases in mercapturic acid biosynthesis. Adv. Enzymol. 1969;32: 173-219

27. Picket CB, Lu AYH, Glutathione S-transferases: gene structure, regulation and biological function. Annu. Rev. Biochem. 1989;58: 743–764

28. Kanaoka Y, Fujimori K, Kikuno R, Sakaguchi, Y, Urade, Y, and Hyaishi, O. Structure and chromosomal localization of human and mouse genes for hematopoeitic prostaglandin D synthase. Eur. J. Biochem. 2000.;267: 3315–3322

29. Meyer DJ, Muimo R, Thomas M, Coates D. and Isaac, R.E. Purification and characterization of prostaglandin-H E-isomerase, a sigma-class glutathione S-transferase, from Ascaridia galli. Biochem. J. 1996;313: 223–227

30. Meyer DJ, Thomas M. Characterization of rat spleen prostaglandin H D-isomerase as a sigma-class GSH transferse. Biochem. J. 1995;311: 739–742

31. Board PG, Taylor MC, Coggan M, Parke MW, Lantum HB, Anders MW. Clarification of the role of key active site residues of glutathione transferase zeta/maleylacetoacetate isomerase by a new spectrometric technique. Biochem. J. 2003;374: 731–737

32. Jakoby WB, Habig WH, 1980. Glutathione S-transferases. In Enzymatic Basis of Detoxificaton Vol II. Academic Press Inc.

33. Chen H, Juchau MR. Recombinant human glutathione S-transferases catalyse enzymic isomerization of 13-cis-retinoic acid to all-trans-retinoic acid in vitro. Biochem. J. 1998;336: 223–226

34. Esterbauer H, Cheeseman KH, Dianzani MU, Poli G, and Slater TF. Separation and characterisation of the aldehydic products of lipid peroxidation stimulated by ADP- $Fe2+$ in rat liver microsomes. Biochem. J. 1982;208: 129–140

35. Ketterer B, Meyer DJ, Taylor JB, Pemble S, Coles B, Fraser G. GSTs and protection against oxidative stress. In : Proceedings of 3rd International GST conference, Scotland : Taylor and Francis, London; 1989. p. 97–109.

36. Cheng JZ, Yang Y, Singh SP, Singhal SS, Awasthi S, Pan SS, Singh SV, Awasthi YC. Two distinct 4-hydroxynonenal metabolising glutathione S-transferase isozymes are differentially expressed in human tissues. Biochem. Biophys. Res. Commun. 2001;282: 1268–1274

37. Hubatsch I, Ridderström M, Mannervik B. Human glutathione transferase A4-4: an alpha class enzyme with high catalytic efficiency in the conjugation of 4-hydroxynonenal and other genotoxic products of lipid peroxidation. Biochem. J. 1998;330: 175– 179

38. Singh SP, Janecki AJ, Srivastava SK, Awasthi S, Awasthi YC, Xia SJ, Zimniak P. Membrane association of glutathione S-transferase mGSTA4-4, and enzyme

that metabolizes lipid peroxidation products. J. Biol. Chem. 2002;277: 4232–4239

39. Jensson H, Guthenber, C, Ålin P, Mannervik B. Rat glutathione transferase 8-8, an enzyme efficiently detoxifying 4-hydroxyalk-2-enals. FEBS Lett. 1986;203: 207–209

40. Stenberg G, Ridderström M, Engström Å, Pemble SE, Mannervik B. Cloning and heterologous expression of cDNA encoding class alpha rat glutathione transferase 8-8, an enzyme with high catalytic activity towards genotoxic α,β-unsaturated carbonyl compounds. Biochem. J. 1992;284: 313–319

41. Cheng JZ, Singhal SS, Sharma A, Saini M, Yang Y, Awasthi S, Zimniak P, Awasthi YC. Transfection of mGSTA4 in HL-60 cells protects against 4-hydroxynonenal-induced apoptosis by inhibiting JNK-mediated signalling. Arch. Biochem. Biophy. 2001;392:197–207

42. Cheng JZ, Singhal SS, Saini M, Singhal J, Piper JT, van Kuijk FJ, , Zimniak GM, Awasthi, YC, Awasthi S. Effects of mGST A4 transfection on 4-hydroxynonenal-mediated apoptosis and differentiation of K562 human erythroleukimia cells. Arch. Biochem. Biophy. 1999;372: 29–36

43. Berhane K, Widersten M, Engstrom A, Kozarich JW, Mannervik B. Detoxication of base propenals and other αβ-unsaturated aldehyde products of radical reaction and lipid peroxidation by human GSTs. Proc. Natl. Acad. Sci. U. S. A. 1994;91: 1480–1484

44. Tan KH, Meyer DJ, Ketterer B. Thymine hydroperoxide, a substrate for rat Se-dependent glutathione peroxidase and glutathione transferase isoenzymes. FEBS Lett. 1986;207: 231–233

45. Storey KB.. Oxidative Stress : Animal adaptations in nature. Braz. J. Med. Biol. Res. 1996;29: 1715-1733

46. Wilce MCJ, Parker MW. Structure and function of glutathione S-transferses. Biochem. Biophys. Acta. 1994;205:1–18

47. Prabhu KS, Reddy PV, Gumpricht E, Hildenbrandt GR, Scholz RW, Sordillo LM, Reddy CC. Microsomal glutathione S-transferase A1-1 with glutathione peroxidase activity from sheep liver: molecular cloning, expression and characterisation. Biochem. J. 2001;360: 345–354

48. Yang Y, Cheng JZ, Singhal SS, Sain, M, Pandya U, Awasthi S, Awasthi YC. Role of glutathione S-transferases in protection against lipid peroxidation. J. Biol. Chem. 2001;276: 19220–19230

49. Zhou ZH, Syvanen M. A complex glutathione transferase gene family in the housefly Musca domestica. Mol. Gen. Genet. 1997;256: 187–194

50. Hurst R, Bao Y, Jemth P, Mannervik B, Williamson G. Phospholipid hydroperoxide glutathione peroxidase activity of human glutathione transferase. Biochem. J. 1998;332: 97–100

51. Vontas JG, Small GJ, Hemingway J. Glutathione S-transferases as antioxidant defence agents confer pyrethroid resistance in Nilaparvata lugens. Biocem. J. 2001;357: 65–72

52. Adler V, Yin Z, Fuchs SY, Benezra M, Rosario L, Tew KD, Pincus MR, Sardana M, Henderson CJ, Wolf CR, Davis RJ, Ronai Z. Regulation of JNK signalling by GSTp. EMBO J. 1999;18: 1321–1334

53. Wang T, Arifoglu P, Ronai Z, Tew KD. Glutathione S-transferase P1-1 (GSTP1-1) inhibits c-Jun N-terminal Kinase (JNK1) signaling through interaction with C-terminus. J. Biol. Chem. 2001;276: 20999–21003

54. Ainbinder E, Bergelson S, Pinkus R, Daniel V. Regulatory mechanisms involved in activator-protein-1 (AP-1)-mediated activation of glutathione S-transferase gene expression by chemical agents. Eur. J. Biochem. 1997;243: 49–57

55. Cho SG, Lee YH, Park HS, Ryoo K, Kang KW, Park J, Eom SJ, Kim MJ, Chang TS, Choi, SY, Shim J, Kim Y, Dong MS, Lee MJ, Kim SG, Ichijo H, Choi EJ. Glutathione S- transferase Mu modulates the stress-activated signals by suppressing apoptosis signal-regulating kinase 1. J. Biol. Chem. 2001;276: 12749–12755

56. Gilot D, Loyer P, Corlu A, Glaise D, Lagadic-Gossmann D, Atfi A, Morel F, Ichijo H, Guguen-Guillouzo C. Liver protection from apoptosis requires both blockage of initiator caspase activities and inhibition of ASK1/JNK pathway via glutathione S-transferase regulation. J. Biol. Chem. 2002;277: 49220–49229

57. Dulhunty A, Gage P, Curtis S, Chelvanayagam G, Board P. The glutathione transferase structural family includes a nuclear chloride channel and a rynodine receptor calcium release channel modulator. J. Biol. Chem. 2001;276: 3319–3323

58. Kampranis SC, Damianova R, Atallah M, Toby G, Kondi G, Tsichlis PN, Makris AM. A novel plant glutathione S-transferase/peroxidase suppresses Bax lethality in yeast. J. Biol. Chem. 2000;275: 29207–29297

59. Lo Bello M, Nuccetelli M, Caccuri AM, Stella L, Parker MW, Rossjohn J, McKinstry WJ, Mozzi AF, Federici G, Polizio F, Pederson JZ, Ricci G. Human glutathione transferase P1-1 and nitric oxide carriers. J. Biol. Chem. 2001;276: 42138–42145

60. Piredda L, Farrace MG, Belloo ML, Malorni W, Melino G, Petruzzeli R, Piacetini M. Identification of 'tissue' transglutaminase binding proteins in neural cells commited to apoptosis. FASEB J. 1999;13: 335–364

61. Fournier D, Bride JM, Poirie M, Berge JB, Flapp Jr. FW. Insect glutathione S-transferases: biochemical characteristics of the major forms from houseflies susceptible and resistant to insecticides. J. Biol. Chem. 1991;267: 1840–1845

62. Chelvanayagam G, Parker MW, Board PG. Fly fishing for GSTs: a unified nomenclature for mammalian and insect glutathione transferases. Chem. Biol. Interact. 2001;133: 256–260

63. Toung YPS, Hsieh TS, Tu CPD. The glutathione S-transferase D genes: a divergently organised, intronless gene family in Drosophila melanogaster. J. Biol. Chem. 1993;268: 9737–9746

64. Ranson HR, Cornel AJ, Fournier D, Vaughan A, Collins FH, Hemingway J. Cloning and localization of a glutathione S-transferase class I gene Anopheles gambiae. J. Biol. Chem. 1997;272: 5464–5468

65. Ranson H, Collins J, Hemingway J. The role of alternative mRNA splicing in generating heterogeneity within the Anapholes gambiae class I glutathione S-transferase family. Proc. Natl. Acad. Sci. U. S. A. 1998;95: 14284–14289

66. Beall C, Fryberg C, Song S, Fryberg E. Isolation of Drosophila gene encoding glutathione S-transferase. Biochem. Genet. 1992;30: 515–527

67. Reiss RA, James AA. A glutathione S-transferase gene of the vector mosquito, Anopheles gambiae. Insect Mol. Biol. 1993;2: 25–32

68. Ranson H, Rossiter L, Ortelli F, Jensen B, Wang X, Rothe CW, Collins FH, Hemingway J. Identification of a novel class of insect glutathione S-tranferases involved in resistance to DDT in the malaria vector Anopheles gambiae Biochem J. 2001;359: 295- 304

69. Enayati AA, Ranson H, Hemingway J. Insect glutathione transferases and insecticides resistance. Insect Mol. Biol. 2005;14: 3–8

70. Oakley AJ, Harnnoi T, Udomsinprasert R, Jirajaroenrat K, Ketterman AJ, Wilce MC. The crystal structures of glutathion S-transferases isoenzymes 1–3 and 1–4 from Anopheles dirus species B. Protein Sci. 2001;10: 2176–2185

71. Udomsinprasert R, Pongjaroenkit S, Wongsantichon J, Oakley AJ, Prapantadhara LA, Wilce MCJ, Ketterman AJ. Identification, characterization and structure of a new Delta class glutathione transferase isoenzyme. Biochem. J. 2005; 388: 763–771

72. Agianian B, Clayton JD, Leanord K, Tucker P, Bullard B, Gros P. Crystallization and preliminary X-ray analysis of Drosophila glutathione S-transferase-2. Acta Crystallogr. D. Biol. Crystallogr. 2000;57: 725–727

73. Agianian B, Tucker PA, Schouten A, Leonard K, Bullard B, Gros P. Structure of a Drosophila sigma class glutathione S-transferase reveals a novel active

site topography suited for lipid peroxidation products. J. Mol. Biol. 2003;326: 151–165

74. Motoyama N, Kulkarni AP, Hodgson E, Dauterman WC. Endogenous inhibitors of glutathione S-transferase in houseflies. Pestic. Biochem. Physiol. 1978;9: 255–262

75. Kulkarni AP, Motoyama N, Dauterman WC, Hodgson E. Inhibition of housefly glutathione S-transferase by catecholamines and quinones. Bull. Environ. Contam. Toxicol. 1978;20: 221–232

76. Clark AG, Letoa M, Ting WS. The purification by affinity chromatography of a glutathione S-transferase from larvae of Galleria mellonella. Life Sci. 1977;20: 141–148

77. Clark, A.G. and Dauterman, W.C. The characterisation by affinity chromatography of glutathine S-transferases from different strains of housefly. Pestic. Biochem. Physiol. 1982;117: 307-314

78. Baker WL, Clark AG, Faulds G, Nielson JS. Multiple glutathione S-transferases in Galleria mellonella ; their detection with fluorigenic substrates. Insect Biochem. Molec. Biol. 1994;24: 301–307

79. Clark AG, Dick GL, Martindal SM, Smith JN. Glutathione S-transferases from New Zealand grass crub, Costelytera zealandica. Insect Biochem. 1985;15: 35–44

80. Clark AG, Shamaan NA. Evidence that DDT-dehydrochlorinase from the housefly is a glutathione S-transferase. Pestic. Biochem. Physiol. 1984;22: 249–261

81. Clark AG, Marshall, SN, Qureshi AR. Synthesis and use of an isoform-specific affinity matrix in the purification of glutathione S-transferases from the house fly, Musca domestica (L.). Protein Expr. Purif. 1990;1: 121–126.

82. Alias Z, Clark AG. Studies on the glutathione S-transferase proteome of adult Drosophila melanogaster: responsiveness to chemical challenge. Proteomics. 2007;7: 3618– 3628.

83. Clark AG, Drake B. Purification and properties of glutathione S-transferases from larvae of Wiseana cervinata. Biochem. J. 1984;217: 41–50

84. Simon PC, van der Jagt DL. Purification of glutathione S-transferases from human liver by glutathione-affinity chromatography. Anal. Biochem. 1977;82: 334–341

85. Cochrane BJ, Morrissey JJ, LeBlanc GA. The genetics of xenobiotic metabolism in Drosophila. IV. Purification and characterisation of the major glutathione S-transferase. Insect Biochem. 1987;17: 731–738

86. Toung YP, Hsieh TS, Tu CPD. Drosophila glutathione S-transferase 1-1 shares a region of sequence homology with the maize glutathione S-transferases III. Proc. Natl. Acad. Sci. U. S. A. 1990;87: 31–35

87. Mannervik B, Guthenberg C. Glutathione transferase (human placenta). Methods Enzymol. 1981;77: 231–235

88. Clark AG, Smith JN, Speir TW. Cross specificity in some vertebrate and insect glutathione S-transferases with methylparathion (dimethyl p-nitrophenyl phophorothionate), 1-chloro-2,4-dinitrobenzene and N-crotonyl-N-acetyl cysteamine as substrates. Biochem. J. 1973;135: 285–392

89. Yawetz A, Koren B. Purification and properties of the Mediterranean fruit fly Ceratitis capitata W. glutathione S-transferase. Insect Biochem. 1984;14: 663–670

90. Keeran WS, Lee RF. The purification and characterisation of glutathione S-transferase from the hepatopancreas of the Blue Crab, Callinectes sapidus. Arch.Biochem. Biophys. 1975;25: 8670–8673

91. Clark, A.G. and Dauterman, W.C. The characterisation by affinity chromatography of glutathine S-transferases from different strains of housefly. Pestic. Biochem. Physiol. 1982;117: 307–314.

92. Grant DF, Dietze EC, Hammock BD. Glutathione S-transferases isoenzymes in: purification, characterisation and isoenzyme-specific regulation. Insect Biochem. 1991;21: 421–433

93. Chiang FM, Sun CN. Glutathione transferase isoenzymes of diamondback moth larvae and their role in the degradation of some organophosphorus insecticides. Pestic. Biochem. Physiol. 1993;45: 7–14

94. Kostaropoulos I, Mantzari AE, Papadoupolos AI. Alterations of some glutathione S- transferase characteristics during the development of Tenebrio molitor (Insecta: Coleoptera). Insect Biochem. Mol. Biol. 1996;26: 963–969

95. Wilson TG. Resistance of Drosophila to toxins. Annu. Rev. Entomol. 2001;46: 545–571.

96. Morton RA. Evolution of Drosophila insecticide resistance. Genome. 1992;36: 1–7

97. Oppenoorth FJ. Biochemistry of insecticide resistance. Pestic. Biochem. Physiol. 14;22: 187–193

98. Oppenoorth FJ, Rupes V, El-Bashir S, Houx NWH, Voerman S. Glutathione dependent degradation of parathion and its significance for resistance in housefly. Pestic. Biochem. Physiol. 1972;2: 262–269

99. Motoyama N, Dauterman WC. Interstrain comparison of glutathione-dependent reactions in susceptible and resistant houseflies. Pestic. Biochem. Physiol. 1975;5: 489– 495

100. Motoyama N., Dauterman WC. In vitro metabolism of azinphosmethyl in susceptible and resistant house flies. Pestic. Biochem. Physiol. 1972;2: 113–122

101. Papadoplus AI., Boukouvala E, Kakaliouras G, Kostaropoulos J, Papadopoulou-Mourkidou. Effect of organophosphate and pyrethroid insecticides on the expression of GST from Tenebrio molitor pupae. Pestic. Biochem. Physiol. 2000;68: 26–33

102. Oppenoorth FJ, van der Pas LJT, Houx NWH. Glutathione S-transferase and hydrolytic activity in a tetrachlorvinphos-resistant strain of housefly and their influence on resistance. Pestic. Biochem. Physiol. 1979;11: 176-188

103. Niwa Y, Miyata T, Saito T. In vitro metabolism of malathion by malathion resistant and susceptible strains of houseflies, Musca domestica. J. Pestic. Sci. 1977;2: 151-157

104. Houpt DR, Pursey JC, Morton RA. Genes controlling malathion resistance in laboratory-selected population of Drosophila melanogaster. Genome. 1988;30: 844–853

105. Cochrane JB, LeBlanc GA. Genetic of metabolism in Drosophila. I. Genetic and environmental factors affecting glutathione S-transferase in larvae. Biochem. Pharmacol. 1986;35: 1679–1684

106. Shepanski MC, Glover TJ, Kuhr RJ. Resistance of Drosophila to DDT. J. Econ. Entmol. 1977;70: 539–543

107. Wilson JA, Clark AG. The role of E3 esterase, glutathione S-transferases and other nonoxidative mechanisms in resistance to diazinon and other organophosphate insecticides in Lucilia cuprina. Pestic. Biochem. Physiol. 1996;54: 85–95

108. Motoyama N, Dauterman WC. Glutathione S-transferases: Their role in the metabolism of organophosphorus insecticides. Rev. Biochem. Toxicol. 1980;2: 49–69

109. Lumjuan N, McCarroll L, Prapanthadara L, Hemingway J, Ranson H. Elevated activity of an Epsilon class glutathione transferase confers DDT resistance in the dengue vector, Aedes aegypti. Insect Biochem. Mol. Biol. 2005;35: 861–871

110. Ding Y, Hawkes N, Meredith J, Eggleston P, Hemingway, J, Ranson, H. Characterization of the promoters of Epsilon glutathione transferases in the mosquito Anopheles gambiae and their response to oxidative stress. Biochem J. 2005;387: 879–888.

111. Sawicki R , Singh S P, Mondal, A K, Benes H, Zimniak P. Cloning, expression and biochemical characterization of one Epsilon-class (GST-3)and ten Delta-class (GST-1) glutathione S-transferases from Drosophila melanogaster, and identification of additional nine members of the Epsilon class. Biochem. J. 2003;370: 661–669

112. Singh SP, Coronella JA, Benes H, Cochrane BJ, Zimniak, P. Catalytic function of Drosophila melanogaster glutathione S-transferase DmGSTS1-1 (GST-2) in conjugation of lipid peroxidation end products. Eur. J. Biochem. 2001;268: 2912–2923

113. Udomsinprasert R, Ketterman AJ. Expression and characterisation of a novel class of glutathione S-transferase from Anopheles dirus. Insect Biochem. Mol. Biol. 2002;32: 425–433

114. Jirajaroenrat K, Pongjaroenkit S, Krittanai C, Prapanthadara L, Ketterman AJ. Heterologous expression and characterization of alternatively spliced glutathione-S-transferases from a single Anopheles gene. Insect Biochem. Mol. Biol. 2001;31: 867–875

115. Wilson TG. Drosophila melanogaster (Diptera: Drosophilidae): a model insect for insecticide resistance studies. J. Econ. Entomol. 1988;81: 22–27

116. Scott JG, Warren WC, Beukeboom LW, Bopp D, Clark AG, Giers SD, Hediger M, Jones AK, Kasai S, Leichter CA, Li M, Meisel RP, Minx P, Murphy TD, Nelson, DR, Reid WR, Rinkevich FD, Robertson, HM, Sackton TB, Sattelle DB, Thibaud-Nissen, F, Tomlinson C, van de Zande L, Walden KKO, Wilson RK, Liu N. Genome of the house fly, Musca domestica L., a global vector of diseases with adaptations to a septic environment. Genome Biol. 2014;15: 466–482

17

Chapter

BIOLOGICAL AND BIOCHEMICAL FOUNDATIONS OF PESTICIDE RESISTANCE IN RHIPICEPHALUS (BOOPHILUS) MICROPLUS

1. INTRODUCTION: AN OVERVIEW

Insects, ticks, and mites constitute the primary categories of arthropods, encompassing species that may act as pests and transmitters of disease. Historically, the issue of arthropod pests and disease vectors impacting public health, agricultural output, and livestock production has been addressed through extensive pesticide application. The extensive application of pesticides has imposed significant selective pressure, resulting in resistance among some economically vital arthropod species to presently utilized insecticides.

Pesticide resistance in arthropod species significant to public health, agriculture, and veterinary medicine has emerged as a critical issue. This condition poses a global problem as numerous arthropod species damage crops, so competing directly with humans for food, while other species serve as significant vectors for infectious agents that cause diseases in humans, cattle, wildlife, and plants. Resistance originates from evolutionary genetic alterations that modify the molecular structure of the target site or induce changes in multigenic metabolizing enzyme systems, leading to increased hydrolysis rates or sequestration of pesticides, along with diminished ability to penetrate the outer chitin protective layer.

Here we review the known molecular and biochemical mechanisms of pesticide resistance and do a gap analysis of processes involved in the evolution of insensitivity to pesticides among arthropods that remain to be fully understood. Significant advances have been achieved in our knowledge of the processes involved in resistance to pyrethroids. More research is required to decrease the knowledge gaps regarding mechanisms of resistance to organophosphates (OP). Less is still known about resistance to amitraz, macrocyclic lactones, and fipronil, whose mechanisms of action appear to involve complex and multifactorial evolutionary mechanisms.

An example of the problem with pesticide-resistant pests is the cattle tick Rhipicephalus (Boophilus) microplus, which is regarded as the most economically important ectoparasite of livestock globally and ranks sixth among the most pesticide resistant pests globally. This cattle tick affects animal health and production in tropical and subtropical regions of the world directly through its obligate hematophagous habit and indirectly by serving as vector of pathogens like Babesia bovis, B. bigemina and Anaplasma marginale, which cause the deadly bovine diseases babesiosis and anaplasmosis, [1, 2]. The control of cattle ticks, bovine babesiosis and anaplasmosisis a costly problem that prevents the livestock industry in Latin America and other parts of the world achieving its full potential [1]. The almost complete reliance on pesticides to control cattle ticks and associated diseases has been a strong selective force for tick populations that now are resistant to multiple classes of acaricides [3]. These advancements underscore the remarkable genetic flexibility and exceptional adaptation of arthropod pests and vectors, which have facilitated their resistance to the majority of synthetic pesticide classes that humans have identified and commercialized.

2. RESISTANCE, ADAPTATION AND COEVOLUTION

Resistance to synthetic pesticides is a genetic condition that confers an arthropod population the capability to get adapted to a toxic environment through a selection process driven by human activity [4]. Based on this concept, the phenomenon of pesticide resistance can be understood as an important model based on natural selection processes [5]. Considering pesticide resistance as a biological model facilitates the study of evolutionary adaptations by arthropods living under selection pressure through constant exposure to pesticides used by humans [6]. However, the natural history of pest-host interactions in a way preadapted arthropods to become resistant to synthetic pesticides.

As a naturally occurring process, the coevolution between plants and arthropods enabled biological and chemical interactions over hundreds of millions of years [7]. As a result of the interactions between these two groups of organisms [8], arthropod species conserved within their genomes genes conferring an evolutionary advantage, or traits that have allowed them to survive. These types of interactions called allelochemical interactions have allowed plants and arthropods to develop very sophisticated molecular mechanisms to maximize their chances of survival. Plants evolved to be a factory of potent allelochemicals such as natural poisons, toxins, and repellents that function as a natural mechanism of defense against predators. As an example of coevolution, herbivore arthropods adapted by evolving a machinery of metabolic hydrolyzing enzymes to prevent poisoning by allelochemicals acquired during plant feeding [9–11]. This resulted in what could be described as natural resistance in arthropods selected through exposure to plant allelochemicals, which took millions of years. Arthropods were in a way preadapted to become resistant to synthetic pesticides rapidly in terms of evolutionary time [12]. Another predisposing factor for this rapid adaptation is the structural analogy between secondary metabolites produced by plants and synthetic pesticides designed by humans to control pests [13, 14].

The development of resistance to chemicals depends on the evolutionary forces exerting selection pressure and the consequent adaptive processes involving selection of genetic variations caused by random mutations, or genetic rearrangements resulting from exposure to pesticides [7], which is reflected in the selection and reproduction of individuals with resistant phenotypes capable of surviving pesticide concentrations that are lethal to the wild- type populations.

Resistance occurs when the selection of low frequency naturally occurring mutations within the genetic pool of a population allows a small fraction of individuals to survive the toxic effect of compounds used as pesticides [15]. Prior to initial exposure of the population to pesticides, there are few individuals with resistant genotypes and most of them are susceptible. When a pesticide is applied, those few survivors have an adaptive advantage. Therefore, they survive initial exposure and reproduce, and the frequency of resistant genotypes in the progeny increases. If repetitive and sustained applications of the pesticide on the population occur, the susceptible individuals are eliminated, which drives the selection of resistant genotypes in the population. At this point, the diverse genetic traits conferring an advantage to survive pesticide exposure are transferred to the progeny and the efficacy of the pesticides decreases.

3. MOLECULAR BASIS OF PESTICIDE RESISTANCE

In general, acaricide resistance can be the result of increased metabolic detoxication or target site modification [4]. Metabolic detoxication is frequently the result of increased enzymatic activity by isozymes encoded by multigene families such as cytochrome P450, glutathione-S-transferase, and carboxylesterase [16, 17]; all these enzymes hydrolyze or sequester different kinds of pesticides. Exposure to a pesticide can exert enough pressure to select an enzymatic system or a specific isozyme within each family. Esterase isozyme overexpression is generally accepted as a mechanism involved in OP resistance. However, in the Coatzacoalcos laboratory strain of R. microplus, designated as such to reflect the name of the village in Mexico where the original tick population sample was obtained, metabolic detoxication has been identified by its efficient esterase activity resulting from enzyme overexpression as a resistance mechanism for permethrin that belongs to the pyrethroid chemical class of pesticides. This strain has been toxicologically characterized using the larval packet test (LPT) [18], which helped to elucidate the esterase-based mechanism of resistance to permethrin. The R. microplus

Coatzacoalcos strain exhibits a significant enhanced capacity to hydrolyze permethrin as well as an increased esterase activity. This suggests an esterase based metabolic mechanism as a main component of permethrin resistance [19]. The esterase gene responsible for permethrin resistance was identified and named CzEST9. It is also known that the overexpression mechanism of this isozyme is the result of CzEST9 duplication in the Coatzacoalcos strain that leads to metabolic detoxication through the overexpression of esterase 9 activity in R. microplus [20]. The sequence of CzEST9 gene has been determined and the recombinant product yielded a 62.8 kDa protein [19]. Since the Coatzacoalcos strain does not include the Kdr variation in the sodium channel gene found in other Mexican strains of R. microplus, it is suggested that there are two independent mechanisms of acaricide resistance to pyrethroids. However, common mechanisms of acaricide resistance to pyrethroids in Mexico apparently involve the presence of sequence variation in the sodium channel gene [21].

The sodium channel is the known target site of pyrethroids. Sequence variation in the sodium channel prevents pyrethroids from attaching to the target site due to an alteration in the sodium channel stereochemical structure. For this reason, the process is described as target site modification mechanism, or Kdr-type resistance (Knock down resistance). This is one of the mechanisms of pesticide resistance in insects that is better understood.

Two important allele variants occurring in the sodium channel gene associated with pyrethroid resistance in the cattle tick R. microplus are the variation occurring in domain III segment 6 (III-S6) [22] and the variation occurring at the bridge joining segments 4–5 in domain II (II- S4-5) [23]. The former is a Phe-Ile substitution produced by a nucleotide variation at domain III-S6 that was first reported in Mexican tick strains. Its role and contribution to pyrethroid resistance has been confirmed [21]. The other is a Le-Ile substitution found thus far only in Australian tick strains; this variation is very similar to a variation found in the crop insect pest Bemicia tabaci [23].

Findings on the diversity of allele variants occurring in the sodium channel gene associated with pyrethroid resistance in the cattle tick R. microplus have been confirmed by experiments based on differences in melting temperatures (Tm) of sodium channel allele specific gene fragments obtained with single larvae DNA from México and Australia. These experiments revealed that substitution III-S6 (Phe-Ile) only occurs in Mexican tick populations whereas substitution II-S4-5 (Le-Ile) only occurs in Australian tick strains [24]. The information available suggests that there are at least two different and independent mechanisms involved based on the different amino acid substitutions (Phe-Ile and Le-Ile) residing in different positions of the sodium channel protein sequence.

4. BIOLOGICAL BASIS OF PESTICIDES RESISTANCE

Allelochemical interactions are defensive processes or are involved in food competition mechanisms that different species employ to inhibit the action of natural enemies. Plant– arthropod coevolution is a natural selection mechanism driven by allelochemical interactions between plants and arthropods over millions of years [7]. As a result of the reciprocal interactions between these two groups of organisms [8], arthropods have conserved within their genomes all those traits conferring them the ability to inhibit or avoid toxicants produced by plants that function as defensive mechanisms against herbivorous insects [9, 10]. Sophisticated metabolic detoxication mechanisms have been developed by herbivorous arthropods in order to survive the exposure to toxic plant metabolites [11], which represents a natural process of resistance to plant toxicants. The preservation of these components in the genome of arthropods provides the foundation of molecular systems that allow them to get adapted and become resistant against pesticides currently used.

Insects affect the survival and reproduction of plants. Secondary metabolites like phenolic acids, flavonoids, terpenoids, steroids, alkaloids, and organic cyanides are produced by plants as part of defensive mechanisms. A coevolutionary race is established through these interactions where mutual selection pressure has led to the process of speciation to preserve natural equilibrium.

Adaptation to new environmental conditions requires the development of defensive processes through natural selection. Plants are biological engines producing a wide variety of natural defensive chemicals including repellents, antifeeding molecules, and poisons, some of which have been used as natural strategies to protect cultivars from plagues or to control vector-borne diseases. These mechanisms evolved separately in different herbivorous insects driven by diverse insect–plant interactions. In general, herbivorous insects feed on few plant species and plant species are attacked by pests specialized to overcome natural defensive substances. Herbivorous insects have developed the molecular machinery to metabolize most of the toxic material produced by plants, but not all toxicants are metabolized by all pest species.

Specialized insects have an adaptive advantage because their biochemical systems evolved to detoxify one or few potentially harmful substrates. The metabolic system of a polyphagous species reflects more diverse detoxication mechanisms against a wide variety of chemically defined plants. Thus, polyphagous insects have a "higher metabolic load". The activity of mixed function oxidases in the intestines of moths and butterfly larvae is higher in polyphagous species than in species restricted to a single family of plants. Pyrethrins are part of the wide variety of allelochemicals metabolized by this family of enzymes [25].

5. METABOLIC DETOXICATION OF PESTICIDES

Plant–insect coevolutionary interactions drive species diversification and the set of genetic traits that allow pest species to survive exposure to a wide variety of secondary metabolites produced by plants. Such genetic traits evolved through evolutionary history, involving several highly specialized multigene families that are responsible for detoxication mechanisms of biotic and xenobiotic compounds. Examples of these multigene families include the glutathione transferases, mixed function oxidases, and carboxilesterases superfamilies [26]. These supergene families are capable of metabolizing a large amount of chemicals, some of which are currently used as pesticides.

Some mechanisms of resistance have been identified for several important arthropod vectors. Increased esterase activity is a major component of organophosphates resistance in Culex mosquitoes [27]. The enhancement of mix function oxidases also plays an important role in OP and pyrethroid resistance mechanisms [28], and the combination of mixed function oxidases and esterases in high concentrations has been detected in permethrin-resistant mosquitoes [29]. Resistance mechanisms in cockroaches include metabolic detoxication and Kdr-type resistance. However, detoxication mechanisms mediated by esterases and oxidases have been identified as the most frequent mechanisms of resistance [30].

6. INSENSITIVE TARGET SITE

Modifications of target sites as a result of point mutations on gene sequences have been also identified as mechanisms of resistance. Variations on genes encoding GABA receptors [31, 32], acetylcholinesterase [33], some detoxifying esterases [34, 35], and sodium channel gene sequences (22) have been discovered in different arthropods. The latter has been identified as Kdr-type resistance; this variation alters the molecular structure of the sodium channel, which is the target site of pyrethroid pesticides [36, 37].

Kdr-type resistance was firstly documented in R. microplus by He et al. [22]. It was shown that there is a variation in the sodium channel gene sequence at position 2134 where the base substitution of thymine by adenine (T2134A) results in an amino acid change from phenylalanine to isoleucine on the transmembrane segment 6 (S6), which is located on domain III of the para-type sodium channel gene [22].

Pyrethroid resistance in arthropods has been associated with nonsynonymous mutation on domains I, II, III, and IV of sodium channel genes [38, 39]. As already mentioned, two important variations have been previously identified in R. microplus sodium channel gene, a domain II variation (C190A) and the domain III variation (T2134A). The latter only found in ticks from Texas and Mexico [22,

40–42] and the former reported in Australia, Africa, and South America [23, 41, 42]. Although pyrethroid resistance in Mexican cattle tick populations has been mostly attributed to the domain III variation T2134A [22, 36, 39, 43], some authors have suggested that additional resistance mechanisms to the sodium channel variations must be present, since genotype frequencies from screened populations do not account for the level of phenotypic resistance observed in field [22, 39, 41].

Previous studies have documented the occurrence of variations on an esterase gene associated with pyrethroid resistance; however, this phenomenon seems to be linked to a pyrethroid metabolic detoxication mechanism since it has been consistently found in pyrethroid-resistant ticks [35, 37]. These results suggest that pyrethroid resistance in ticks is the result of genetic traits involved in both metabolic and insensitive target mechanisms, depending on which gene the variation occurs.

7. GENOMIC PERSPECTIVES FOR PEST CONTROL

Applying the genomics approach to pesticide research offers the opportunity to advance our knowledge of the mechanisms of resistance and to find sustainable solutions to problems associated with pesticide resistance diagnosis, prediction, and prevention. This will also expand options to improve integrated pest management programs. In the case of livestock, a more rational use of pesticides could be achieved by combining genomics-based knowledge of acaricide resistance with the use of more efficient anti tick vaccines developed through modern technologies [43, 44].

The use of recombinant DNA technologies and the application of bioinformatics to mine genome databases such as GenBank, are powerful foundations to innovate diagnostic tools based on the identification and amplification of single nucleotide polymorphisms (SNPs) associated to target site insensitivity mechanisms [39, 42, 43]. Recent technologies such as polymerase chain reaction (PCR) is a powerful tool used to amplify or detect SNPs that can be employed as biomarkers of pesticide resistance, which provides an alternative to the time consuming bioassays that mitigates the risk of exposure to pesticides by laboratory personnel [21, 35–39, 44].

Genomics approaches are also enabling the design of new target antigens through in silico analysis of transcriptomic and genomic data to develop vaccines against ticks [43, 45], mosquitoes, other biting flies, and parasitic worms, as well as markers for pesticide resistance detection [46, 47]. The integration of molecular methods for pesticide resistance detection, prediction, and vaccine development efforts against hematophagous arthropods is an exciting alternative to manage the emergence of pesticide resistance and to improve vector and vector- borne disease control technologies.

8. CONCISE SUMMARY

Pesticide resistance is a genetic phenomenon that indicates the existence and selection of a set of genes within an arthropod population. A primary role of this gene suite is to neutralize the pesticides employed in pest management. The application of pesticides exerts significant selective pressure for resistance within pest populations.

Genetic plasticity in arthropods facilitates the development of resistance to presently utilized pesticides. Due to the natural host defense mechanisms developed over millions of years, arthropod pests are likely preadapted to develop resistance to new pesticides, particularly if the current extensive pesticide application persists.

The escalating issue of multiple resistance in arthropods globally necessitates research to enhance our comprehension of pesticide resistance mechanisms. This knowledge can be converted into enhanced diagnostic and predictive instruments to alleviate the effects of pesticide resistance. The integration of genomic methods for detecting pesticide resistance and developing vaccines against hematophagous arthropods will enhance strategies to prevent and predict the emergence of pesticide resistance, thereby reducing the impact of pests and vector-borne diseases on humans, livestock, and wildlife.

REFERENCES

1. de la Fuente J, Estrada-Pena A, Venzal JM, Kocan KM, Sonenshine DE. Overview: ticks as vectors of pathogens that cause disease in humans and animals. Frontiers in Bioscience. 2008; 13: 6938–6946.
2. Estrada-Pena A, Garcia Z, Sanchez HF. The distribution and ecological preferences of Boophilus microplus (Acari: Ixodidae) in Mexico. Experimental and Applied Acarology. 2006; 38 (4): 307–316. DOI: 10.1007/s10493-006-7251-2.
3. George JE, Pound JM, Davey RB. Chemical control of ticks on cattle and the resistance of these parasites to acaricides. Parasitology. 2004; 129: S353–366.
4. Rosario-Cruz R, Almazan C, Miller RJ, Dominguez-Garcia DI, Hernandez-Ortiz R, de la Fuente J. Genetic basis and impact of tick acaricide resistance. Frontiers in Biosciences. 2009; 14: 2657–2665.
5. Ffrench-Constant RH, Daborn PJ, Le Goff G. The genetics and genomics of insecticide resistance. Trends in Genetics. 2004; 20: 163–170. DOI: http://dx.doi.org/10.1016/ j.tig.2004.01.003.
6. Palumbi SR. Humans as the world's greatest evolutionary force. Science. 2001; 293: 1786–1790. DOI: 120.1126/science.293.5536.1786.
7. Wheat CW, Vogel H, Wittstock U, Braby MF, Underwood D, Mitchell-Olds T. The genetic basis of a plant–insect coevolutionary key innovation. Proceedings of

the National Academy of Science of the United States of America. 2007; 104 (51): 20427- 20431. DOI:10.1073/pnas.0706229104.

8. Rausher MD. Co-evolution and plant resistance to natural enemies. Nature. 2001; 411: 857–864. DOI: 10.1038/35081193.

9. Kessler A, Baldwin IT. Plant responses to insect herbivory: the emerging molecular analysis. Annual Review of Plant Physiology and Plant Molecular Biology. 2002; 53: 299–328. DOI: 10.1146/annurev.arplant.53.100301.135207.

10. Wittstock, U., Gershenzon, J. Constitutive plant toxins and their role in plant defense. Current opinions in Plant Biology. 2002; 5: 300–307. DOI: 10.1016/S1369-5266(02)00264-9.

11. Ranson H, Claudianos C, Ortelli F, Abgrall C, Hemingway J, Sharakhova MV, Unger MF, Collins FH, Feyereisen R. Evolution of supergene families associated with insecticide resistance. Science. 2002; 298:179–181. DOI: 10.1126/science.1076781.

12. Howe GA, Jander G. Plant immunity to insect herbivores. Annual Review of Plant Biology. 2008; 59: 41–66. DOI: 120.1146/annurev.arplant.59.032607.092825.

13. Isman MB. Botanical insecticides, deterrents and repellents in modern agriculture and an increasingly regulated world. Annual review of Entomology. 2006; 51: 45–66. DOI: 10.1146/annurev.ento.51.110104.151146.

14. Mencke N. Acaricidal and repellent properties of permethrin, its role in reducing transmission of vector-borne pathogens. Parassitologia. 2006; 48: 139–140.

15. Roush RT. Occurrence, genetics and management of insecticide resistance. Parasitology Today. 1993; 9: 174–179. DOI: 10.1016/0169-4758(93)90141-2.

16. Miller RJ, Li AY, Tijerina M, Davey RB, George JE. Differential response to diazinon and coumaphos in a strain of Boophilus microplus (Acari: Ixodidae) collected in Mexico. Journal of Medical Entomology. 2008; 45: 905–911.

17. Baffi MA, De Souza GR, De Sousa CS, Ceron CR, Bonetti AM. Esterase enzymes involved in pyrethroid and organophosphate resistance in a Brazilian population of Riphicephallus (Boophilus) microplus (Acari: Ixodidae). Molecular and Biochemical Parasitology. 2008; 160(1): 70–73. DOI: 10.1016/j.molbiopara.2008.03.009.

18. Ortiz EM, Santamaría VM, Ortiz NA, Soberanes CN, Osório MJ, Franco BR, Martinez IF, Quezada DR, Fragoso SH. Characterization of Boophilus microplus resistance to ixodicides in México. In: Seminario internacional de Parasitología Animal Acapulco, Gro. México. 1995; 58–66.

19. Pruett JH, Guerrero FD, Hernandez R. Isolation and identification of an esterase from a Mexican strain of Boophilus microplus (Acari: Ixodidae). Journal of Economical Entomology. 2002; 95: 1001–1007.

20. Jamroz RC, Guerrero FD, Pruett JH, Oehler DD, Miller RJ. Molecular and biochemical survey of acaricide resistance mechanisms in larvae from Mexican strains of the southern cattle tick, Boophilus microplus. Journal of Insect Physiology. 2000; 46: 685– 695. DOI: 10.1016/S0022-1910(99)00157-2.

21. Rosario-Cruz R, Guerrero FD, Miller RJ, Rodriguez-Vivas RI, Domínguez-García DI, Cornel AJ, Hernandez-Ortiz R, George JE. Roles played by esterase activity and by a sodium channel mutation involved in pyrethroid resistance in populations of Boophilus microplus (Acari: Ixodidae) collected from Yucatan, Mexico. Journal of Medical Entomology. 2005; 42: 1020–1025.

22. He H, Chen AC, Davey RD, Ivie GW, George JE. Identification of a point mutation in the para-type sodium channel gene from a pyrethroid-resistant cattle tick. Biochemistry and Biophysics Research Communications. 1999; 261: 558–561. DOI: 10.1006/bbrc. 1999.1076.

23. Morgan JA, Corley SW, Jackson LA, Lew-Tabor AE, Moolhuijzen PM, Jonsson, NN. Identification of a mutation in the para-sodium channel gene of the cattle tick Rhipicephalus (Boophilus) microplus associated with resistance to synthetic pyrethroid acaricides International Journal of Parasitology. 2009; 39: 775–779.

24. Chen AC, He H, Temeyer KB, Jones S, Green P, Barker S.C. A survey of Rhipicephalus microplus populations for mutations associated with pyrethroid resistance Journal of Economical Entomology. 2009; 102: 373–380.

25. Whittaker RH, Feeny PP. Allelochemics: chemical interactions between species. Science. 1971; 171: 757–770. DOI: 10.1126/science.171.3973.757.

26. Devonshire AL and Field LM. Gene amplification and insecticide resistance. Annual Review of Entomology. 1991; 36: 1–23. DOI: 10.1146/annurev. en.36.010191.000245.

27. Hemingway J, Karunaratne SH. Mosquito carboxylesterases: a review of the molecular biology and biochemistry of a major insecticide resistance mechanism. Medical and Veterinary Entomology. 1998; 12: 1–12. DOI: 10.1046/j.1365-2915.1998.00082.x.

28. Bisset J, Rodriguez MM, Diaz C, Soca A. Resistance in a strain of Culex quinquefasciatus coming from Medellin, Colombia. Revista Cubana de Medicina Tropical. 1998; 50: 133–137.

29. Vulule JM, Beach RF, Atieli FK, McAllister JC, Brogdon WG, Roberts JM, Mwangi RW, Hawley WA. Elevated oxidase and esterase levels associated

with permethrin tolerance in Anopheles gambiae from Kenyan villages using permethrin-impregnated nets. Medical and Veterinary Entomology. 1999; 13: 239–244. DOI: 10.1046/ j1365-23915.1999.00177.x.

30. Hemingway J, Dunbar SJ, Monro AG, Small GJ. Pyrethroid resistance in German cockroaches (Dictyoptera: Blattelidae): resistance levels and underlying mechanisms. Journal of Economical Entomology. 1993; 86: 1631–1638.

31. Zhang HG, Ffrench-Constant RH, Jackson MB. A unique amino acid of the Drosophila GABA receptor with influence on drug sensitivity by two mechanisms. Journal Physiology. 1994; 479: 65–75. DOI: 10.1113/jphysiol.1994.sp020278.

32. Ffrench-Constant RH, Steichen JC, Rocheleau TA, Aronstein K, Roush RT. A single- amino acid substitution in a gamma aminobutiric acid subtype a receptor locus is associated with cyclodiene insecticide resistance in Drosophila populations. Proceedings of the National Academy of Sciences U.S.A. 1993; 90: 1957–1961.

33. Mutero A, Pralavorio M, Bride JM, Fournier D. Resistance associated point mutations in insecticide-insensitive acetylcholinesterase. Proceedings of the National Academy of Sciences USA. 1994; 91: 5922–5926.

34. Campbell PM, Newcomb RD, Russell RJ, Oakeshott JG. Two different amino acid substitutions in the ali-esterase, E3, confer alternative types of organophosphorus insecticide resistance in the sheep blowfly, Lucilia cuprina. Insect Biochemistry and Molecular Biology. 1998; 28: 139–150.

35. Hernandez R, He H, Chen AC, Waghela SD, Ivie GW, George JE, Wagner GG. Identification of a point mutation in an esterase gene in different populations of the southern cattle tick, Boophilus microplus. Insect Biochemistry and Molecular Biology. 2000; 30: 969- 977. DOI: 10.1016/S0965-1748(00)00069-2.

36. Guerrero FD, Davey RB, Miller RJ. Use of an allele-specific polymerase chain reaction assay to genotype pyrethroid resistant strains of Boophilus microplus. Journal of Medical Entomology. 2001; 38(1): 44–50.

37. Guerrero FD, Pruett JH, Li AY. Molecular and biochemical diagnosis of esterase-mediated pyrethroid resistance in a Mexican strain of Boophilus microplus (Acari: Ixodidae). Experimental and Applied Acarology. 2002; 28: 257–264.

38. Dong K. Insect sodium channels and insecticide resistance. Invertebrate Neuroscience. 2007; 7(1): 17–30. DOI: 10.1007/s10158-006-0036-9.

39. Rosario-Cruz R, Guerrero FD, Miller RJ, Rodriguez-Vivas RI, Tijerina M, Dominguez-Garcia DI, Hernandez-Ortiz R, Cornel AJ, McAbee RD, Alonso-Diaz MA. Molecular survey of pyrethroid resistance mechanisms in Mexican field

populations of Rhipicephalus (Boophilus) microplus. Parasitology Research. 2009; 105(4): 1145–1153. DOI: 10.1007/s00436-009-1539-1.

40. Miller RJ, Davey RB, George JE. First report of permethrin-resistant Boophilus microplus (Acari: Ixodidae) collected within the United States. Journal of Medical Entomology. 2007; 44(2): 308–315.

41. Li AY, Davey RB, Miller RJ, Guerrero FD, George JE. Genetics and mechanisms of permethrin resistance in the Santa Luiza strain of Boophilus microplus (Acari: Ixodidae). Journal of Medical Entomology. 2008; 45(3): 427–438.

42. Lovis L, Guerrero FD, Miller RJ, Bodine DM, Betschart B, Sager H. Distribution patterns of three sodium channel mutations associated with pyrethroid resistance in Rhipicephalus (Boophilus) microplus populations from North and South America, South Africa and Australia. International Journal of Parasitology. 2012; 2: 216–224. DOI: 10.1016/j.ijpddr.2012.08.001.

43. Temeyer KB, Chen AC, Davey RB, Guerrero F, Freeman JM, Kammlah DM, Li AY, Lohmeyer K H, Olafson P, Pérez de León AA, Phillips P L, Pound JM, Welch JB. Novel approaches for control of Rhipicephalus (Boophilus) microplus. Revista Mexicana de Ciencias Pecuarias. 2012; 3(Supl 1): 25–40.

44. Guerrero FD, Jamroz RD, Kammlah D, Kunz SE. Toxicological and molecular characterization of pyrethroid-resistant horn flies, Haematobia irritans: identification of kdr and super-kdr point mutations. Insect Biochemistry and Molecular Biology. 1997; 27: 745–755. DOI: 10.1016/S0965-1748(97)00057-X.

45. Miller R, Estrada-Peña A, Almazan C, Yeater KM, Messenger M, Ellis D, and Pérez de León, AA. Exploring the use of an anti-tick vaccine as a tool for the integrated eradication of the cattle fever tick, Rhipicephalus (Boophilus) annulatus. Vaccine. 2012; 30: 5682–5687. DOI: 10.1016/j.vaccine.2012.05.061.

46. Williamson MS, Martinez-Torres D, Hick CA, Devonshire A.L. Identification of mutations in house-fly para-sodium channel gene associated with knockdown resistance (Kdr) to pyrethroid insecticides. Molecular and General Genetics. 1996; 252: 51–60.

47. Temeyer KB, Pruett JH, Chen AC. R86Q a mutation in BmAChE3 yielding a Rhipicephalus microplus organophosphate insensitive acetylcholinesterase. Journal of Medical Entomology. 2007; 1013–1018.

18 Chapter BIOCHEMICAL RESISTANCE OF INSECTICIDES IN TEA PESTS

1. INTRODUCTION: AN OVERVIEW

1.1. Tea, the economically valuable plantation crop: From shrub to cup

Tea is derived from the young leaves and buds of the Camellia sinensis (L.) plant. O. Kuntze hails from China, where its utilization is believed to have been discovered approximately 4,700 years ago. Shen Nung, a Chinese ruler from approximately 4,700 years ago, is credited with the discovery that tea leaves immersed in boiling water produce a delicious beverage. Tea subsequently gained popularity in China due to its flavor and therapeutic properties. Ultimately, the practice of consuming tea proliferated across Asia and subsequently across the globe. At present, tea ranks as the second most favored and widely used beverage following water. The term tea originated from t'e (pronounced 'tay') in the Amoy dialect, but in Cantonese it is referred to as ch'a ('chah'). This is the designation of this remarkable beverage in Japan, Iran, Russia, Indonesia, Malaysia, Vietnam, and India. Tea was brought to Japan around AD 800 and was considered a medicinal substance for almost 500 years. Tea was brought to Europe in the early 17th century with the commencement of trade between Europe and Southeast Asia.

Now tea is produced in almost every region of the globe. The tea plant is predominantly grown in Asia followed by in Africa and to a very small extent in Europe, South America, Australia and New Zealand. Now, tea is grown in 36 tropical and subtropical countries. Major tea producing countries are India, China, Sri Lanka, Kenya, Japan, Indonesia, Thailand, Bangladesh, Nepal, Vietnam, Turkey and Argentina.

Tea plants are indigenous to East and South Asia, perhaps originating near the junction of northeast India, northern Burma, southwestern China, and Tibet. The commercially grown tea plants originate from small-leaved China plants, C. sinensis, Assam plants, C. assamica (Masters), Cambod plants, C. assamica ssp. lasiocalyx (Planchon ex Watt) Wight, and many hybrids among these species. Tea plants necessitate a warm, humid climate, evenly distributed rainfall, and extended hours of sunlight. Shoots, consisting of two or three delicate leaves and a bud, are collected and processed in factories to produce various types of tea.

2. INSECT PEST OCCURRENCES IN TEA

Being a widespread perennial monoculture crop, tea plantations provide the most congenial microclimate as well as continuous food supply to a number of arthropods. Every part of the tea plant, i.e. leaf, stem, root, flower and seed is subjected to attack by at least one arthropod pest species. According to recent estimates, globally, more than a thousand (1,034 species) arthropods and 82 species of nematodes are associated with tea plantations [1]. Among the insect pests, 32% are from order Lepidoptera, followed by 27% pest species from Hemiptera [2]. In India, only 300 species of arthropods are recorded as pest and about 167 species are from tea-growing Northeast India (Table 1) [3]. The dynamic adaptations of insects have facilitated them to exploit every part of the tea plant and the maximum numbers of pests occur on the foliage.

Among the arthropods that attack tea plant, insect and mite pests are the most damaging [2], causing on average a 5%–55% yield loss [4–6]. Insect pests alone can cause on an average 11%– 55% yield loss, if left unrestrained [1].

Tea mosquito bug, thrips, jassids, tea caterpillars (looper, redslug and bunch), aphids, termites, cockchafers and red spider mite are among the various insects and mite pests that cause severe loss in tea production. The damage caused by the tea pests frequently leads to a significant impact on quality and yield, while the degree of pest infestation differs in different tea-growing areas depending on altitude, climate, forest cover and local cultural practices. Tea pests can be broadly classified into the different categories based on their feeding nature, time of occurrence and severity. The major pests are mite pests: red spider mite (Oligonychus coffeae), scarlet mite (Brevipalpus phoenicis), pink mite (Acaphylla theae), purple mite (Calacarus carinatus); sucking pests: tea mosquito bug (Helopeltis theivora), tea thrips (Scirtothrips dorsalis and Myctrothrips setiventris), tea jassids (Empoasca flavescens), aphids, scale insects; leaf eaters: looper caterpillars (Hyposidra talaca, H. infixaria, Buzura suppressaria), red slug caterpillar (Eterusia magnifica), bunch caterpillar (Andraca bipunctata); and soil-borne pests: termites (Microcerotermes sp. Odontotermes sp.), cockchafers, nematodes, weevils, cricket [8].

Table 1. Major insect and mite pests that occur on tea in India.

Sl. no.	Name of the pests		Nature of damage	Family: Order
1.	*Andraca bipunctata* Walk.	Bunch caterpillar	Folivores	Bombycidae: Lep.
2.	*Eterusia magnifica* Butl.	Red slug caterpillar	Folivores	Zygaenidae: Lep.
3.	*Buzura suppressaria* Guen.	Looper caterpillar	Folivores	Geometridae: Lep.
4.	*Hyposidra talaca, H. infixaria* Walk.	Black inch worm	Folivores	Geometridae: Lep.

5.	*Lymantria albulunata* Mre.	Sungma caterpillar	Folivores	Lymantridae: Lep.
6.	*Gracillaria theivora* Walsh.	Tea leaf roller	Folivores	Gracilariidae: Lep.
7.	*Homona coffearia* Nietner	Tea tortrix	Folivores	Torticidae:Lep.
8	*Laspeyresia leucostoma* Mayer.	Flushworm	Folivores	Eucosmidae: Lep.
9.	*Holotrichia impressa* Burm.	Cockchafer grubs	Root of young tea	Scarabaeidae: Col.
10.	*Serica assamensis* Brenske	Leaf-eating cockchafer	Leaf eater	Scarabaeidae: Col.
11.	*Asticus chrysochlorus* Wied.	Large green weevil	Leaf eater	Curculionidae: Col.
12.	*Agromyzidae* (Bigot) Meij	Tea leaf miner	Leaf eater	Agromyzidae: Dip.
13.	*Microtermes* spp.	Live wood-eating termite	Stem and root eater	Termitidae: Iso.
14.	*Odontotermes* sp.	Scavenging termite	Stem and root eater	Termitidae: Iso.
15.	*Helopeltis theivora* Waterhouse	Tea mosquito bug	Leaf sucker	Miridae: Hem.
16.	*Empoasca flavescens* Fabr.	Tea greenfly/tea jassid	Leaf sucker	Jassidae: Hem.
17.	*Toxoptera aurantii* Boyer de Fons.	Tea aphid	Leaf sucker	Aphididae: Hem.
18.	*Scirtothrips dorsalis* Hood	Yellow tea thrips	Leaf and bud sucker	Thripidae: Thy.
19.	Mycterothrips (Teaniothrips) setiventris Bagnall	Common thrips	Leaf and bud sucker	Thripidae: Thy.
20.	*Oligonychus coffeae* Nietner	Red spider mite	Leaf sucker	Tetranychidae: Aca.
21	*Brevipalpus phoenicis* Geijskes	Scarlet mite	Leaf sucker	Tenuipalpidae: Aca.

3. INSECT PESTS OF TEA AND MANAGEMENT PROBLEMS

Tea is produced from the young foliage, i.e. young leaves and a bud, and foliage production is increased by seasonal pruning which enhances the leaf cover. The major pests of the crop are those associated with the young foliage. The most important insect pest groups are the folivores (chewing) and sap suckers of the young tender leaves, buds and stems (sucking pest), which damage the most economic part of tea plant that is processed in tea industry for making the tea. These pests cause substantial loss in yield to the tea industry.

In India, different management practices are followed to protect the tea crop against different insect pest groups. Most of the plantations are managed conventionally i.e. using different organosynthetic insecticides, whereas some organic plantations use plant- and animal-based herbal and microbial insecticides. In conventional tea plantations, organo-synthetic insecticides of different functional groups such as organochlorines, organophosphates, synthetic pyrethroids (SPs) and neonicotinoids (NNs) are regularly used throughout the year to control the invasion of different insect pest groups (sucking, folivores and others) [8]. The use of insecticide is cost-effective to planters and a major concern for the environmental degradation due to contamination as well as in resurgence of primary pests [6], outbreak of secondary pests [9], development of insecticide resistance [10, 11], including undesirable residues in made tea [12]. Regular spraying of insecticides leads to the development of higher level of tolerance or resistance to insecticides in many insects [11, 13].

From the early forties onwards, dichlorodiphenyltrichloroethane (DDT) (organochlorine) was regularly used to manage the infestation of H. theivora, the major sucking pest in Northeast India [14]. In 1968, endosulfan (cyclodiene: organochlorine) was introduced in the tea plantations of the Dooars region of West Bengal, India, in the form of thiodan 35 EC [15]. Currently, in different conventional tea plantations of tea-growing regions of India, cypermethrin, deltamethrin, quinalphos, monocrotophos, chlorpyriphos, imidacloprid, etc. are extensively used during cropping season to control insect pests [16–18].

Recently, a number of insecticides have been found to be ineffective in controlling the insect and mite pests in different tea-growing regions of India [8]. The development of resistance to different classes of insecticides is one of the causes for persistence and resurgence of insect pests on tea crop [8, 19–21]. A major concern in managing the major insect pests of tea is its high potential to develop resistance rapidly to regularly used insecticides [11]. Continuous and repeated exposure to different classes of insecticides for many years, in addition to their high reproductive potential, short life cycle and numerous annual generations, has limited the management of major pests of tea [11]. Recently, there are reports on the development of resistance to many commonly used synthetic insecticides and consequent failure in controlling many tea pests [10, 22–26]. Such failures are already known in case of organochlorines (OCs), organophosphorus (OP) and synthetic pyrethroid (SP) insecticides and more recently for the newer compound such as neonicotinoids [19–21, 26]. The development of resistance in H. theivora populations to different classes of insecticides has been in the range of 1.47–62.99-fold for males and 1.25–62.82-fold for females in Northeast India [19]. Relative toxicity to commonly used insecticides has been observed to vary in H. theivora populations from Jorhat, Assam [20], Darjeeling [27], and from sub-Himalayan Dooars region of Northeast India [28]. LC50 values of insecticides, when compared with the field dose against H. theivora recommended by TRA (Tea Research Association, Tocklai, Assam, India), revealed a pronounced shift in the level of susceptibility of H. theivora to all insecticides except acephate [20–21].

For the management of other sucking insects such as yellow tea thrips, S. dorsalis, insecticides are also used in conventional tea plantations. In tea ecosystem, control failure and the development of biochemical resistance have been reported in S. dorsalis [11, 24]. In India, S. dorsalis populations have developed a high degree of resistance to a range of organochlorine (DDT, BHC and endosulfan), organophosphate (acephate, dimethoate, phosalone, methyl-O- demeton and triazophos) and carbamate insecticide (carbaryl) in chili ecosystem [31]. S. dorsalis has also developed a high degree of resistance to various insecticides, viz. monocrotophos, acephate, dimethoate, phosalone, carbaryl and triazophos [32]. Recently, several insecticides have been tested on S. dorsalis in chili ecosystem in USA and found limited success with chlorfenpyr, spinosad and imidacloprid [33, 34]. The performance of novaluron, abamectin, spiromesifen, cyfluthrin, methiocarb and azadirachtin failed to provide effective control of this pest [35].

Similarly, in another emerging sucking insect pest of tea, tea greenfly, E. flavescens, repeated management failure and biochemical insecticide resistance in tea ecosystem from Northeast India have been reported [11, 24]. In China, chemical insecticides including fenvalerate, cyfluthin, cypermethrin and imidacloprid are sprayed to control the leafhoppers as frequently as seven times annually or even more frequently [36–37]. A high level of resistance against many insecticides has been reported in related species, E. vitis [38]. The resistance to thiamethoxam was highest and to cyperemethrin was lowest in E. vitis. Recently, in Fujian province of China, a regional diversity of resistance to eight insecticides in E. vitis has been reported in tea ecosystem with higher resistance level to bifenthrin, acetamiprid, imidacloprid, cartap and chlorfenapyr [39].

A high level of insecticide resistance in folivores, such as black hairy caterpillar, bunch caterpillar, looper pest complex (Hyposidra talaca, H. infixaria, Buzura suppressaria, Eturesia magnifica) and in termite of tea ecosystem, has been reported with reduced susceptibility against different insecticides.

Detoxification of insecticides is an important toxicokinetic mechanism for insect pests to tolerate regularly applied insecticides [8, 40–42]. Susceptibility levels against insecticides change mainly due to metabolic detoxification of the insecticides through the induction of some detoxifying enzymes under the stress of different management practices [43–45].

Generally, three principal enzymes, general esterases (GEs), glutathione S-transferases (GSTs) and cytochrome P450-mediated monooxygenases (CYP450s), are involved in the process of metabolic detoxification of insecticides [41]. Estimation of the activities of these metabolic defence-related detoxifying enzymes gives information on the level of tolerance/resistance of the insect pest population to insecticides and is a useful tool in monitoring the tolerance/ resistance to insecticides at population level of the pest. The early detection of metabolic threats related to tolerance/resistance to insecticides in pest specimens is of crucial importance for devising pest control techniques that would minimize the development of tolerant/resistant forms and prevent any undesirable wastage of insecticide, money and manpower.

4. INSECTICIDE RESISTANCE MECHANISMS IN INSECT PESTS OF TEA

Insects come across with numerous toxins as they go through their life cycle. Some of these toxins are naturally produced by plants (plant allelochemical) and others by humans (synthetic insecticides). To protect themselves against the natural toxins, insects have evolved various detoxification mechanisms [41] These mechanisms also cross-protect insect pests when they are exposed to synthetic insecticides [25]. Herbivorous insect groups (agricultural pests) are significantly more diverse than their non-herbivorous sister groups [46]. The role of plant in promoting diversification in insects has occurred through co-evolutionary 'defence strategies' among them [47]. This diversification could also have been a result of insects 'tracking' plant phylogenies, with minor chemical changes in plants allowing the evolving populations of insects to change and speciate accordingly, which probably has occurred long after chemical changes in plants [48]. Evolution to herbivory preceded via mixed feeding on reproductive parts or spores, dead tissues of plants and animals and fungi. This progression implies that omnivory preceded generalized herbivory and the evolution of specialization on specific plant taxa was a later accomplishment [49].

Among sucking insect herbivores, the actual food used, i.e. digested whole tissue particularly parenchyma (as in Helopeltis sp.), cell content (thrips) and phloem flow (Empoasca sp.), influences both the feeding mechanism and feeding behaviour [48]. While the chewing insects (looper caterpillar complex) cause extensive damage, the sucking insects cause modest to barely perceptible damage. However, sucking insects, particularly phloem and digested tissue feeder,impose an additional challenge to the plants as they deplete photosynthates, act as vector of viruses and introduce chemical and protein effectors that alter plant defence mechanisms (signalling) and development [50]. When these attributes are combined with a broad host range, breeding strategies that promote invasiveness, highly evolved feeding strategies, the ability to adapt to a wide range plant habitats and the emergence of insecticide resistance, it is not surprising that sucking insects cause heavy losses in agriculture and horticulture [51].

Insecticide resistance is a genetic change in response to selection pressure of toxicants that impair pest control in the field [52]. Insecticide resistance does not occur unless a structural genetic change occurs that is heritable. Therefore, insecticide resistance is an evolutionary phenomenon that results under the selection pressure of a new toxicant in the environment [8].Thus insecticide resistance is different from insecticide tolerance. Insecticide tolerance is the natural ability of a population to withstand the toxic effect of a particular insecticide. It can develop within one generation as a result of physiological adaptation, i.e. induction of xenobiotic detoxifying enzymes. Hence, variation within a population may include individuals with genetic traits that make them better adapted to survive in exposure to an insecticide. If these individuals survive the insecticide exposure, then the tolerance traits can be passed on to the next generation, thereby enriching the gene pool with those genes. The mechanisms of development of insecticide tolerance can be divided into four levels:

Altered behaviour - avoidance of contact with the insecticide

Development of barrier tissues - reduced penetration of the insecticide though the integument
Enhanced detoxification - higher metabolism of the insecticide

Alteration of receptor - at the target site for the insecticide.

The first level, at which insecticide tolerance can develop, is when the insect encounters an insecticide. An altered behaviour helps the insect to avoid coming into contact with the insecticide. Once the insect comes in contact with the insecticide, a reduced and delayed penetration through the cuticle will reduce the effect of the insecticide at the target site; this is yet another level of resistance. Within the insect's body, the insecticide may be enzymatically metabolized and thereby inactivated. At the third level of resistance mechanism, three systems of xenobiotic detoxification enzymes operate: esterases, glutathione S-transferases and cytochrome P450-dependent monooxygenases. The increased activity of one of these enzyme systems in metabolizing insecticides will result in insecticide tolerance. Alterations at the target site for the insecticide are the last level of insecticide resistance mechanisms. Different classes of insecticides bind to specific target sites and reduced binding at the target site, or increased number of target site molecules may confer insecticide resistance.

4.1 Behavioural resistance

Behavioural resistance mechanisms are the least studied resistance mechanisms in insects, but this is not to say that behavioural resistance is the least significant. Behavioural resistance can be defined as 'evolved behaviours that will reduce an insect's exposure to toxic compounds or that allows an insect to survive in what would otherwise be a toxic and fatal environment' [53]. Behavioural resistance has been observed in more than 30 species of insects [53]. Avoidance is the first step in the evolution of behavioural resistance [54]. In H. theivora, this kind of resistance has been seen [55]. H. theivora shows a different egg-laying strategy to avoid insecticide exposure. Even E. flavescens avoids exposure to direct sunlight and therefore prefers to stay on the underside of the tea leaves. This behaviour cross-protects it from the direct insecticide exposure in conventional tea plantations during spraying [8]. The same has been found in S. dorsalis which resides inside the leaf bud during development and underside of the leaf during adult stage, thereby avoiding direct exposure to insecticides [8].

4.2 Reduced penetration

Reduced penetration of insecticides through barrier tissues of insects is another way in which an insect can modify the effective dose of insecticide at the target site. The mechanism may not prevent the insecticide from eventually entering the insect, but it can reduce the rate at which the insecticide reaches the target site. Reduced penetration has been shown to function as a resistance mechanism to many different insecticides, and, by the nature of this mechanism, cross-resistance is often found [56]. The rate of penetration of insecticides through the insect cuticle or other barriers (peritrophic membrane) depends on the physicochemical properties of the insecticide and the barrier. A reduced penetration contributes to DDT resistance in the tobacco budworm, Heliothis virescens F. (Lepidoptera: Noctuidae). DDT-resistant larvae had an altered composition of the cuticle. The protein and lipid contents are greater in the cuticle of resistant larvae and, furthermore, the cuticle of the resistant larvae probably had a higher degree of sclerotization [57]. In M. domestica, two resistant strains, with reduced penetration as one of the resistance mechanisms, also showed increased cuticular lipid content; more total lipids, mono-glycerides, fatty acids, sterols and phospholipids were present in the resistant strains compared to a susceptible strain [58]. Reduced penetration has been documented as a resistance mechanism only at the level of the insect cuticle, but any biological membrane may serve as a barrier and thereby give resistance [59]. As a single resistance mechanism it usually only confers low levels (less than threefold) of resistance [59]. Reduced penetration has been shown to function as a resistance mechanism to many different insecticides, including insecticides of the three major classes, OPs, carbamates and pyrethroids. Reduced penetration of OPs through the cuticular barrier has been reported, for example, diazinon in M. domestica [60], azinphos-methyl in the pear psylla, Psylla pyricola Foester (Hemiptera: Psyllidae) [61] and profenofos in H. virescens [62]. However, by slowing the penetration rate of insecticides, this mechanism reduces the risk that the insects' detoxification systems become overloaded, and the dose of insecticide reaches to a lethal level at the target site. In female H. theivora, a higher level of body lipid has been found which effectively reduces the penetration of insecticide to the target site [63]. No studies on resistance due to reduced penetration in E. flavescens and S. dorsalis or any other tea pests have been reported to date. The studies had shown that when different resistance mechanisms are combined in the same individuals, a synergistic effect, resulting in a high level of resistance, may arise [64, 65]. Therefore, even a small degree of reduced penetration can contribute significantly to the overall insecticide resistance of the insect pests.

4.3 Metabolic detoxification

Metabolic detoxification of insecticides is an important toxicokinetic mechanism for insects to tolerate the toxic effects of insecticides. Generally, lipophilic (hydrophobic) insecticides are rapidly detoxified. Organophosphates, organochlorines, carbamates and pyrethroids are lipophilic compounds, and detoxification enzymes transform these insecticides to more hydrophilic and less biologically active compounds so that can be eliminated more easily by excretion. Increased detoxification of insecticides has often been reported in many resistant populations [40]. Three enzyme systems are generally recognized as the major detoxification systems involved in insecticide resistance in insects. These are carboxylesterases, cytochrome P450-dependent monooxygenases and glutathione S-transferases [40, 41].

4.4 Alteration at the target site for insecticide (target site insensitivity)

The biochemical sites for insecticide action differ for different insecticides and are a potential field of research for developing insecticides, which can act specifically or more efficiently on insect biochemical sites compared to mammals. The target site receptor for action of organo-phosphates, carbamates, organochlorines and pyrethroids is in the nervous system. The enzyme acetylcholinesterases (AChEs) (EC 3.1.1.7) are the target sites for organophosphates and carbamates, and voltage-gated sodium channel of the nerve membrane is the target of pyrethroids and DDT. Neurotoxic insecticides such as cyclodienes (e.g. dieldrin and endosul-fan), y-HCH (lindane) and fipronil target gamma-aminobutyric acid (GABA)-receptor [68, 69] and nicotinyl insecticides (imidacloprid and nicotine) target the nicotinic acetylcholine receptor (nACHR) [70]. Alteration at the target site, to less a sensitive target for neurotoxic insecticides, is an important toxicodynamic resistance mechanism in insects [71].

In insects, the potent inhibitors of AChE are organophosphates and carbamates. These compounds inhibit the activity of AChE by forming a stable covalent intermediate, preventing the enzyme to hydrolyse acetylcholine. An accumulation of acetylcholine keeps the ion channel of the receptor permanently open, which eventually kills insect. OPs and carbamates are quasi- irreversible inhibitors of AChE. The organophosphates and carbamates phosphorylate and carbamylate the active site serine of AChE, respectively [72]. Generally, the reactivation time of phosphorylated or carbamylated AChE is long. However, the half-lives of reactivation vary considerably, from minutes to several days, depending on the compound interacting with AChE [73]. Carbamylated AChE generally reactivates faster than phosphorylated AChE. Reduced sensitivity of AChE to inhibition by OPs and carbamates is an important resistance mechanism in insects and is often referred to as altered or insensitive AChE [74]. The presence of insensitive AChE conferring resistance was first noticed in OP-resistant mites, Tetranychus urticae Koch (Acari: Tetranycidae) [74] and also found in several insect populations resistant to these compounds [75–77]. Insecticide susceptible and resistant insect pest populations differ in the level of AChE activity [78–80]. A higher level of AChE activity has been reported in H. theivora sampled from conventional tea plantations than from organic garden indicating the presence of resistance to insecticides in conventional tea ecosystems [81, 82]. There is no such report on S. dorsalis and E. flavescens in conventional tea ecosystems to date.

5. MAJOR METABOLIC DETOXIFYING ENZYMES IN INSECTS

Carboxylesterases, glutathione S-transferases and cytochrome P450-mediated monooxygenases are the three principal enzymes that facilitate the insects to metabolize different kind of toxins. These large enzyme families contain multiple forms with overlapping substrate specificities. Knowledge of insecticide detoxification helps in understanding the mechanism of insecticide resistance, hence the development of a sound resistance management strategy. Detoxification can be divided into phase I (primary) and phase II (secondary) processes (Figure 1).

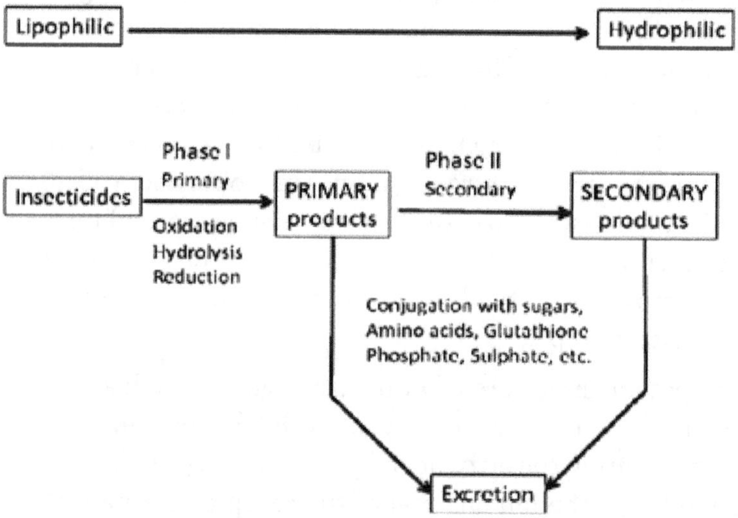

Figure 1. Insecticide detoxification pathways.

Phase I reactions consist of oxidation, hydrolysis and reduction. The phase I metabolites are sometimes polar enough to be excreted but are usually further converted by phase II reactions. In phase II reactions, the polar products are conjugated with a variety of endogenous compounds such as sugars, sulphate, phosphate, amino acids or glutathione and subsequently excreted. Phase I reactions are usually responsible for decreasing the biological activity of toxins, and therefore the enzymes involved are rate limiting with respect to toxicity. The most important function of biotransformation is to decrease the lipophilicity of insecticides, so that they can be excreted quickly [83].

5.1 Phase I reactions

5.1.1 Cytochrome P450 monooxygenases (E.C. 1.14.-.-)

Oxidation is considered the most important among phase I reactions. The oxidative reactions are carried out mainly by a group of enzymes called cytochrome P450 monooxygenases [also known as mixed function oxidases (MFO) or polysubstrate monooxygenases (PSMO), microsomal oxidase, P450 enzymes]. Cytochrome P450, or CYP genes, constitutes one of the largest family of genes, with representatives in virtually all living organisms, from bacteria to protists, plants, fungi and animals [84].

In insects, P450 monooxygenases are involved in many processes including roles in the metabolism of plant allelochemicals by herbivores and in detoxification of insecticides. The human genome carries about 57 CYP genes, and insect genomes can carry from 36 CYP genes in the body louse Pediculus humanus [85] to 170 in a mosquito [86]. Each P450 protein is the product of a distinct CYP gene, and P450 diversity is the result of successive gene (or genome) duplications followed by sequence divergence [84].

The typically 45–55-kDa P450 proteins are heme-thiolate enzymes. Their essential common feature is the absorbance peak near 450 nm of their FeII–CO complex for which they are named [87]. P450 enzymes are best known for their monooxygenase role, catalysing the transfer of one atom of molecular oxygen to a substrate and reducing the other to water. The simple stoichiometry commonly describes the monooxygenase or mixed function oxidase reaction of P450:

$$RH+O_2+NADPH+H^+ ® ROH+H_2O+NADP^+$$

However, oxygen atom transfer is not the only catalytic function of P450 enzymes. They also show activities such as oxidases, reductases, desaturases, isomerases, etc. and collectively are known to catalyse at least 60 chemically distinct reactions (Table 2).

The first insect P450 cloned and sequenced was CYP6A1 from Musca domestica in 1989 [88]. The P450 gene complement (CYPome) size of an insect genome is not a definite number [89]. Insects can survive with small CYPomes even in toxic environments. The human body louse Pediculus humanus, with 36 CYP genes, is known to become highly resistant to many classes of insecticides [85], and the honeybee, with just 46 CYP genes [90], is not more sensitive than other species in a comparison to the toxicity of 62 insecticides [91]. The main driver of CYPome evolution is of course gene duplication, followed by divergence (by neofunctionalization or subfunctionalization) or death (pseudogenization or deletion) [84].

Table 2. Enzymatic reactions catalysed by insect P450 enzymes (adapted from Feyereisen, 2005).

Reaction catalysed	P450
Oxidase activity O2 to H2O, H2O2, O2 -	CYP6A1 (and probably most P450 enzymes)
Aliphatic hydroxylation C–H hydroxylation	CYP4C7, CYP6A1, CYP6A2, CYP6A8, CYP6G1, CYP6M2, CYP6CM1vQ, CYP9T2, CYP12A1, CYP18A1, CYP302A1 CYP306A1, CYP312A1, CYP314A1, CYP315A1
O-dealkylation	CYP6A1, CYP6D1, CYP6A5, CYP6B4, CYP6B17, CYP6B21, CYP6G1, CYP6Z2, CYP6CM1vQ, CYP9A12, CYP9A14, CYP12A1, CYP321A1,
Dehalogenation	CYP6G1
Epoxidation	CYP6A1, CYP6A2, CYP6B8, CYP6B27, CYP6AB3, CYP6 CYP6AB11, CYP9E1, CYP12A1, CYP15A1, CYP321A1
Aromatic hydroxylation	CYP6D1, CYP6G1, CYP6M2
Heteroatom oxidation and dealkylation Phosphorothioate ester oxidation	CYP6A1, CYP6A2, CYP6D1, CYP12A1
N-dealkylation	CYP6A5, CYP12A1
N-oxidation	+(Nicotine)
S-oxidation	+(Phorate)
Aldehyde oxidation	CYP18A1
Complex and atypical reactions	
Cyanogenic glucoside biosynthesis:	CYP40
Val/Ile to oximes	5A2
Oximes to cyanohydrins	CYP6M2
Aryl ether cleavage	CYP33
Carbon–carbon cleavage	2A3
Decarbonylation with C–C cleavage	M2
Aromatization	+?(Sterols, ecdysteroid)
Reduction	CYP4G1
Endoperoxide isomerisation	+(Defensive steroids) - -

5.1.2 Carboxylesterases (EC 3.1.1.1)

Carboxylesterase or esterase is a collective term for the enzymes that hydrolyse carboxylic esters [92]. Classification of these enzymes is difficult because of their overlapping substrate specificity [93]. However, the esterase classification of Aldridge [94] is generally recognized. According to that classification, esterases inhibited by paraoxon in a progressive and temperature-dependent manner are called B-esterases and those which are not inhibited are A- esterases [94]. Some A-esterases can hydrolyse OPs, through an acylated cysteine in their active site, and are termed phosphoric triester hydrolases (EC 3.1.8.) [95, 96]. The term carboxylesterase is now mainly attributed to B-esterases [95, 96]. These enzymes have an active site serine residue, hence the terms B-esterase and serine hydrolase are synonymous. Insecticides such as organophosphates, carbamates, pyrethroids and some juvenoids, which contain ester linkages, are susceptible to hydrolysis. Esterases are hydrolases that split ester compounds by the addition of water to yield an acid and alcohol.

Esterases that metabolize organophosphates can be divided into three groups: A-esterases which are not inhibited by organophosphates but hydrolyse them; B-esterases, which are susceptible to organophosphate inhibition; and C-esterases which are uninhibited by organo-phosphates and do not degrade them [41].

There are two types of esterases that are important in metabolizing insecticides, namely, carboxylesterases and phosphatases (also called phosphorotriester hydrolases or phosphor- triesterases). Carboxyl esterases, which are B-esterases, play a significant role in degrading organophosphates, carbamates, pyrethroids, and some juvenoids in insects. The best example is malathion hydrolysis, which yields both α- and β-monoacids and ethanol [41]. Phosphatases are A-esterases that detoxify many organophosphorous insecticides especially phosphates in insects. In houseflies, paraoxon can be hydrolysed to diethyl phosphoric acid and p-nitrophenol. Phosphatases also hydrolyse the alkyl groups of organophosphates. Paraoxon is hydrolysed by the enzyme in houseflies. Several amides containing organophosphorous insecticides such as dimethoate and acephate have been shown to be hydrolysed by carboxylamidases to their corresponding carboxylic acid derivatives [41].

5.2 Phase II reactions

Phase I reactions with xenobiotics result in the addition of functional groups such as hydroxyl, carboxyl and epoxide. These phase I products can further undergo conjugation reactions with endogenous molecules. These conjugations are called phase II reactions. The endogenous molecules include sugars, amino acids, glutathione, phosphate and sulphate. Conjugation products are usually more polar, less toxic and more readily excreted than their parent compounds. Thus, the process with only a few exceptions results in detoxifications.

Three types of conjugation reactions occur in insects. Type I requires an activated conjugating agent that then combines with the substrate to form the conjugated product. Type II involves the activation of the substrate to form an activated donor that then combines with an endogenous molecule to yield a conjugated product. In Type III, conjugation can proceed directly between the substrate and the conjugating agent without involving activation. Thus, Type I and II require information of high-energy intermediates before the conjugation reactions proceed. The chemical groups required for Type I are –OH, NH2, COOH and SH (glucose conjugation, sulphate conjugation and phosphate conjugation); for Type II COOH (amino acid conjugation); and for Type III, halogens, alkenes, NO2, epoxides, ethers and esters (glutathione conjugation).

5.2.1 Glutathione S-transferases (EC 2.5.1.18.)

Glutathione conjugations are performed by a group of multifunctional enzymes known as glutathione S-transferases and are involved in detoxification mechanisms of many molecules. GSTs are involved in the transport of physiologically important lipophilic compounds. These enzymes catalyse reactions in which the sulphur atom of glutathione provides electron for nucleophilic attack on a second electrophilic substrate; the latter can be endogenous natural substrates such as epoxides, organic hydroperoxides, or activated alkenals resulting from oxidative metabolism. These enzymes catalyse the conjugation of reduced glutathione (GSH) with electrophilic substrates. Glutathione S-transferases perform a variety of reactions including:

The S-alkylation of GSH by alkyl halides and related compounds. The replacement of labile aryl halogen or nitro groups by GSH. The replacement of labile arakyl halogen and ester groups by GSH. The addition of GSH to various epoxides.

The addition of GSH to α -, β-unsaturated compounds including aldehydes, ketones, lactones, nitriles and nitro compounds.

The O-alkyl and O-aryl conjugation of phosphorothiotes and phosphates with GSH.

The glutathione conjugate is subsequently transformed to mercapturic acid through the stepwise loss of glutamic acid and glycine to a cysteine conjugate, which is finally acetylated before excretion. Because of their broad substrate specificities, glutathione S-transferases are responsible for the detoxification of numerous xenobiotics [97]. More than 40 GST genes have been identified in insects [98, 99]. Mammalian GSTs have been classified into eight cytosolic classes (alpha, mu, pi, theta, sigma, zeta, kappa and omega) and a microsomal class on the basis of their amino acid sequence, immunological properties and substrate specificity. Each class shares 40% or higher amino acid in common. The classification of insect GST is not clear. The majority of insect GSTs do not belong to mammalian classes. Insect glutathione S- transferases consist of two subunits (homodimers and heterodimers) of molecular weight between 19 and 35 kD. Two classes of insect GSTs (Class I and

Class II) were reported [100], which have been referred to as the Delta and Sigma class, respectively. Recently, a new class of insect GSTs, referred to as Epsilon has been described in several species of insects including Anopheles gambiae [101]. Purified cytosolic and microsomal glutathione S-transferase isozymes from fall armyworm larvae possessed cumenehydrperoxide peroxidase [102]. A Delta class GST purified from German cockroaches also showed high peroxidase activity [103]. The name of each GST is composed of the initials of the species scientific name, followed by the acronym GST, a capital letter to designate the class name and an Arabic number for the individual protein, such as AgGSTD2.

Glutathione S-transferases are important in the metabolism of organophosphorous insecticides resulting in detoxification [99, 104]. For example, methyl parathion is dealkylated by glutathione S-tranferases to form desmethyl parathion and methyl glutathione [41]. On the contrary, parathion can be de-arylated by glutathione S-transferases to produce diethyl phosphorothioic acid and S-(p-nitrophenyl) glutathione [41]. Interestingly, a glutathione S- transferase isozyme from the housefly exhibits DDT-dehydrochlorinase activity, showing that DDT-dehydrochlorinase (DDTase) is one of the glutathione S-tranferases [104]. DDT-dehydrochlorinase converts DDT to DDE, resulting in detoxification [41].

6. DETOXIFYING ENZYMES AND INSECTICIDE METABOLISM

The metabolism of insecticides by P450 enzymes is very often a key factor in determining toxicity to insects and to non-target species. The importance of monooxygenases in insecticide resistance became evident in the early 1960s, when it was shown that resistance to carbaryl could be abolished by the P450 inhibitor sesame [106]. Additional evidence of monooxygenase- based resistance quickly amassed [107, 108]. Monooxygenase-mediated detoxification is frequently found as a major mechanism of resistance, and unlike target site resistance, detoxification has the potential to confer cross-resistance to toxins independent of their target sites [109, 110]. Most cases of monooxygenase-mediated resistance result from an increase in detoxification (Table 3). However, in cases where the parent insecticide must undergo monooxygenase-mediated bioactivation, as is the case for many organophosphates, it is also possible that resistance could be achieved through decreased activation [111]. Although this has been reported once, it does not appear to be a common mechanism of resistance. This may explain why esterases are relatively more common than monooxygenases in resistance to some organophosphates [110, 112]. The classical example is probably the metabolism of phosphorothioate insecticides. In many cases, the active ingredients of organophosphorus insecticides are phosphorothioate (P=S) compounds (also known as phosphorothionates), whereas the molecule active at the acetylcholinesterase target site is the corresponding phosphate (P=O) (Figure 2).

P450 enzymes that metabolize OPs can metabolize other insecticides as well, and this sometimes leads to potentially useful interactions. Thus, enhanced detoxification of dicofol in spider mites can lead to enhanced chlorpyriphos activation, and hence negative cross- resistance [113]. Similarly, permethrin resistance in horn flies is suppressible by piperonyl butoxide, and negatively related to diazinon toxicity [114]. In H. armigera populations from West Africa, triazophos shows negative cross-resistance with pyrethroids, and in this case, the synergism shown by the OP towards the pyrethroid appears due to an enhanced activation to the oxon form [115]. Organophosphorus compounds (disulfoton and fenthion) are also activated by thioether oxidation (formation of sulphoxide and sulfone). The metabolism of pyrethroid by P450 enzymes is well studied in insects. Hydroxylations and further metabolism make pyrethroid metabolism and has been noticed for the single enantiomer of deltamethrin (Figure 3).

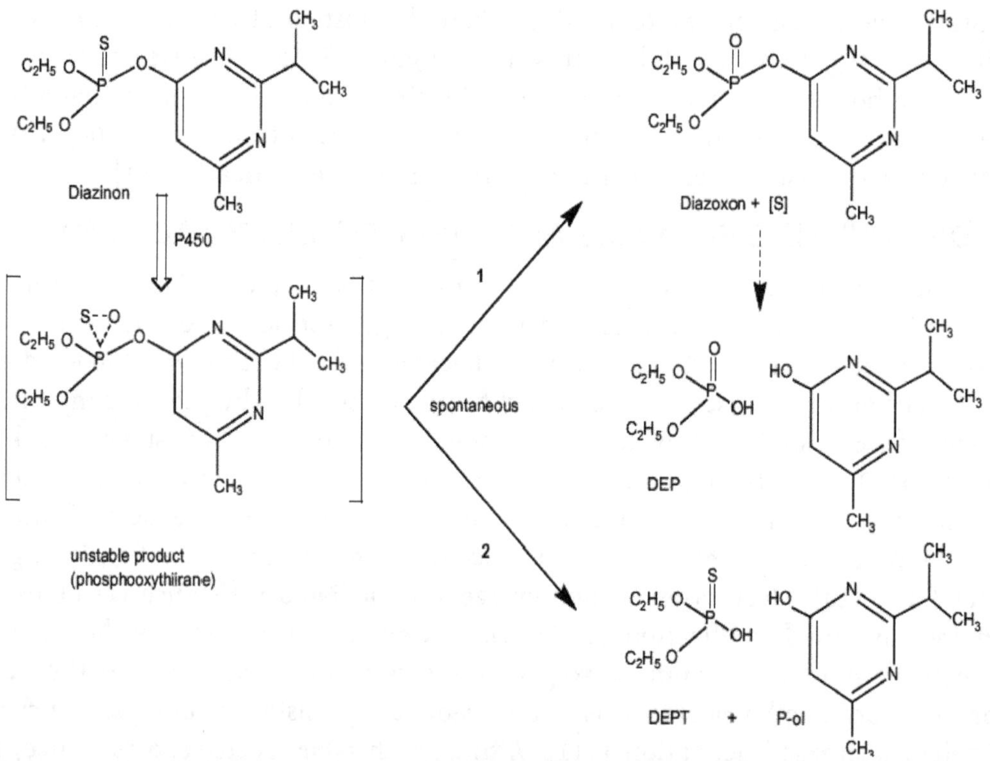

Figure 2. Metabolism of diazinon by cytochrome P450. Following an insertion of oxygen into the substrate, a reactive intermediate collapses (1) by desulphuration or (2) by cleavage of the ester linkage. DEP, diethylphosphate; DEPT, diethylphosphorothioate; P-ol, 2-isopropoxy-4-methyl-6-hydroxypyrimidine; [S], reactive form of sulphur released during the reaction. Adopted from Feyereisen, 2012.

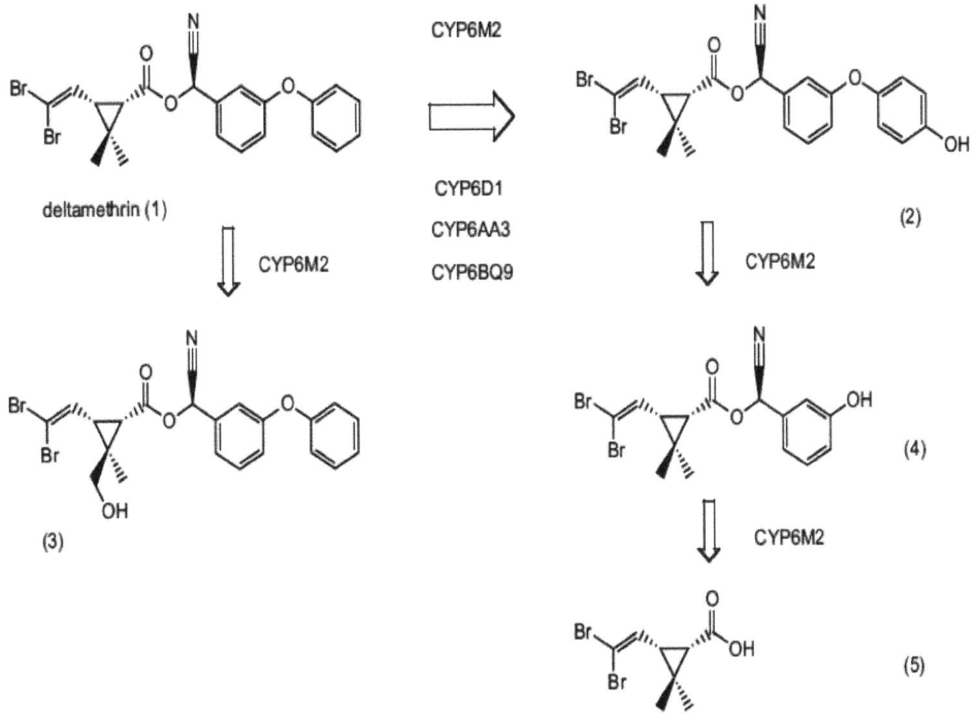

Figure 3. Metabolism of deltamethrin by insect P450 enzymes: (1) deltamethrin; (2) 4′ hydroxydeltamethrin; (3) trans- hydroxymethyl-deltamethrin; (4) cyano (3-hydroxyphenyl) methyl deltamethrate; (5) deltamethric acid. Adopted from Feyereisen, 2012.

The currently banned cyclodiene insecticides aldrin, heptachlor and isodrin are epoxidized by P450 enzymes to the environmentally stable toxic epoxides dieldrin, heptachlor epoxide and endrin, respectively [116]. Recombinant CYP6A1, -A2, -A8, -B8 and -B27; CYP12A1; and CYP321A1 can catalyse these epoxidations. Examples of pro-insecticide metabolism include the activation of chlorfenapyr by N-dealkylation [117] and diafenthiuron by S-oxidation [118]. In each case, the insect P450-dependent activation is a key in the selective toxicity of these pro-insecticides that target mitochondrial respiration. Recombinant housefly CYP6A1 catalyses the activation of chlorfenapyr (Figure 4).

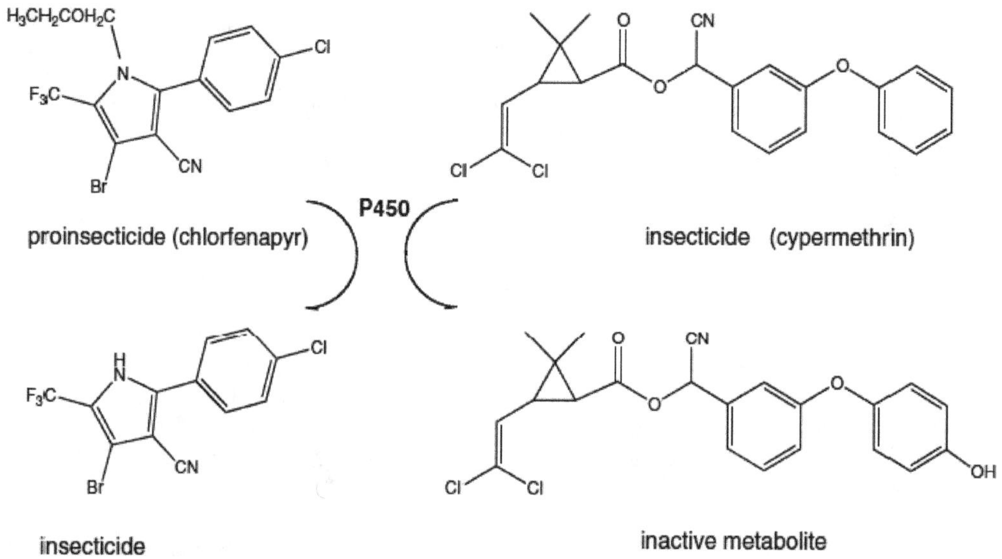

Figure 4. Chlorfenapyr and cypermethrin metabolism. The same P450 in Heliothis virescens probably activates the pyrrole and inactivates the pyrethroid, resulting in negative cross-resistance. Adopted from Feyereisen, 2012.

In H. virescens, the toxicity of chlorfenapyr is negatively correlated with cypermethrin toxicity [119]. The metabolism of imidacloprid is also of interest, particularly in relation to resistance. Piperonylbutoxide can synergize the toxicity of imidacloprid, and two P450 enzymes, CYP6G1 of D. melanogaster and CYP-6CM1vQ of Bemisia tabaci, have been shown to metabolize this neonicotinoid [120, 121] (Figure 5). Hydroxylations at the 4 and 5 positions can lead to the olefinic metabolite or to the dihydroxylated metabolite. In the whitefly, the 5-hydroxy metabolite is not toxic, but the 4-hydroxy metabolite is as toxic as the parent compound, so region selectivity may be of importance.

Despite the continuous use of insecticides, there are repeated failures in controlling the sucking insect pest species in recent years [11, 21] in different conventional tea plantations of Terai, the Dooars and Darjeeling foothill regions. Such a failure occurs due to changes in the susceptibility level of the pest species to the applied insecticides. Susceptibility level changes mainly due to metabolic detoxification of the insecticides through higher level of activity of some insecticide detoxifying enzymes under the stress of different management practices [8, 10, 11, 22]. In another mirid pest, Lygus lineolaris, metabolic resistance to insecticides due to enhanced level of activity has been reported by many authors [177, 178].

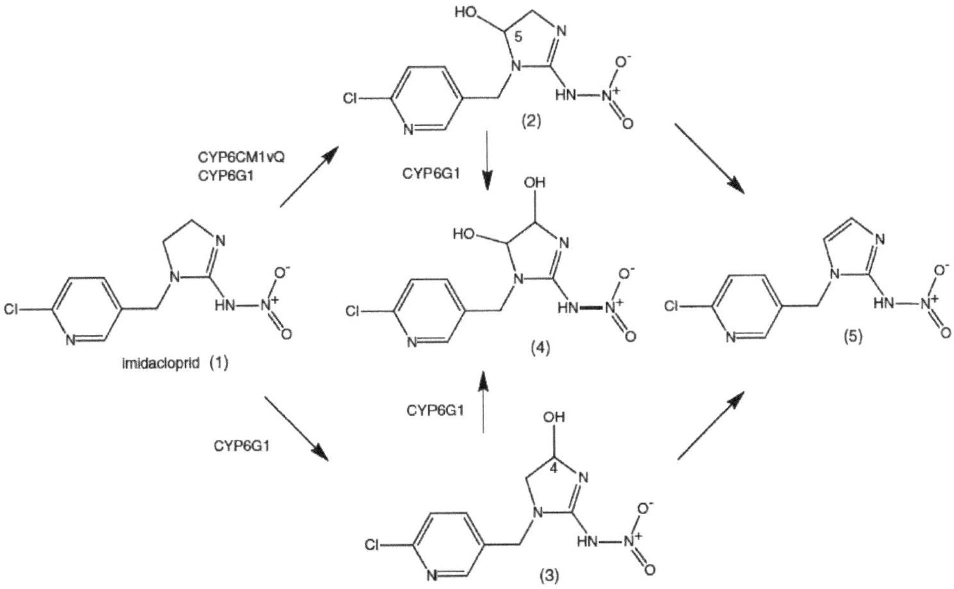

Figure 5. Metabolism of imidacloprid by insect P450 enzymes: (1) imidacloprid; (2) 5-hydroxyimidacloprid; (3) 4-hydroxyimidacloprid; (4) dihydroxyimidacloprid; (5) Non-enzymatically derived dehydroimidacloprid. Adopted from Feyereisen, 2005.

Table 3. Over-expressed CYP genes in insecticide-resistant strains.

Species	P450 Over-expressed	Resistance pattern
Musca domestica	CYP6A1	OP, carbametes [122]
		IGR [123, 124]
	CYP6A5v2, CYP6A36	Pyrethroids [125–127]
	CYP6D1, CYP6D3	Pyrethroids [128–130]
	CYP6D1	Pyrethroids [131]
	CYP6D1, CYP6D3v2	Pyrethroids [132]
	CYP6A24	Pyrethroids [133]
	CYP12A1	Pyrethroids [134]
Drosophila melanogaster	CYP6A2	DDT, malathion [135–137]
	CYP6A8	Malathion [137, 138]
		Malathion [137]
		DDT [139]
	CYP6G1	DDT [140, 141]
		DDT [142]
		Lufenuron, propoxur [141

Species	P450 Over-expressed	Resistance pattern
		Imidacloprid [141]
		Imidacloprid [142]
		DDT [143]
		Diazinon [144]
	CYP12D1/2	DDT [143]
		DDT [139]
	CYP12A4	Lufenuron [145]
Drosophila simultans	CYP6G1	DDT, imidacloprid, Malathion [139]
Anopheles gambiae	CYP6Z1	Pyrethroids [146]
	CYP325A3	Pyrethroids [147]
	CYP6M2, CYP6P3	Pyrethroids [148]
	CYP6P3	Permethrin [149]
	CYP6M2, CYP6Z2	Permethrin [150]
	CYP4C27, CYP4H15	DDT [151]
	CYP6Z1,2, CYP12F1, CYP314A1	DDT [147]
A. stephensi	CYP325C1	Pyrethroids [152]
A. funetus	CYP6P4, CYP6P9	Pyrethroids [153, 154]
Aedesaegypti	CYP9J10,27,32	Pyrethroids [155]
Culex pipien squinquefasciatus	CYP9M10	Permethrin [156]
	CYP6F1	Permethrin [157]
	CYP4H34, CYP6Z10, CYP9M10	Permethrin [158]
	CYP4H21, H22, H23, CYP4J4, CYP4J6	Deltamethrin [159]
Heliothis virescens	CYP9A1	Thiodicarb [160]
Helicoverpa zea	CYP6B8,B9	Cypermethrin [161]
	CYP4G8	Pyrethroids [162]
	CYP6B7	Pyrethroids [163]
H. armigera	CYP6B7, CYP9A12, CYP9A14	Pyrethroids [164]
	CYP4S1, CYP337B1	Fenvalerate [165]
	CYP4L5,11, CYP4M6,7, CYP6AE11, CYP9A14, CYP332A1, CYP337B1	
Plutella xylostella	CYP6BG1	Cypermethrin [167]

Species	P450 Over-expressed	Resistance pattern
	CYP4M20	Cypermethrin [168]
Lygus lineolaris	CYP6X1	Permethrin [169]
Bemisia tabaci	CYP6CM1vQ	Imidacloprid [170]
Nilaparvata lugens	CYP6ER1	Imidacloprid [171]
Myzus persicae	CYP6CY3	Neonicotinoids [172]
Diabrotica virgifera	CYP4	Me-parathion [173]
Tribolium castaneum	CYP6BQ8,9,10, CYP436B1, B2	Deltamethrin [174]
Blattella germanica	P450MA,	Chlorpyriphos [175]
	CYP4G19	Pyrethroids [176]

In Western Flower Thrips, Frankliniella occidentalis, metabolic detoxification of insecticides has been reported by many authors [179]. In Bemisia tabaci, metabolic resistance due to enhanced activity of insecticide resistance-related enzymes has also been reported [180].

7. HOST ALLELOCHEMICALS, INDUCTION OF DETOXIFYING ENZYMES AND INSECTICIDE RESISTANCE

Understanding the diversity of insect responses to chemical pressures (plant allelochemicals and insecticides) in their local ecological context represents a key challenge in developing sustainable pest control strategies. Plants and insects have had co-existing relationships for a long time. Insects were suppressed either by other insects or toxins or by plant defence mechanisms in order to create a balance between the insect pest population and host. Each plant species has a unique set of defence traits ranging from morphological to phytochemical parameters that have behavioural and physiological ramifications for a potential herbivore consumer [181, 182]. Therefore, the resistance mechanisms evolved by insects to deal with the chemical defences of plants are similar to those mechanisms that have evolved to resist synthetic insecticides. The chemical structure of some synthetic insecticides is comparable to that of some plant-produced compounds (e.g. pyrethroids and nicotinoids). Insect resistance to plant allelochemicals interferes with their resistance to synthetic insecticides [183]. From the evolutionary perspective, despite the key role of the chemical 'arms race' in driving the co- evolution of plants and insects, much research has focused so far on describing the diversity of plant chemicals and their effects on herbivores. Hence, the understanding of insecticide resistance mechanisms as well as taking into account other ecological parameters is important in predicting the spread of insecticide resistance in natural pest populations and in choosing the optimum strategy for managing insect pest populations. Less is known about the multiple mechanisms evolved by insects to overcome these chemical defences (Table 4). These mechanisms include contact and ingestion avoidance, excretion, sequestration, degradation of the toxin and target site mutation.

Biotransformation of plant toxins is one of the major weapons that insects have evolved in their co-evolutionary arms race with plants [204]. To date, metabolic resistance to plant chemicals has been identified not only in herbivorous insects [194] but also in detrivorous insects such as mosquito larvae feeding plant debris [205]. Metabolic resistance often results from the overproduction of 'detoxification enzymes' that can metabolize plant xenobiotics (allelochemicals). This mechanism is often associated with phenotypic plasticity, as the production of detoxification enzymes is usually induced by the presence of plant xenobiotics in the diet of the insect.

Induction of insect detoxifying enzyme activities by plant allelochemicals is a clear manifestation of biochemical phenotypic plasticity and has been documented in several instances [206, 207]. Many of the theories and some of the experiments implicitly or explicitly deal with the insect's ability to metabolize plant secondary substances by P450 and other enzymes. In those studies, a 'higher activity of midgut microsomal oxidase enzymes in polyphagous than in monophagous species indicates that the natural function of these enzymes is to detoxify natural insecticides present in the larval food plants'. The estimation of aldrin epoxidation in gut homogenates of last instar larvae from 35 species of Lepidoptera showed that polyphagous species had on average a 15 times higher activity than monophagous species. This trend was seen in sucking insects as well. A 20-fold lower aldrin epoxidase activity was found in the oleander aphid Aphis nerii (specialist feeder on two plant families, Asclepiadaceae and Apocynaceae) when compared to the potato aphid Myzus euphorbiae or to the green peach aphid Myzus persicae (both are generalists found on 30–72 plant families) [208]. A similar type of observation was made for other detoxification enzymes. In mites, predatory mite has a five times lower aldrin epoxidase activity than its herbivorous prey [209]. The toxicity of the natural phototoxin α-terthienyl is inversely proportional to the level of its metabolism in Lepidoptera and is related to diet breadth. Metabolism is highest in Ostrinia nubilalis, which feeds on numerous phototoxic Asteraceae; lower in Helicoverpa virescens, which has a broad diet, including some Asteraceae that are non-phototoxic; and lowest in Manduca sexta, a specialist of Solanaceae [210].

Table 4. Plant allelochemicals and associated resistance mechanisms in insects.

Plant allelochemicals	Target (mechanism of effect)	Resistance mechanisms	Species
Alkaloids	Neuroreceptors (inhibition), ion	Modification of nicotine	*Helicoverpa zea* (Lep.) [184]
	channels (antagonists), nucleic acids (disruption of DNA synthesis), feeding (deterrent owing to bitterness), enzymes (inhibition)	synthesis by salivary glucose oxidase	
Cardenolides	Nervous system (depressing activity); Na^+, K^+-ATPase (specific inhibitor)	Canal trenching behaviour, target site mutation	*Danaus plexippus* (Lep.) [185] *Chrysochus* sp. (Col.) [186]
Cyanogenic glycosides	Electron transport (inhibition of mitochondrial cytochrome oxidase)	Ingestion avoidance, sequestration and detoxification	*Schistocerca americana* (Ort.), *Hypera brunneipennis* (Col), *Zygaena* sp. (Lep.), *Clossiana euphrosyne* (Lep.), *Heliconius sara* (Lep.) [187]
Glucosinolates	Respiration (inhibition)	Detoxification by GSTs, detoxification by glucosinolate sulphatase, formation of nitriles instead of isothiocyanate detoxification by P450s, detoxification by N- oxidation and sequestration	*Myzus persicae* (Hem.) [188] Plutella xylostella (Lep.) [189] Pieris rapae (Lep.) [190] Drosophila melanogaster [191] Estigmene acrea (Lep.) [192] Tyria jacobaeae (Lep.) [193]
Flavonoids and phenolic acids	Respiration (inhibition), growth (inhibition)	Ingestion avoidance, decrease of toxin levels in gall tissue, glycosylation by UDP-glycosyl transferase, sequestration and/ or excretion	Manduca sexta (Lep.) [194] Potania sp. (Hym) [195] Bombyx mori (Lep.) [196]

Iridoid glycosides	Feeding (deterrent owing to bitterness), nucleic acids (inhibition of DNA polymerase), proteins (denaturant and cross-linking activities)	Sequestration	Longitarsus sp. (Col.) [197]
Coumarins and furanocoumarins	Nucleic acids (photoactivie DNA bonding), pro-oxidant activity	Detoxification by P450s, detoxification by GSTs	Papilio polyxenes (Lep.) [198] Depressaria pastinacella [199] Spodoptera frugiperda (Lep.) [200]
Protease inhibitors	Digestive system (inhibition of protease)	Over-expression of insensitive protease	Callosobruchus maculatus (Col.) [201]
Terpinoids	Nervous system (inhibition of acetylcholine esterases); feeding (deterrent owing to physical barrier and bitterness); growth and development inhibitor (pheromone analogue)	Repression of genes involved in biosynthetic pathways	Spodoptera exigua (Lep.) [202]
Tannins	Feeding (complexation of salivary and gut proteins); pro-oxidant activity	Synthesis of anti-oxidant compounds	Orgyia leucostigma (Lep.) [203]
Lep., Lepidoptera; Col., Coleoptera; Ort., Orthoptera; Hym., Hymenoptera; Dip., Diptera; Hem., Hemiptera.			

In addition to insecticides, insect carboxylesterases also metabolize many glycosides. Β - glucosidase enzyme is active towards a variety of glucosides in fall armyworms, corn ear-worms, cabbage loopers and valvet bean caterpillars. The p-nitrophenyl β-D-glucoside, 4- methyl umbelliferyl β -D-glucoside, D (+)-cellobiose, D-amygdalin and helicon were preferred substrates whereas sinigrin, phloridzin, α-solanine, tomatine and linamarin were poor substrates for these insects and many other insects reported to date [211]. β-Glycosidases have been shown to play important roles in the survival of certain phytophagous insects [211]. The ability of peach tree borer, Synanthedon exitiosa, larvae to survive well on prunasin-containing peach tree is because they can metabolize cyanogenic glycosides through β-glucosidase and detoxify the released cyanide by β-cyanoalanine synthase, thereby allowing them to utilize peach trees [212]. Another example is the larvae of the tiger swallowtail, Papilio glaucus, which feeds on quacking aspen, which contains various phenolics glycosides (e.g. salicortin). These larvae hydrolyse the glycosides by β-glucosidase and detoxify the released phenolics aglycone by a highly active esterase, thereby allowing them to survive on aspen [213, 214].

Glutathione S-transferases are also involved in the metabolism of many toxic plant allelo-chemicals. These plant allelochemicals may be of many diverse groups including α, β- unsaturated carbonyl compounds (e.g. trans-cinnamaldehyde, trans-2-hexanal), isothiocyanates (e.g. allylisothiocyanate, benzyl isothiocyanate) and organothiocyanates (e.g. benzyl thiocyanate) as have been documented in some instances [41]. The glutathione S- transferase activities are lower in the specialist insects than in the generalists. In the crucifer- adapted cabbage looper for the metabolism of isothiocyanates (plant allelochemical), the activity of this enzyme wasfound to be two- to sixfold higher than that in the fall armyworm [215, 216]. These findings strongly advocate that glutathione S-transferases play an important role in developing resistance towards plant allelochemicals in phytophagous lepidopteran insects [215–217]. Many plant allelochemicals are potent inhibitors of glutathione S-transferases in many insects [218]. Many flavonoids, other phenols and α-, β-unsaturated carbonyl compounds are also found to be potent inhibitors of the enzymes.

H. theivora, E. flavescens and S. dorsalis all are polyphagous in nature. H. theivora known to feed on at least sixteen different plant families reported till date [11]. Similarly, E flavescens is also polyphagous [11]. S. dorsalis has been documented to attack more than 150 hosts from at least 40 different plant families [219]. Hence, these pests are exposed to a wide variety of plant allelochemicals of diverse groups having the potential to induce the activity of these resistance- related enzymes. A higher level of detoxifying enzyme activity in H. theivora has been reported when reared on two alternative hosts, i.e. Mikania micrantha (Asteraceae) and Psidium guajava (Myrtaceae), than on tea [220]. Over the four hundred million years of co-evolution with plants, phytophagous insects have developed diverse resistance mechanisms to cope with plant chemical defences. Because insects face a geographical mosaic of chemical environments, from non-toxic to highly toxic plants, the costs associated with resistance traits vary with the probability of encountering a toxin. Moreover, other selection pressures, such as the presence or absence of competitors and predators, can also influence the costs and selection of particular resistance traits. Thus, the complexity of the local community composition is a key factor in maintaining the diversity of adaptive mechanisms to plant xenobiotics. These mechanisms are more plastic and complex compared with those involved in resistance to insecticides, perhaps because environments in which insecticides are heavily used also tend to have communities of low diversity and complexity. However, because some detoxification enzymes are involved in plant toxins and insecticides metabolism, cross-resistance mechanisms can be predicted to be observed under specific environmental conditions. Deciphering the impact of allelochemicals in cross-resistance mechanisms with insecticides at a local scale, and comparing the molecular and evolutionary mechanisms of resistance to phytotoxins and synthetic insecticides, represent promising areas of research for developing long-term sustainable insect control strategies for the effective management of pest concern [220].

8. GENETICS AND INSECTICIDE RESISTANCE IN TEA PESTS

Earlier, common visible markers including mophometrices, eye colour, body spots or bands and hairs or spines, wing venation were used as phenotypic markers in studying the pattern of dispersal, mating behaviour, population variability and inheritance of genetic traits in insects [221, 222]. Although the phenotypic markers are found at all times of life span of the organism and can be readily used for studies in field conditions, they suffer from many practical limitations. The major drawback is that these visible phenotypes are relatively infrequent and often hard to score. Because the phenotype markers are rare, use of these markers in mapping a trait is difficult. For all such difficulties and with the concurrent advancement in biochemical methodologies, protein markers then became more popular. Protein markers made a significant contribution in the early periods when DNA technologies were not so much advanced, as it is now [223]. A diverse range of novel molecular (DNA) markers are now available for entomological investigations. Currently, both DNA and protein markers have revolutionized the biological sciences and have enhanced many fields of insect study, especially agricultural entomology [224].

Insecticide resistance is the result of an increase in the ability of individuals of an insect species to survive insecticide application and is an important example of man-driven evolution [225]. Alleles conferring resistance may arise and spread in populations and to other populations with variable success, depending on factors such as selective forces, genetic variability, gene flow, population size and environmental conditions [220]. Studies that map the population structure of pest insects, as well as the potential for gene flow between populations, are needed to understand the development of resistance and prevention of its spread [226, 227]. Development of resistance is often rapid in isolated populations that have been treated by insecticides [228]. The rate of development of insecticide resistance may, however, be influenced by gene flow between treated and untreated populations by maintaining the frequency of resistance alleles at a low level [229]. Contaminant exposure was a poor predictor of population structure and the level of gene flow was a better predictor of relatedness [230]. Gene flow may balance divergence by opposing the effect of selection pressures [229]. Population genetic patterns should therefore be investigated with reference to geographical variability, as well as selection pressure. Detoxification resistance occurs when enhanced levels or modified activities of biotransformation enzymes prevent the insecticide from acting on its site of action because the metabolites produced have little or no activity compared with the original substance [231]. These alterations may stem from mutations that produce a protein with slightly modified characteristics or altered expression levels.

Due to the regular application of chemical control to prevent economic loss, sucking insects have seen significant selection pressure. Insecticides are likely to remain the primary control strategy in the foreseeable future; thus, it is essential to examine the composition of sucking insect populations and variations in pesticide susceptibility. Various methodologies exist for assessing genetic diversity, including randomly amplified polymorphic DNA analysis, microsatellite analysis, minisatellite analysis, restriction fragment length polymorphism analysis, and amplified fragment length polymorphism (AFLP) study. DNA markers are appropriate for use with little quantities of insect material and can be utilized with preserved, desiccated, or aged materials. Certain entities exhibit intricate multi-locus banding patterns, perhaps of a non-Mendelian character (e.g., randomly amplified polymorphic DNAs (RAPDs)). They possess a growing array of applications, many of which entail intra- and interspecific distinctions.

REFERENCES

1. Chen Z. M. and Chen X. F. (1989) An analysis of world tea fauna. Journal of Tea Science 9: 73–88.
2. Muraleedharan N. (1992) Pest control in Asia. In Tea: Cultivation to Consumption. Wilson, K.C. and Clifford, M.N. (edited by N. Muraleedharan). Chapmann & Hall, London, pp. 375–412.
3. Das G. M. (1965) Pests of tea in North East India and their control. Memorandom No. 27, pp. 169–173, Tocklai Experimental Station, Tea Research Association, Jorhat, Assam.
4. Muraleedharan N. (2007) Tea insects: ecology and control, pp. 672–674. In Encyclopedia of Pest Management. CRC Press, London.
5. Rattan P. S. (1992) Pest and disease control in Africa, pp. 331–352. In Tea: Cultivation to Consumption (edited by K. C. Wilson and M. N. Clifford). Chapman & Hall, London.

6. Sivapalan P. (1999) Pest management in tea, pp. 625–646. In Global Advances in Tea Science (edited by N. K. Jain). Aravali Books, New Delhi.
7. Hazarika L. K., Bhuyan M. and Hazarika B. N. (2009) Insect pests of tea and their management. Annual Review of Entomology 54: 267–284
8. Saha D. and Mukhopadhyay A. (2013) Insecticide resistance mechanisms in three sucking insect pests of tea with reference to North East India - an appraisal. International Journal of Tropical Insect Science 33(1): 46–70.
9. Cranham J. E. (1966) Tea pests and their control. Annual Review of Entomology 11: 491– 514.
10. Saha D., Roy S. and Mukhopadhyay A. (2012) Insecticide susceptibility and activity of major detoxifying enzymes in female Helopeltis theivora Waterhouse (Heteroptera: Miridae) from sub-Himalayan tea plantations of North Bengal, India. International Journal of Tropical Insect Science 32: 85–93.
11. Saha, D. (2014) Assessment of Population Variability at Subcellular Level of Some Common Sucking Tea Pests from Darjeeling Hill and its Adjoining Plain. A Ph.D. Thesis. University of North Bengal, Sliguri-734013, District - Darjeeling, West Bengal, India.
12. Chaudhuri T. C. (1999) Pesticide residues in tea, pp. 369–378. In Global Advances in Tea Science (edited by N. K. Jain). Aravali Books, New Delhi.
13. Komagata O., Kasai S. and Tomita T. (2010) Overexpression of cytochrome P450 genes in pyrethroid- resistant Culex quinquefasciatus. Insect Biochemistry and Molecular Biology 40: 146–152.
14. Das G. M. (1963) Some important pests of tea. Two and a Bud 10: 4–8.
15. Mukerjea T. D. (1968) Thiodan: a broad spectrum new insecticide. Two and a Bud 15: 6–10.
16. Barbora B. C. and Biswas A. K. (1996) Use pattern of pesticides in tea estates of North East India. Two and a Bud 47: 19–21.
17. Sannigrahi S. and Talukdar T. (2003) Pesticide use patterns in Dooars tea industry. Two and a Bud 50: 35–38.
18. Gurusubramanian G., Rahman A., Sarmah M., Roy S. and Bora S. (2008) Pesticide usage pattern in tea ecosystem, their retrospects and alternative measures. Journal of Environmental Biology 29: 813–826.
19. Gurusubramanian G. and Bora S. (2007) Relative toxicity of some commonly used insecticides against adults of Helopeltis theivora Waterhouse (Miridae: Hemiptera) collected from Jorhat area tea plantations, South Assam, India. Resistant Pest Management Newsletter 17: 8–12.

20. Gurusubramanian G. and Bora S. (2008) Insecticidal resistance to tea mosquito bug, Helopeltis theivora Waterhouse (Miridae: Heteroptera) in North East India. Journal of Environmental Research Development 2: 560–567.

21. Gurusubramanian G., Senthilkumar N., Bora S., Roy S. and Mukhopadhyay A. (2008) Change in susceptibility in male Helopeltis theivora Waterhouse (Jorhat population, Assam, India) to different classes of insecticides. Resistant Pest Management Newsletter 18: 36–39.

22. Saha D., Mukhopadhyay A. and Mahadur M. (2010) Variation in detoxifying enzymes of Assam thrips, Scirtothrips dorsalis Hood (Thysanoptera: Thripidae) from organically and insecticide managed tea plantations. North Bengal University. Journal of Animal Science 4: 45–52.

23. Saha D., Mukhopadhyay A. and Mahadur M. (2012) Detoxifying enzymes activity in tea greenfly, Empoasca flavescens Fabricius (Homoptera: Jassidae) from sub Himalayan tea plantations, pp. 257–266. In Proceedings of National Symposium on Biodiversity Status and Conservation Strategies with special Reference to North East India (edited by R. Varatharajan). Centre for Advanced Studies, Department of Life Sciences, Manipur University, Manipur, India.

24. Saha D., Roy S. and Mukhopadhyay A. (2012) Seasonal incidence and enzyme-based susceptibility to synthetic insecticides in two upcoming sucking insect pests of tea. Phytoparasitica 40: 105–115.

25. Saha D., Mukhopadhyay A. and Bahadur M. (2012) Variation in the activity of three principal detoxifying enzymes in major sucking pest of tea, Helopeltis theivora Waterhouse (Heteroptera: Miridae) from sub- Himalayan tea plantations of West Bengal, India. Proceedings of Zoological Society. (July-Dec 2013) 66(2): 92–99.

26. Kumar, B., Saha, D. and Mukhopadhyay, A. (2014) Enhancement of resistant ratio vis-a-vis defence-enzymes activity in tea mosquito bug, Helopeltis theivora Water house (Hemiptera: Miridae) selected through exposure to sub-lethal dose of monocrotophos. Proceedings of Zoological Society. DOI: 10.1007/s12595-014-0124-5 (Published online on 11th September, 2014)

27. Bora S., Rahman A., Sarmah M. and Gurusubramanian G. (2007) Relative toxicity of pyrethroid and nonpyrethroid insecticides against male and female tea mosquito bug (Darjeeling strain). Journal of Entomological Research 37: 37–41.

28. Roy S., Gurusubramanian G. and Mukhopadhyay A. (2008) Insecticide persistence and residual toxicity monitoring in tea mosquito bug, Helopeltis theivora Waterhouse (Hemiptera: Heteroptera: Miridae). Resistant Pest Management Newsletter 17: 9–14.

29. Rahman A., Sarmah M., Phukan A. K., Roy S., Sannigrahi S., Borthakur M. and Gurusubramanian G. (2005) Approaches for the management of tea mosquito bug, Helopeltis theivora Waterhouse (Miridae: Heteroptera), pp. 146–161. In Proceedings of the 34th Tocklai Conference 'Strategies for Quality in the Digital Era' (edited by A. K. Barooah, M. Borthakur and J. N. Kalita). Tocklai Experimental Station, TRA, Jorhat, Assam.

30. Roy S. (2010) Evaluation of the levels of insecticide susceptibility of Helopeltis theivora Waterhouse (Heteroptera: Miridae) and development of an efficacious strategy for management of the pest in Dooars tea plantation of North Bengal, PhD thesis, University of North Bengal, India.

31. Reddy G. P. V., Prasad V. D. and Rao R. S. (1992) Relative resistance in chilli thrips, Scirtothrips dorsalis (Hood) populations in Andhra Pradesh to some conventional insecticides. Indian Journal of Plant Protection 20: 218–222.

32. Vanisree K., Upendhar S. and Rajasekhar P. (2011) Toxicity of certain novel insecticides against chilli thrips, Scirtothrips Dorsalis (Hood). Resistant Pest Management Newsletter 21: 17–21.

33. Seal D. R., Ciomperlik M., Richards M. L. and Klassen W. (2006) Comparative effectiveness of chemical insecticides against the chilli thrips, Scirtothrips dorsalis Hood (Thysanoptera: Thripidae), on pepper and their compatibility with natural enemies. Crop Protection 25: 949–955.

34. Seal D. R., Klassen W. and Sabines C. (2007) Management of chilli thrips, Scirtothrips dorsalis (Thysanoptera: Thripidae): effectiveness of neonicotinoids and spinosyns and ineffectiveness of pyrethroids. Proceedings of the Caribbean Food Crops Society 43: 39–48.

35. Dog˘ramaci M., Arthurs S. P., Chen J., McKenzie C., Irrizary F. and Osborne L. (2011) Management of chilli thrips Scirtothrips dorsalis (Thysanoptera: Thripidae) on peppers by Amblyseius swirskii (Acari: Phytoseiidae) and Orius insidiosus (Hemiptera: Anthocoridae). Biological Control 59: 340–347.

36. Shi C. H., Shang J. N. and Chen Y. F. (2001) Dynamic residues of imidacloprid in tea products and its application to control of tea insect pests, pp. 165–166. In Integrated Pest Management in Relation to Safe Agricultural Products (edited by Y. J. Yu and Q. H. Zhang). China Agriculture Press, Beijing.

37. Tao T., Hu K. M., She Y. P., Zhu Q. Z. and Luo Z. (1996) Studies on the integrated management techniques of smaller green leafhopper in tea plantation of south Yunnan. Acta Phytophysiologica Sinica 23: 310–314.

38. Nian-wu W., Jin-han X. U., Zheng C., Hui W., Ling-ling Z. and Xiong G. (2004) Resistance level of Empoasca vitis (Gothe) in different tea plantations [J]. Journal of Tea Science, 2-.

39. Jia-Xiang Z., Jian-Wei F. U., Qing-Quan S. U., Jian-Yu L. I. and Zhi-Xiong Z. (2009) The regional diversity of resistance of Tea Green Leafhopper (Gothe), to insecticides in Fujian Province. Journal of Tea Science 2-.

40. Soderland D. M. and Bloomquist J. R. (1990) Molecular mechanisms of insecticide resistance, pp. 58–96. In Pesticide Resistance in Arthropods (edited by R. T. Roush and B. E. Tabashnik). Chapman and Hall, New York.

41. Yu S. J. (2008) The Toxicology and Biochemistry of Insecticides. CRC Press, Florida. 296 p.

42. Wu S., Yang Y., Yuan G., Campbell P. M., Teese M. G., Russell R. J., Oakeshott J. G. and Wu Y. (2011) Overexpressed esterases in a fenvalerate resistant strain of the cotton bollworm, Helicoverpa armigera. Insect Biochemistry and Molecular Biology 41: 14–21.

43. Buès R., Bouvier J. C. and Boudinhon L. (2005) Insecticide resistance and mechanisms of resistance to selected strains of Helicoverpa armigera (Lepidoptera: Noctuidae) in the south of France. Crop Protection 24: 814–820.

44. Cao C.W., Zhang J. Gao X.W. Liang P. and Guo H.L. (2008) Overexpression of carboxylesterase gene associated with organophosphorous insecticide resistance in cotton aphids, Aphis gossypii (Glover). Pesticide Biochemistry and Physiology 90: 175–180.

45. Cao C.W., Zhang J., Gao X.W., Liang P. and Guo H.L. (2008) Differential mRNA expression levels and gene sequences of carboxylesterase in both deltamethrin resistant and susceptible strains of the cotton aphid, Aphis gossypii. Insect Science 15: 209–216.

46. Mitter C., Farrell B. D. and Futuyma D. J. (1991) Phylogenetic studies of insect/plant interactions: Insights into the genesis of diversity. Trends in Ecology and Evolution 6: 290–293.

47. Ehrlich P. R. and Raven P. H. (1964) Butterflies and plants: a study in coevolution. Evolution 18: 586–608.

48. Bernays E. A. (1998) Evolution of feeding behavior in insect herbivores: success seen as different ways to eat without being eaten. Bioscience 48: 35–44.

49. Dethier V. G. (1954) Evolution of feeding preferences in phytophagous insects. Evolution 8: 33–54.

50. Kaloshian I. and Walling L. L. (2005) Hemipterans as plant pathogens. Annual Review of Plant Biology 43: 491–521.

51. Goggin F. L. (2007) Plant-aphid interactions: molecular and ecological perspectives. Current Opinion in Plant Biology 10: 399–408.

52. Sawicki, R. M. (1987) Definition, detection and documentation of insecticide resistance, pp.105-112. In Combating resistance to xenobiotics: Biological and chemical approaches (edited by E. Hoorwood). Chichester, England,

53. Sparks T. C., Lockwood J. A., Byford R. L., Graves J. B. and Leonard B. R. (1989) The role of behavior in insecticide resistance. Journal of Pesticide Science 26: 383–399.

54. Brattsten L. B. (1988) Potential role of plant allelochemicals in the development of insecticide resistance, pp. 313–348. In Novel Aspects of Insect–Plant Interactions (editd by P. Barbosa and D. K. Letourneau). John Willey & Sons, New York.

55. Roy S. and Mukhopadhyay A. (2011) Insecticide-Induced Change in Egg-Laying Strategy of Helopeltis theivora (Hemiptera: Miridae) on Tea Shoot (Camellia sinensis). Proceedings of Zoological Society 64 (1): 54–56.

56. Price N. R. (1991) Insect resistance to insecticides: mechanisms and diagnosis. Comparative Biochemistry and Physiology 100C, 319–326.

57. Vinson S. B. and Law P. K. (1971) Cuticular composition and DDT resistance in the tobacco budworm. Journal of Economic Entomology 64: 1387–1390.

58. Patil V. L. and Guthrie F. E. (1979) Cuticular lipids of two resistant and a susceptible strain of houseflies. Pesticide Science 10: 399–406.

59. Scott J. G. (1990) Investigating mechanisms of insecticide resistance: methods, strategies, and pitfalls, pp. 39–57. In Pesticide Resistance in Arthropods (edited by R. T. Roush and B. E. Tabashnik). Chapman and Hall, New York.

60. Forgash A. J., Cook B. J. and Riley R. C. (1962) Mechanisms of resistance in diazinon- selected multi resistant Musca domestica. Journal of Mechanisms of Economic Entomology 55: 544–551.

61. Van de Baan H. E. and Croft B. A. (1991) Resistance to insecticides in winter- and summer-forms of pear psylla, Psylla pyricola. Pesticide Science 32: 225–233.

62. Kanga L. H. B. and Plapp F. W. J. (1995) Target-site insensitivity as the mechanism of resistance to organophosphorus, carbamate, and cyclodiene insecticides in tobacco budworm adults. Journal of Economic Entomology 88: 1150–1157.

63. Roy S., Gurusubramanian G. and Mukhopadhyay A. (2008) Variation in endosulfan susceptibility and body lipid content of Helopeltis theivora Waterhouse (Heteroptera: Miridae) in relation to the use pattern of the insecticide, in sub-Himalayan Dooars tea plantations. Journal of Plantation Crops 36: 388–392.

64. Martinez-Torres D., Foster S. P., Field L. M., Devonshire A. L. and Williamson M. S. (1999) A sodium channel point mutation is associated with resistance to DDT and pyrethroid insecticides in the peach-potato aphid, Myzus persicae (Sulzer) (Hemiptera: Aphididae). Insect Molecular Biology 8: 339–346.

65. Mutero A., Pralavorio M., Bride J. M. and Fournier D. (1994) Resistance-associated point mutations in insecticide-insensitive acetylcholinesterase. Proceedings of the National Academy of Sciences USA 91: 5922–5926.

66. Brown T. M. and Brogdon W. G. (1987) Improved detection of insecticide resistance through conventional and molecular techniques. Annual Review of Entomology 32: 145–162.

67. Scharf M. E., Neal J. J. and Bennett W. G. (1998) Changes of insecticide resistance levels and detoxication enzymes, following insecticide selection in the German cockroach, Blattella germanica (L.). Pesticide Biochemistry and Physiology 59: 67–79.

68. Casida J. E. (1993) Insecticide action at the GABA-gated chloride channel: recognition, progress, and prospects. Archives of Insect Biochemistry and Physiology 22: 13–23.

69. Narahashi, T. (1996) Neuronal ion channels as the target sites of insecticides. Pharmacology and Toxicology 79: 1–14.

70. Sattelle D. B. (1985) Acetylcholine receptors, pp. 395–434. In Comprehensive Insect Physiology, Biochemistry, and Pharmacology (edited by G. A. Kerkut and L. I. Gilbert). Vol. 11. Pergamon, Oxford.

71. Byrne F.J. and Toscano N. C. (2001) An Insensitive Acetylcholinesterase Confers Resistance to Methomyl in the Beet Armyworm Spodoptera exigua (Lepidoptera: Noctuidae) Journal of Economic Entomology 94(2): 524–528.

72. Eldefrawi A. T. (1985) Acetylcholinesterases and anticholinesterases, pp. 115–130. In Comprehensive Insect Physiology, Biochemistry, and Pharmacology (edited by G. A. Kerkut and L. I. Gilbert). Vol. 12. Pergamon Press, Oxford.

73. Aldridge W. N. and Reiner E. (1972). Enzyme inhibitors as substrates. North-Holland Publishing Co., Amsterdam.

74. Smissaert, H. R. (1964) Cholinesterase inhibition in spider mite susceptible and resistant to organophosphate. Science 143: 129–131.

75. Fournier D. and Mutero A. (1994) Modification of acetylcholinesterase as a mechanism of resistance to insecticides. Comparative Biochemistry and Physiology 108C, 19–31.

76. Hama H. (1983) Resistance to insecticides due to reduced sensitivity of acetylcholinesterase, pp. 299–331. In Pest Resistance to Pesticides (edited by G. P. Georghiou and T. Saito). Plenum Press, New York.

77. Oppenoorth F. J. (1985) Biochemistry and genetics of insecticide resistance, pp. 731– 773. In Comparative Insect Physiology, Biochemistry and Pharmacology (edited by G. A. Kerkut and L. I. Gilbert). Vol. 12. Pergamon Press, Oxford.

78. Guedes R. N. C., Zhu K. Y., Kambhampati S. and Dover B. A. (1997) An altered acetylcholinesterase conferring negative cross-insensitivity to different insecticidal inhibitors in organophosphate-resistant lesser grain borer, Rhyzopertha dominica. Pesticide Biochemistry and Physiology 58: 55–62.

79. Harold J. A. and Ottea J. A. (1997) Toxicological significance of enzyme activities in profenofos resistant tobacco budworms. Heliothis virescens (F.). Pesticide Biochemistry and Physiology 58: 23–33.

80. Levitin E. and Cohen E. (1998) The involvement of ace-tylcholinesterase in resistance of the California red scale, Aonidiella aurantii to organophosphorus pesticides. Entomologia Experimentalis et Appllicata 88: 115–121.

81. Sarker M. and Mukhopadhyay A. (2006a) Studies on salivary and midgut enzymes of a major sucking pest of tea, Helopeltis theivora (Heteroptera: Miridae) from Darjeeling plains, India. Journal of the Entomological Research Society 8: 27–36.

82. Sarker M. and Mukhopadhyay A. (2006b) Studies on some enzymes related to insecticide resistance in Helopeltis theivora Waterhouse (Insecta: Heteroptera: Miridae) from Darjeeling foothills and plains. Journal of Plantation Crops 34: 423–428.

83. Li X., Schuler M. A. and Berenbaum M. R. (2007) Molecular mechanisms of metabolic resistance to synthetic and natural xenobiotics. Annual Review of Entomology 52: 231– 253.

84. Feyereisen R. (2012) Insect CYP genes and P450 enzymes, pp. 236–316. In Insect Molecular Biology and Biochemistry (edited by L. I. Gilbert). Academic Press, London.

85. Lee S. H., Kang J. S., Min J. S., Yoon K. S., Strycharz J. P., Johnson, R., Mittapalli, O., Margam, V.M., Sun, W., Li, H.M., Xie, J., Wu, J,, Kirkness, E.F., Berenbaum, M.R., Pittendrigh, B.R. and Clark, J.M. (2010) Decreased detoxification genes and genome size make the human body louse an efficient model to study xenobiotic metabolism. Insect Molecular Biology 19: 599–615.

86. Arensburger P., Megy K., Waterhouse R. M., Abrudan J., Amedeo P., Antelo, B., Bartholomay L., Bidwell S., Caler E., Camara F., Campbell C.L., Campbell

K.S., Casola C., Castro M.T., Chandramouliswaran I., Chapman S.B., Christley S., Costas J., Eisenstadt E., Feschotte C., Frasher-Liggett C., Guigo R., Haas B., Hammond M., Hansson B.S., Hemingway J. Hill S.R., Howarth C., Ignell R., Kennedy R.C., Kodira C.D., Lobo N.F., Mao C., Mayhew G., Michel K., Mori A., Liu N., Naveira H., Nene V., Nguyen N., Pearson M.D., Pritham E.J., Puiu D., Qi Y., Ranson H., Ribeiro J.M. Roberston H.M., Severson D.W., Shumway M., Stanke M., Strausberg R.L., Sun C., Sutton G., Tu Z.J., Tubio J.M., Unger M.F., Van Lan Dingham D.L., Vilella A.J., White O., White J.R., Wondji C.S., Wortman J., Zdobnov E.M., Birren B., Christensen B.M., Collins F.H., Cornel A., Dimopoulos G., Hannick L.I., Higgs S., Lanzaro G.C., Lawson D., Lee N.H., Muskavitch M.A., Raikhel A.S., Atkinson P.W., (2010) Sequencing of Culex quinquefasciatus establishes a platform for mosquito comparative genomics. Science 330: 86–88.

87. Omura T. and Sato R. (1964) The carbon monoxide-binding pigment of liver microsomes I. Evidence for its hemoprotein nature. Journal of Biological Chemistry 239: 2370– 2378.

88. Feyereisen R., Koener J. F., Farnsworth D. E. and Nebert D. W. (1989) Isolation and sequence of cDNA encoding a cytochrome P-450 from an insecticide-resistant strain of the house fly, Musca domestica. Proceedings of the National Academy of Science USA 86: 1465–1469.

89. Feyereisen R. (2011) Arthropod CYPomes illustrate the tempo and mode in P450 evolution. Biochimica et Biophysica Acta 1814: 19–28.

90. Claudianos C., Ranson H., Johnson R. M., Biswas S., Schuler M. A., Berenbaum M. A., Feyereisen R. and Oakeshott J. G. (2005) A deficit of detoxification enzymes: pesticide sensitivity and environmental response in the honeybee. Insect Molecular Biology 15, 615–636.

91. Hardstone M. C. and Scott J. G. (2010) Is Apis mellifera more sensitive to insecticides than other insects? Pest Management Science 66: 1171–1180.

92. Hemingway J. and Karunaratne S. H. P. P. (1998) Mosquito carboxylesterases: a review of the molecular biology and biochemistry of a major insecticide resistance mechanism. Medical and Veterinary Entomology 12: 1–12.

93. Heymann E. and Jakoby W.B. (1980) Carboxylesterases and amidases, pp. 291–323. In Enzymatic basis of detoxication (edited by E. Heymann and W. B. Jokoby). Academic Press, New York,. Aldridge W.N. (1993) The esterases: perspectives and problems. Chemico-Biological Ineractions 87: 5–13.

94. Reiner E. (1993) Recommendations of the IUBMB nomenclature committee – comments concerning classification and nomenclature of esterases

hydrolysing organo-phosphorus compounds. Chemico-Biological Interactions 87: 15, 16.

95. Walker C.H. (1993) The classification of esterases which hydrolyse organophosphates–recent developments. Chemico-Biological Interactions 87: 17–24.

96. Yu S.J. (1996) Insect glutathione S-transferases. Zoological Studies 35: 9–19.

97. Ranson H., Claudianos C., Ortelli F., Abgrall C., Hemingway J., Sharakhova M.V., Unger M.F., Collins F.H. and Feyereisen R. (2002) Evolution of supergene families associated with insecticide resistance. Science 298: 179–181.

98. Ranson, H. and Hemingway, J. (2005) Mosquito glutathione transferases. Review. Methods in Enzymology 401: 226–241.

99. Fournier D., Bride J.B., Poirie M., Berge J.P. and Plapp F.W. Jr. (1992) Insect Glutathione S-transferases: Biochemical characteristics of the major forms from house flies susceptible and resistant to insecticides. Journal of Biological Chemistry 267: 1840–1845.

100. Ranson H., Rossiter L., Ortelli F., Jensen B., Wang X., Roth C.W., Collins F.H. and Hemingway J. (2001) Identification of a novel class of insect glutathione S-transferases involved in DDT resistance in the malaria vector, Anopheles gambiae. Biochemical Journal 359: 295–304.

101. Yu S.J. (2002) Biochemical characteristics of microsomal and cytosolic glutathione S- transferases in larvae of the fall armyworm, Spodoptera frugiperda (J. E. Smith). Pesticide Biochemistry and Physiology 72: 100–110.

102. Ma B. and Chang F.N. (2007) Purification and cloning of a delta class glutathione s- transferase displaying high peroxidase activity isolated from German cockroach, Blattella germanica. FEBS Journal 274: 1793–1803.

103. Che-Mendoza A., Penilla R.P. and Rodríguez D.A. (2009) Insecticide resistance and glutathione S-transferases in mosquitoes: A review. African Journal of Biotechnology 8 (8): 1386–1397

104. Clark A.G. and Shamaan N.A. (1984) Evidence that DDT-dehydrochlorinase from the house fly is a glutathione S-transferase. Pesticide Biochemistry and Physiology 22: 249– 261.

105. Eldefrawi M.E., Miskus R. and Sutcher V. (1960) Methylenedioxyphenyl derivatives as synergists for carbamate insecticides on susceptible, DDT-and parathion-resistant house flies. Journal of Economic Entomology 53: 231–234.

106. Georghiou G. P. and Metcalf R. L. (1961) The adsorption and metabolism of 3-isopropylphenyl-n-methylcarbamate by susceptible and carbamate selected strains of house flies. Journal of Economic Entomology 54(2): 231–233.

107. Schonbrod R.D., Phillco W.W. and Terriere L.C. (1965) Hydroxylation as a factor in resistance in house flies and blow flies. Journal of Economic Entomology 58(1): 74–77.

108. Agosin M. (1985) Role of microsomal oxidations in insecticide degradation, pp. 647- 712. In Comparative Insect Physiology, Biochemistry and Pharmacology (edited by L. I. Gilbert and G. A. Kerkut). Vol. 12. Pergamon, Oxford.

109. Scott J. G. (1991) Insecticide resistance in insects, pp. 663–677. In Handbook of Pest Management (edited by D. Pimental). CRC Press, Boca Raton.

110. Konno T. E. H. and Dauterman W. C. (1989) Studies on methyl parathion resistance in Heliothis virescens. Pesticide Biochemistry and Physiology 33: 189–199.

111. Oppenoorth F.J. (1985) Biochemistry and genetics of insecticide resistance, pp. 731- 773. In Comparative insect physiology, biochemistry and pharmacology (edited by G. A. Kerkut and L. I. Gilbert). Vol. 12. Pergamon Press Oxford, UK.

112. Hatano R., Scott J. G. and Dennehy T. J. (1992) Enhanced activation is the mechanism of negative crossresistance to chlorpyrifos in the dicofol-IR strain of Tetranychus urticae (Acari: Tetranychidae). Journal of Economic Entomology 85: 1088–1091.

113. Cilek J. E., Dahlman D. L. and Knapp F.W. (1995) Possible mechanism of diazinon negative cross-resistance in pyrethroid-resistant horn flies (Diptera: Muscidae). Journal of Economic Entomology 88: 520–524.

114. Martin T., Chandre F., Ochou O. G., Vaissayre M. and Fournier D. (2003) Oxidases responsible for resistance to pyrethroids sensitize Helicoverpa armigera (Hubner) to triazophos in West Africa. Insect Biochemistry and Molecular Biology 33: 883–887.

115. Drabek J. and Neumann R. (1985) Proinsecticides, pp. 35–86. In Insecticides (edited by D. H. Huston and T. R. Roberts). Wiley, London.

116. Black B. C., Hollingworth R. M., Ahammadsahib K. I., Kukel C. D. and Donovan S. (1994) Insecticidal action and mitochondrial uncoupling activity of AC303, 630 and related halogenated pyrroles. Pesticide Biochemistry and Physiology 50: 115–128.

117. Kayser H. and Eilinger P. (2001) Metabolism of diafenthiuron by microsomal oxidation: Procide activation and inactivation as mechanisms contributing to selectivity. Pest Management Science 57: 975–980.

118. Pimprale S. S., Besco C. L., Bryson P. K. and Brown T. M. (1997) Increased susceptibility of pyrethroid-resistant tobacco budworm (Lepidoptera: Noctuidae) to chlorfenapyr. Journal of Economic Entomology 90: 49–54.

119. Joussen N., Heckel D. G., Haas M., Schuphan I. and Schmidt B. (2008) Metabolism of imidacloprid and DDT by P450 CYP6G1 expressed in cell cultures of Nicotiana tabacum suggests detoxification of these insecticides in Cyp6g1-overexpressing strains of Drosophila melanogaster, leading to resistance. Pest Management Science 64: 65–73.

120. Karunker I., Morou E., Nikou D., Nauen R., Sertchook R., Stevenson B. J., Paine M. J. I., Morin S. and Vontas J. (2009) Structural model and functional characterization of the Bemisia tabaci CYP6CM1vQ, a cytochrome P450 associated with high levels of imidacloprid resistance. Insect Biochemistry and Molecular Biology 39: 697–706.

121. Feyereisen R., Koener J. F., Farnsworth D. E. and Nebert D. W. (1989) Isolation and sequence of cDNA encoding a cytochrome P-450 from an insecticide-resistant strain of the house fly, Musca domestica. Proceedings of the National Academy of Science USA 86: 1465–1469.

122. Cariño F., Koener J. F., Plapp F. W., Jr. and Feyereisen R. (1992) Expression of the cytochrome P450 gene CYP6A1 in the housefly, Musca domestica. ACS Symposium Series 505: 31–40.

123. Sabourault C., Guzov V.M., Koener J.F., Claudianos C., Plapp F.W. and Feyereisen R., (2001) Overproduction of a P450 that metabolizes diazinon is linked to a loss-of- function in the chromosome 2 ali-esterase (Mda E7) gene in resistant house flies. Insect Molecular Biology 10: 609–618

124. Zhu F., Liu N. (2008a) Differential expression of CYP6A5 and CYP6A5v2 in pyrethroid resistant house flies, Musca domestica. Archives of Insect Biochemistry and Physiology 67: 107–119.

125. Zhu F., Feng J., Zhang L., Liu N. (2008b) Characterization of two novel cytochrome P450 genes in insecticide-resistant house-flies. Insect Molecular Biology 17: 27–37.

126. Zhu F., Li T., Zhang L., Liu N. (2008c) Co-up-regulation of three P450 genes in response to permethrin exposure in permethrin resistant house flies, Musca domestica. BMC Physiology 8: 18.

127. Liu N. and Scott J. G. (1996) Genetic analysis of factors controlling high-level expression of cytochrome P450, CYP6D1, cytochrome b5, P450 reductase, and monooxygenase activities in LPR house flies, Musca domestica. Biochemical Genetics 34: 133–148.

128. Kasai S. and Scott J. G. (2001a) Cytochrome P450s CYP6D3 and CYP6D1 are part of a P450 gene cluster on autosome 1 in the house fly. Insect Molecular Biology 10: 191–196.

129. Kasai S. and Scott J. G. (2001b) Expression and regulation of CYP6D3 in the house fly, Musca domestica (L.). Insect Biochemistry and Molecular Biology 32: 1–8.

130. Kasai S. and Scott J. G. (2000) Overexpression of cytochrome P450 CYP6D1 is associated with monooxygenase-mediated pyrethroid resistance in house flies from Georgia. Pesticide Biochemistry and Physiology 68: 34–41.

131. Kamiya E., Yamakawa M., Shono T. and Kono Y. (2001) Molecular cloning, nucleotide sequences and gene expression of new cytochrome P450s (CYP6A24, CYP6D3v2) from the pyrethroid resistant housefly, Musca domestica L. (Diptera: Muscidae). Applied Entomology and Zoology 36: 225–229.

132. Shono T., Kasai S., Kamiya E., Kono Y. and Scott J. G. (2002) Genetics and mechanisms of permethrin resistance in the YPER strain of house fly. Pesticide Biochemistry and Physiology 73: 27–36.

133. Guzov V. M., Unnithan G. C., Chernogolov A. A. and Feyereisen R. (1998) CYP12A1, a mitochondrial cytochrome P450 from the house fly. Archives of Biochemistry and Biophysics 359: 231–240.

134. Waters L. C., Zelhof A. C., Shaw B. J. and Ch'ang L. Y. (1992) Possible involvement of the long terminal repeat of transposable element 17.6 in regulating expression of an insecticide resistance-associated P450 gene in Drosophila. Proceedings of the National Academy of Science USA 89: 4855–4859.

135. Maitra S., Dombrowski S. M., Waters L. C. and Ganguly R. (1996) Three second chromosome-linked clustered Cyp6 genes show differential constitutive and barbital-induced expression in DDT-resistant and susceptible strains of Drosophila melanogaster. Gene 180: 165–171.

136. Maitra S., Dombrowski S. M., Basu M., Raustol O., Waters L. C. and Ganguly R. (2000) Factors on the third chromosome affect the level of cyp6a2 and cyp6a8 expression in Drosophila melanogaster. Gene 248: 147–156.

137. Maitra S., Price C. and Ganguly R. (2002) Cyp6a8 of Drosophila melanogaster: Gene structure, and sequence and functional analysis of the upstream DNA. Insect Biochemistry and Molecular Biology 32: 859–870.

138. Le Goff G., Boundy S., Daborn P. J., Yen J. L., Sofer L. Lind R., Sabourault C., Madi- Ravazzi L. and ffrench-Constant R.H. (2003) Microarray analysis of cytochrome P450 mediated insecticide resistance in Drosophila. Insect Biochemistry and Molecular Biology 33: 701–708.

139. Daborn P., Boundy S., Yen J., Pittendrigh B. and ffrench-Constant R. (2001) DDT resistance in Drosophila correlates with Cyp6g1 over-expression and confers crossresistance to the neonicotinoid imidacloprid. Molecular Genetics and Genomics 266: 556–563.

140. Daborn P. J., Yen J. L., Bogwitz M. R., Le Goff G., Feil E., Jeffers S., Tijet N., Perry T., Heckel D., Batterham P., Feyereisen R., Wilson T. G. and ffrench-Constant R. H.

141. (2002) A single P450 allele associated with insecticide resistance in Drosophila. Science 297: 2253–2256.

142. Catania F., Kauer M. O., Daborn P. J., Yen J. L., ffrench-Constant R. H. and Schlotterer C. (2004) World-wide survey of an Accord insertion and its association with DDT resistance in Drosophila melanogaster. Molecular Ecology 13: 2491– 2504.

143. Brandt A., Scharf M., Pedra J. H., Holmes G., Dean A., Kreitman M. and Pittendrigh B. R. (2002) Differential expression and induction of two Drosophila cytochrome P450 genes near the Rst (2) DDT locus. Insect Molecular Biology 11: 337–341.

144. Pyke F. M., Bogwitz M. R., Perry T., Monk A., Batterham P. and McKenzie J. A. (2004) The genetic basis of resistance to diazinon in natural populations of Drosophila melanogaster. Genetica 121: 13–24.

145. Bogwitz M. R., Chung H., Magoc L., Rigby S., Wong W., O'Keefe M., McKenzie J.A., Batterham P. and Daborn P. J. (2005) Cyp12a4 confers lufenuron resistance in a natural population of Drosophila melanogaster. Proceedings of the National Academy of Science USA 102: 12807–12812.

146. Nikou D., Ranson H. and Hemingway J. (2003) An adult specific CYP6 P450 gene is overexpressed in a pyrethroids resistant strain of the malaria vector, Anopheles gambiae. Gene 318: 91–102.

147. David J. P., Strode C., Vontas J., Nikou D., Vaughan A., Pignatelli P.M., Louis C., Hemingway J. and Ranson H. (2005) The Anopheles gambiae detoxification chip: A highly specific microarray to study metabolic-based insecticide resistance in malaria vectors. Proceedings of the National Academy of Science USA 102: 4080–4084.

148. Djouaka R. F., Bakare A. A., Coulibaly O. N., Akogbeto M. C., Ranson H., Hemingway J. and Strode C. (2008) Expression of the cytochrome P450s, CYP6P3 and CYP6M2 are significantly elevated in multiple pyrethroid resistant populations of Anopheles gambiae s.s. from Southern Benin and Nigeria. BMC Genomics 9: 538.

149. Muller P., Warr E., Stevenson B. J., Pignatelli P. M., Morgan J. C., Steven A., Yawson

150. A. E., Mitchel S. N., Ranson H., Hemingway J. Paine M. J. I. Donnelly M. J. (2008) Field-caught Permithrin-Resistant Anopheles gambiae overexpress CYP6P3, a P450 that metabolises Pyrethroids. PLoS Genetics 4 (11): e1000286.

151. Muller P., Donnelly M. J. and Ranson H. (2007) Transcription profiling of a recently colonised pyrethroid resistant Anopheles gambiae strain from Ghana. BMC Genomics 8: 36.

152. Vontas J., Blass C., Koutsos A. C., David J. P., Kafatos F. C., Louis C., Hemingway J. Christophides G. K. and Ranson H. (2005) Gene expression in insecticide resistant and susceptible Anopheles gambiae strains constitutively or after insecticide exposure. Insect Molecular Biology 14: 509–521.

153. Vontas J., David J. P., Nikou D., Hemingway J., Christophides G. K., Louis C. and Ranson H. (2007) Transcriptional analysis of insecticide resistance in Anopheles stephensi using cross-species microarray hybridization. Insect Molecular Biology 16: 315– 324.

154. Amenya D. A., Naguran R., Lo T. C., Ranson H., Spillings B. L., Wood O.R., Brooke B. D., Coetzee M. and Koekemoer L. L. (2008) Over expression of a cytochrome P450 (CYP6P9) in major African malaria vector, Anopheles funestus, resistant to pyreth roids. Insect Molecular Biology 17: 19–25.

155. Wondji C. S., Irving H., Morgan, J., Lobo N. F., Collins F. H., Hunt R.H., Coetzee M., Hemingway J. and Ranson H. (2009) Two duplicated P450 genes are associated with pyrethroid resistance in Anopheles funestus, a major malaria vector. Genome Research 19: 452–459.

156. Strode C., Steen K., Ortelli F. and Ranson H. (2006) Differential expression of the de-toxification genes in the different life stages of the malaria vector Anopheles gambiae. Insect Molecular Biology 15: 523–530.

157. Hardstone M. C., Komagata O., Kasai S., Tomita T. and Scott J. G. (2010) Use of isogenic strains indicates CYP9M10 is linked to permethrin resistance in Culex pipiens quinquefasciatus. Insect Molecular Biology 19: 717–726.

158. Kasai S., Weerashinghe I. S., Shono T. and Yamakawa M. (2000) Molecular cloning, nucleotide sequence and gene expression of a cytochrome P450 (CYP6F1) from the pyrethroid-resistant mosquito, Culex quinquefasciatus Say. Insect Biochemistry and Molecular Biology 30: 163–171.

159. Komagata O., Kasai S. and Tomita T. (2010) Overexpression of cytochrome P450 genes in pyrethroid-resistant Culex quinquefasciatus. Insect Biochemistry and Molecular Biology 40: 146–152.

160. Shen B., Dong H. Q., Tian H. S., Ma L., Li X. L., Wu G. L. and Zhu C. L. (2003) Cytochrome P450 genes expressed in the deltamethrin- susceptible and -resistant strains of Culex pipiens pallens. Pesticide Biochemistry and Physiology 75: 19–26.

161. Rose R. L., Goh D., Thompson D. M., Verma K. D., Heckel D. G., Gahan L. J., Roe R. M. and Hodgson E. (1997) Cytochrome P450 (CYP) 9A1 in Heliothis virescens: The first member of a new CYP family. Insect Biochemistry and Molecular Biology 27: 605–615.

162. Hopkins B. W., Longnecker M. T. and Pietrantonio P. V. (2010) Transcriptional over expression of CYP6B8/CYP6B28 and CYP6B9 is a mechanism associated with cypermethrin survivorship in field-collected Helicoverpa zea (Lepidoptera: Noctuidae) moths. Pest Management Science 67: 21–25.

163. Pittendrigh B., Aronstein K., Zinkovsky E., Andreev O., Campbell B., Daly J., Trowell S. and ffrench-Constant R. H. (1997) Cytochrome P450 genes from Helicoverpa armigera: Expression in a pyrethroid-susceptible and -resistant strain. Insect Biochemistry and Molecular Biology 27: 507–512.

164. Ranasinghe C. and Hobbs A. A. (1998) Isolation and characterization of two cytochrome P450 cDNA clones for CYP6B6 and CYP6B7 from Helicoverpa armigera (Hubner): Possible involvement of CYP6B7 in pyrethroid resistance. Insect Biochemistry and Molecular Biology 28: 571–580.

165. Yang Y., Chen S., Wu S., Yue L. and Wu Y. (2006) Constitutive overexpression of multiple cytochrome P450 genes associated with pyrethroid resistance in Helicoverpa armigera. Journal of Economic Entomology 99: 1784–1789.

166. Wee C. W., Lee S. F., Robin C. and Heckel D. G. (2008) Identification of candidate genes for fenvalerate resistance in Helicoverpa armigera using cDNA-AFLP. Insect Molecular Biology 17: 351–360.

167. Brun-Barale A., Hema O., Martin T., Suraporn S., Audant P., Sezutsu H. and Feyereisen R. (2010) Multiple P450 genes overexpressed in deltamethrin-resistant strains of Helicoverpa armigera. Pest Management Science 66: 900–909.

168. Bautista M. A., Tanaka T. and Miyata T. (2007) Identification of permethrin-inducible cytochrome P450s from the diamondback moth, Plutella xylostella (L.), and the possibility of involvement in permethrin resistance. Pesticide Biochemistry and Physiology 87: 85–93.

169. Baek J. H., Clark J. M. and Lee S. H. (2010) Cross-strain comparison of cypermethrin- induced cytochrome P450 transcription under different induction conditions in diamondback moth. Pesticide Biochemistry and Physiology 96: 43–50.

170. Zhu Y. C. and Snodgrass G. L. (2003) Cytochrome P450 CYP6X1 cDNAs and mRNA expression levels in three strains of the tarnished plant bug Lygus lineolaris (Hetero ptera: Miridae) having different susceptibilities to pyrethroids insecticide. Insect Molecular Biology 12: 39–49.

171. Karunker I., Benting J., Lueke B., Ponge T., Nauen R., Roditakis E., Vontas J., Gorman K., Denholm I. and Morin S. (2008) Over-expression of cytochrome P450 CYP6CM1 is associated with high resistance to imidacloprid in the B and Q biotypes of Bemisia tabaci (Hemiptera: Aleyrodidae). Insect Biochemistry and Molecular Biology 38: 634–644.

172. Bass C., Carvalho R. A., Oliphant L., Puinean A. M., Field L. M., Nauen R., Williamson M. S., Moores, G. and Gorman, K. (2011) Overexpression of a cytoichrome P450 monooxygenase, CYP6ER1, is associated with resistance to imidacloprid in the brown plant hopper, Nilapatvata lugens. Insect Molecular Biology 20(6): 763–773.

173. Puinean A. M., Foster S. P., Oliphant L., Denholm I., Field L. M. Millar N. S., Williamson M. S. and Bass C. (2010) Amplification of a cytochrome P450 gene is associated with resistance to neonicotinoid insecticides in the aphid Myzus persicae. PLoS Genetics 6(6): e1000999. doi:10.1371/journal.pgen.1000999.

174. Scharf M. E., Parimi S., Meinke L. J., Chandler L. D. and Siegfried B. D. (2001) Expression and induction of three family 4 cytochrome P450 (CYP4)* genes identified from insecticide-resistant and -susceptible western corn rootworms, Diabrotica virgifera virgifera. Insect Molecular Biology 10: 139–146.

175. Zhu F., Parthasarathy R., Bai H., Woithe K., Kaussmann M., Nauen R., Harrison D. A. and Palli S. R. (2010) A brain-specific cytochrome P450 responsible for the majority of deltamethrin resistance in the QTC279 strain of Tribolium castaneum. Proceedings of the National Academy of Sciences USA 107: 8557–8562.

176. Scharf M. E., Lee C. Y., Neal J. J. and Bennett G. W. (1999) Cytochrome P450 MA expression in insecticide-resistant German cockroaches (Dictyoptera: Blattellidae). Journal of Economic Entomology 92: 788–793.

177. Pridgeon J. W., Zhang L. and Liu N. (2003) Overexpression of CYP4G19 associated with a pyrethroid-resistant strain of the German cockroach, Blattella germanica (L.). Gene 314: 157–163.

178. Zhu Y. C., Snodgrass G. L. and Ming Shun Chen M. S. (2004) Enhanced esterase gene expression and activity in a malathion-resistant strain of the tarnished plant bug, Lygus lineolaris. Insect Biochemistry and Molecular Biology 34: 1175–1186.

179. Zhu Y. C., West S., Snodgrass G. and Luttrell R. (2011) Variability in Resistance-relat ed Enzyme Activities in Field Populations of the Tarnished Plant Bug, Lygus lineol ris. Pesticide Biochemistry and Physiology 99: 265–273.

180. Ferrari J. A., Morse J. G., Georghiou G. P., Sun Y (1993) Elevated esterase activity and acetylcholinesterase insensitivity in citrus thrips (Thysanoptera:

Thripsidae) populations from the San Joaquin Valley of California. Journal of Economic Entomology 86: 1645–1650.

181. Vassiliou V., Emmanouilidou M., Perrakis A., Morou E., Vontas J., Tsagkarakou A. and Roditakis E. (2011) Insecticide resistance mechanisms in tea pests 69 Insecticide resistance in Bemisia tabaci from Cyprus. Insect Science 18 30–39.

182. Wink M. and Waterman P. (1999) Chemotaxonomy in relation to molecular phylogeny of plants. Annual Plant Reviews 2, 300–341

183. Wink M. (2003) Evolution of secondary metabolites from an ecological and molecular phylogenetic perspective. Phytochemistry 64: 3–19.

184. Yu S. J. and Ing R. T. (1984) Microsomal biphenyl hydroxylase of fall armyworm larvae and its induction by allelochemicals and host plants. Comparative Biochemistry and Physiology 78C: 145–152.

185. Musser R. O., Hum-Musser S. M., Eichenseer H., Peiffer M., Ervin G., Murphy J. B. and Felton G. W. (2002) Herbivory: caterpillar saliva beats plant defences – a new weapon emerges in the evolutionary arms race between plants and herbivores. Nature 416: 599–600.

186. Helmus M. R. and Dussourd D. E. (2005) Glues or poisons: which triggers vein cutting by monarch caterpillars? Chemoecology 15: 45–49

187. Labeyrie E. and Dobler S. (2004) Molecular adaptation of Chrysochus leaf beetles to toxic compounds in their food plants. Molecular Biology and Evolution 21: 218–221. Zagrobelny M., Bak S., Rasmussen A. V., Jørgensen B., Naumann C. M. and Møller B. L. (2004) Cyanogenic glucosides and plant–insect interactions. Phytochemistry 65: 293–306.

188. Francis F., Vanhaelen N. and Haubruge E. (2005) Glutathione S-transferases in the adaptation to plant secondary metabolites in the Myzus persicae aphid. Archives of Insect Biochemistry and Physiology 58: 166–174.

189. Ratzka A., Vogel H., Kliebenstein D. J. Mitchell-Olds T. and Kroymann J. (2002) Disarming the mustard oil bomb. Proceedings of the National Academy of Science USA 99: 11223–11228.

190. Wittstock U., Agerbirk N., Stauber E. J., Olsen C. E., Hippler M., Mitchell-Olds T., Gershenzon J. and Vogel H. (2004) Successful herbivore attack due to metabolic diversion of a plant chemical defense. Proceedings of the National Academy of Science USA 101: 4859–4864.

191. Fogleman J.C. (2000) Response of Drosophila melanogaster to selection for P450- mediated resistance to isoquinoline alkaloids. Chemico-Biological Interactions 125: 93– 105.

192. Hartmann T., Theuring C., Beuerle T., Klewer N., Schulz S., Singer M. S. and Bernays E. A. (2005) Specific recognition, detoxification and metabolism of pyrrolizidine alkaloids by the polyphagous arctiid Estigmene acrea. Insect Biochemistry and Molecular Biology 35: 391–411.

193. Naumann C., Hartmann T. and Dietrich Ober D. (2002) Evolutionary recruitment of a flavin dependent monooxygenase for the detoxification of host plant acquired pyrrolizidine alkaloids in the alkaloid-defended arctiid moth Tyria jacobaeae. Proceedings of the National Academy of Science USA 99: 6085–6090.

194. Glendinning J. I. (2002) How do herbivorous insects cope with noxious secondary plant compounds in their diet? Entomologia Experimentalis et Applicata 104: 15–25.

195. Nyman T. and Julkunen-Tiitto R. (2000) Manipulation of the phenolic chemistry of willows by gall-inducing sawflies. Proceedings of the National Academy of Science USA 97: 13184–13187.

196. Luque T., Okano K. and O'Reilly D. R. (2002) Characterization of a novel silkworm (Bombyx mori) phenol UDP-glucosyltransferase. European Journal of Biochemistry 269: 819–825.

197. Willinger G. and Dobler S. (2001) Selective sequestration of iridoid glycosides from their host plants in Longitarsus flea beetles. Biochemical Systematics and Ecology 29: 335–346.

198. Petersen R. A., Zangerl A. R., Berenbaum, M. R. and Schuler M. A. (2001) Expression of CYP6B1 and CYP6B3 cytochrome P450 monooxygenases and furanocoumarin metabolism in different tissues of Papilio polyxenes (Lepidoptera: Papilionidae). Insect Biochemistry and Molecular Biology 31: 679–690.

199. Nitao J. K., Berhow M., Duval S. M., Weisleder, D., Vaughn S. F., Zangerl A. and Berenbaum M. R. (2003) Characterization of furanocoumarin metabolites in parsnip webworm, Depressaria pastinacella. Journal of Chemical Ecology 29: 671–682.

200. Yu S. J. (2002) Biochemical characteristics of microsomal and cytosolic glutathione S- transferases in larvae of the fall armyworm, Spodoptera frugiperda (J. E. Smith). Pesticide Biochemistry and Physiology 72: 100–110.

201. Moon J., J., Salzman R. A., Ahn J. E., Koiwa H. and Zhu-Salzman K. (2004) Transcriptional regulation in cowpea bruchid guts during adaptation to a plant defence protease inhibitor. Insect Molecular Biology 13: 283–291.

202. Bede J. C., Musser R. O., Gary W. Felton G. W. and Korth K. L. (2006) Caterpillar herbivory and salivary enzymes decrease transcript levels of Medicago truncatula genes encoding early enzymes in terpenoid biosynthesis. Plant Molecular Biology 60: 519– 531.

203. Barbehenn R. V., Poopat U. and Spencer B. (2003) Semiquinone and ascorbyl radicals in the gut fluids of caterpillars measured with EPR spectrometry. Insect Biochemistry and Molecular Biology 33: 125–130

204. Berenbaum M. R. (2002) Postgenomic chemical ecology: from genetic code to ecological interactions. Journal of Chemical Ecology 28(5): 873–896.

205. Meyran J. C., David J. C., Rey J. P., Cuany D. Bride A. and Amichot J. M. (2002). The biochemical basis of dietary polyphenols detoxification by aquatic detritivorous Arthropoda. Recent Research Developments in Analytical Biochemistry 2: 185–199.

206. Fraenkel G. S. (1959) The raison d'etre of secondary plant substances. Science 129: 1466–1470.

207. Krieger R. I., Feeny P. P. and Wilkinson C. F. (1971) Detoxication enzymes in the guts of caterpillars: an evolutionary answer to plant defenses? Science 172: 579–581.

208. Mullin C. A. (1986) Adaptive divergence of chewing and sucking arthropods to plant allelochemicals, pp. 175–209. In Molecular Aspects of Insect-Plant Associations (edited by L. B. Brattsten and S. Ahmad). Plenum Press, New York.

209. Mullin C. A., Croft B. A., Strickler K., Matsumura F. and Miller J. R. (1982) Detoxification enzyme differences between a herbivorous and predatory mite. Science 217: 1270, 1271.

210. Iyenger S., Arnason J. T., Philogene B. J. R., Werstiuk N. H. and Morand P. (1990) Comparative metabolism of the photoxic allelochemical α-terthienyl in three species of lepidopterans. Pesticide Biochemistry and Physiology 37(2): 154–164.

211. Yu S. J. (1983) Induction of detoxifying enzymes by allelochemicals and host plants in the fall armyworm. Pesticide Biochemistry and Physiology 19: 330–336.

212. Reilly C. C., Gentry C. R. and McKay J. R. (1987) Biochemical evidence for resistance of root stocks to the peach tree borer and species separation of peach tree borer and lesser peach tree borer (Lepidoptera: Sesiidae) on peach trees. Journal of Economic Entomology 80: 338–343.

213. Lindroth R. L. (1989) Host plant alteration of detoxication activity in Papilio glaucus glaucus. Entomologia Experimentalis et Applicata 50: 29–35.

214. Lindroth R. L., Scriber J. M. and Hsia M. T. S. (1988) Chemical ecology of the tiger swallowtail: mediation of host use by phenolic glycosides. Ecology 69, 814–822.

215. Wadleigh R. W. and Yu S. J. (1987) Glutathione transferase activity of fall armyworm larvae toward a, b-unsaturated carbonyl allelochemicals and its induction by allelochemicals. Insect Biochemistry 17: 759–764.

216. Wadleigh R. W. and Yu S. J. (1988) Detoxification of isocyanate allelochemicals by glutathione transferase in three lepidopterous species. Journal of Chemical Ecology 14: 1279–1288.

217. Yu S. J. (1989) Purification and characterization of glutathione transferases from five phytophagous Lepidoptera. Pesticide Biochemistry and Physiology 35: 97–105.

218. Yu S. J. and Abo-Elghar G. E. 2000. Allelochemicals as inhibitors of glutathione S- transferase in the fall armyworm. Pesticide Biochemistry and Physiology 68: 173–183.

219. Mound L. A. and Palmer G. M. (1981) Identification, distribution and host plants of the pest species of Scirtothrips (Thysanoptera: Thripidae). Bulletin of Entomological Research 71: 467–479.

220. Saha D., Mukhopadhyay A. and Bahadur M. (2012) Effect of host plants on fitness traits and detoxifying enzymes activity of major sucking insect pest of tea, Helopeltis theivora (Heteroptera: Miridae). Phytoparasitica 40: 433–444.

221. Bartlett A. C., Wilson N. M. and Mattix E. B. (1968) The fate of genetic markers in populations of boll weevils. Journal of Economic Entomology 61: 808–812.

222. Bartlett A. C. and Butler G. D. Jr. (1975) Genetic control of the cabbage looper by a recessive lethal mutation. Journal of Economic Entomology 68: 331–335.

223. Loxdale H. D. and Lushai G. (1998) Molecular Markers in Entomology. Bulletin of Entomological Research 88: 577–600.

224. Loxdale H. D., Brookes C. P. and De Barro, P. J. (1996) Application of novel molecular markers (DNA) in agricultural entomology. In The Ecology of Agricultural Pests (edited by W. O. C. Symondson and J. E. Liddell). Chapman and Hall, London.

225. Daly J. C. (1993) Ecology and genetics of insecticide resistance in Helicoverpa armigera: interactions between selection and gene flow. Genetica 90: 217–226.

226. Roush R. T. and Daly J. C. (1990) The role of population genetics in resistance research and management, pp. 97–152. In Pesticide Resistance in Arthropods (edited by R. T. Roush and B. E. Tabashnik). Chapman & Hall, New York.

227. Labbe P., Lenormand T. and Raymond M. (2005) On the world wide spread of an insect resistance gene: a role for local selection. Journal of Evolutionary Biology 18: 1471– 1484.

228. Denholm I., Sawicki R. M. and Farnham A. W. (1985) Factors affecting the resistance to insecticides in houseflies, Musca domestica. IV: the population biology of the flies on animal farms in south-eastern England and its implications for the management of resistance. Bulletin of Entomological Research 75: 143–158.

229. Lenormand T. (2002) Gene flow and the limits to natural selection. Trends in Ecology and Evolution 17: 183–189.

230. Whitehead A., Anderson S. L., Kuivila K. M., Roach J. L. and May B. (2003) Genetic variation among interconnected populations of Catostomus occidentalis: implications for distinguishing impacts of contaminants from biogeographical structuring. Molecular Ecology 12: 2817–2833.

231. Brogdon W. G. and McAlliester J. C. (1998) Insecticide resistance and vector control. Emerging Infectious Diseases 4: 605–613.

Index

www.ingramcontent.com/pod-product-compliance
Lightning Source LLC
LaVergne TN
LVHW081941200926
848484LV00009B/2483